Innovations in Crystallography

Innovations in Crystallography

Edited by **Sharon Levine**

CLANRYE
INTERNATIONAL

New Jersey

Published by Clanrye International,
55 Van Reypen Street,
Jersey City, NJ 07306, USA
www.clanryeinternational.com

Innovations in Crystallography
Edited by Sharon Levine

International Standard Book Number: 978-1-63240-310-0 (Hardback)

This book contains information obtained from authentic and highly regarded sources. Copyright for all individual chapters remain with the respective authors as indicated. A wide variety of references are listed. Permission and sources are indicated; for detailed attributions, please refer to the permissions page. Reasonable efforts have been made to publish reliable data and information, but the authors, editors and publisher cannot assume any responsibility for the validity of all materials or the consequences of their use.

The publisher's policy is to use permanent paper from mills that operate a sustainable forestry policy. Furthermore, the publisher ensures that the text paper and cover boards used have met acceptable environmental accreditation standards.

Trademark Notice: Registered trademark of products or corporate names are used only for explanation and identification without intent to infringe.

Printed in the United States of America.

Contents

Preface

Over the recent decade, advancements and applications have progressed exponentially. This has led to the increased interest in this field and projects are being conducted to enhance knowledge. The main objective of this book is to present some of the critical challenges and provide insights into possible solutions. This book will answer the varied questions that arise in the field and also provide an increased scope for furthering studies.

This book presents latest innovations and advances in the field of crystallography. The emergence of X-ray diffraction during the earlier years of twentieth century led to the transformation of crystallography. From a field of scientific inquiry limited to mathematics, physics and mineralogy, it has evolved as a field which encompasses almost all physical, life and material sciences as well as engineering. This book is a compilation of the contributions by eminent researchers and scientists from all over the world. It provides insights into the recent studies and advances in the field of crystallography.

I hope that this book, with its visionary approach, will be a valuable addition and will promote interest among readers. Each of the authors has provided their extraordinary competence in their specific fields by providing different perspectives as they come from diverse nations and regions. I thank them for their contributions.

Editor

History of Crystallography

Histories of Crystallography by Shafranovskii and Schuh

Bart Kahr and Alexander G. Shtukenberg

Additional information is available at the end of the chapter

1. Introduction

There are at least six book length biographies of Herman Melville (1819-1891) and ten histories of the Russian Revolution currently in print in the English language. On the other hand, if you chase after crystals not whales, or believe that the determination of the structure of matter was a historical pivot, you will be disappointed that there does not exist a single narrative history of crystallography in print in English or any other language to the best of our knowledge. By any measure, crystallography now receives scant attention by historians and scholars.

One admirable attempt to fill this chasm is the wonderfully idiosyncratic *Historical Atlas of Crystallography* published by the International Union of Crystallography (Lima-de-Faria, 1990). It is a treasure of timelines, portraits of crystallographers, and fetishistic reproductions of cover pages of classic monographs, accompanied by revealing essays on various aspects of the history of crystallography by acknowledged experts. But, the *Historical Atlas* is not a narrative history written with one strong voice.

Burke's *The Origin of the Science of Crystals* (Burke, 1966) is such a narrative that runs up to the discovery of X-ray diffraction. It is the best source for those interested in an English language analysis of the history of crystallography. But, this book has been long out-of-print. (We are not oblivious to the ironies of lamenting in an open-access journal about access to print media. Google Books may ultimately obviate such lamentations but to date only a limited preview of *The Origin...* is available on-line).

As a remedy, we set out to produce an English language edition of I. I. Shafranovskii's two volume *History of the Science of Crystals* only available in Russian. The first volume was subtitled *From Ancient Times to the Beginning of the 19^h Century* (Shafranovskii, 1978), and the second volume *The 19th Century* (Shafranovskii, 1980, Figure 1). A third volume covering the age of X-rays was planned but never materialized. Shafranovskii was a professor at the Leningrad Mining Institute. He had a long-standing interest in the history of crystallography (For a biographical sketch, see next section) and earlier published *The History of Crystallography in Russia* (Shafranovskii, 1962). In many ways, Shafranovskii's later two volume *History of Crystallography* is the best effort to cover a massive subject spanning centuries, countries, and languages. It is a valuable complement to Burke in that it is Russo-centric. Generally speaking, Russian science historians comfortably read English, French, and German while American science historians comfortably read…English, French, and German. This has naturally created a bias against Cyrillic texts that became apparent to us during excursions in some highly circumscribed aspects of the history of crystallography (Shtukenberg & Punin, 2007). Burke, for instance, made scant use of Russian sources. For this reason, Shafranovskii restores some balance to the history of crystallography, even if he sometimes chauvinistically overemphasizes Russian sources.

We began our translation project more than one year ago. We made some progress but the labor ahead is many times over the labor that is behind.

Very recently, we became aware of several remarkable manuscripts in English that are freely downloadable from achives.org. Their author is Curtis P. Schuh, whose surname is linked with Shafranovskii in the title of this article. Though incomplete and unpublished, Schuh's manuscripts obviate our perceived need for an English language translation of Shafranovskii. In light of Schuh, the rewards of fully translating Shafranovskii are diminished. Herein, we aim to introduce readers first to Shafranovskii's book, and then to Schuh's unpublished manuscripts in the final section.

Our translations of Shafranovskii's introduction, table of contents, and a sample chapter, follow. Here, we can see his strategy and style. In preparing an English language edition of Shafranovskii's book we did not aspire to make a one-to-one translation. While Shafranovskii is a formidable historian, he is a tiresome, repetitive writer. He engages the reader with an old-fashioned, didactic, 'Soviet' style. Our intent was to reduce his two volumes to one and in the process produce a readable *History of Crystallography*. Striking out redundancies, directive phrases such as "It is important to remember that…", and so on, nods to Academicians, and irrelevant minutiae, should have accomplished most of our aim. We aspired to preserve Shafranovskii's organization and style when it did not interfere with driving the narrative forward. At the same time we intended to add material that has since come to light, and insert narrative glue in places, even while scraping off irksome residues in other places. We had planned to eviscerate a few of Shafranovskii's chapters that give the impression that the author 'ran out of gas' during his extensive undertaking. In reviewing the birth of physical crystallography, Shafranovskii summarizes the seventeen experiments in Bartholinus' (1625-1698) book on the discovery of double refraction in Iceland spar (Bartholinus, 1669, 1959) in the order given. The numbing chapter reads as follows: "In the first experiment… In the second experiment… In the sixteenth experiment… In the last experiment." We elected to rewrite this chapter from scratch. Even though we planned to take considerable liberties, a small effort would spare readers from author's weakest efforts.

Shafranovskii reviewed the relevant historical literature in his introduction. Here, we introduce major sources upon which he was most reliant, and those he was most critical of. Shafranovskii's naturally acknowledges Burke's text. Both Shafranovskii and Burke were admirers of Metzger (1889-1944?) a crystallographer turned philosopher of science. In her doctoral dissertation, *La génese de la Science des Cristaux* (Metzger, 1918), she emphasized the separation of crystallography from other disciplines during 17th and 18th Centuries with special attention to French texts. "Unfortunately," says Shafranovskii, "the fates of the author and her interesting book were tragic. The first page of the manuscript, kindly sent [to me] from Paris by Dr. K. I. Kurilenko, bears a foreboding inscription: 'Author and her book disappeared during the German occupation 1940-1944'". It is now known that Metzger was deported from Lyon to Auschwitz and was not among the twenty who survived her transport of 1501 persons (Freudenthal, 1990).

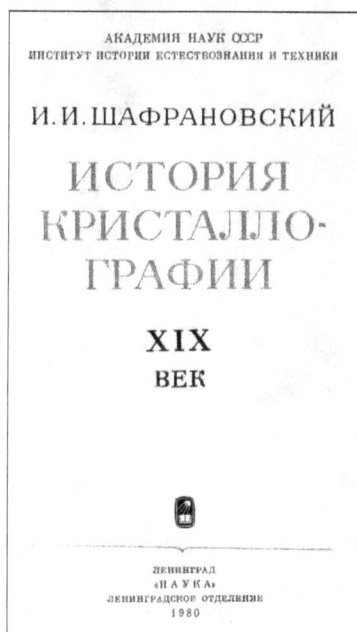

АКАДЕМИЯ НАУК СССР
ИНСТИТУТ ИСТОРИИ ЕСТЕСТВОЗНАНИЯ И ТЕХНИКИ

И. И. ШАФРАНОВСКИЙ

ИСТОРИЯ
КРИСТАЛЛО-
ГРАФИИ

XIX
ВЕК

ЛЕНИНГРАД
«НАУКА»
ЛЕНИНГРАДСКОЕ ОТДЕЛЕНИЕ
1980

Figure 1. I. I. Shafronovski's History of Crystallography, XIXth Century, Volume 2.

German texts dominated the 19th century literature on the history of crystallography, especially those of Marx (1794-1864) and Kobell (1803-1882). Marx's *Geschichte der Kristallkunde* (1825) was valued by Shafranovskii because of its numerous quotations from ancient sources. Kobell's *Geschichte der Mineralogie von 1650-1860*, current at the time of publication (1864, see also Kobell, 1866), contained histories of individual minerals and mineral properties such as magnetism and luminescence. Kobell earned fame as a poet of the upper-Bavarian dialect whose compositions became folk songs. His extended poem *Die Urzeit der Erde* (Kobell, 1856) showcased his knowledge of geoscience in verse.

Groth (1843-1927) published *Entwicklungsgeschichte der mineralogischen Wissenschaften* in 1926. As the founder of the journal *Zeitschrift für Krystallographie*, the author of the collection of crystallographic knowledge *Chemische Kristallographie* (1906-1919), and the source of the crystals that von Laue used for the discovery of X-ray diffraction, his place in the crystallographic firmament is assured. However, according to Shafranovskii, "Despite the prominence of the author, unfortunately the presentation of material [in his history] is sketchy. The review of the second half of the 19th Century is too brief and fragmented for a balanced narrative." Also falling short, according to Shafranovskii, was the Austrian mineralogist Tertsch, whose popular history, *Secrets of the Crystal World. A Romance of Science* (1947), trumpeted hyperbolic language not justified by its contents.

Figure 2. I. I. Shafranovskii age ~ 50.

Naturally, Shafranovskii gave special attention to the Russian literature. Terniaev's (1767-1827) history of mineralogy predated (1819) Marx's comparable work, with a stronger focus on recent events, especially emphasizing the contributions of Haüy (1743-1822). Vernadsky's (1863-1945) *Foundations of Crystallography* (1904) contains a splendid introduction to the history of crystallograpy. He gives affectionate portraits of giants such as Kepler (1571-1630), Steno (1638-1686), Romé de l'Lisle, Haüy (1736-1790), and Bravias (1811-1863), but also acknowledges lesser heroes such as Bernhardi (1770-1850), who helped to conceive the crystallographic systems, and Grassmann Sr. (1779-1852) who developed the stereographic projection, among others. Lemmlein (1901-1962), a specialist in mineral genesis, treated crystallography's past with great respect, especially the work of Lomonosov (Lemmlein, 1940). His brilliant comments to *On Precious Stones* (1989) by the 11th Century Persian scholar Al-Biruni, frame gemology. Shubnikov (1887-1979) posthumously published his brief "Origins of Crystallography" (1972), a popular introduction to the history of crystallography that, like Vernadsky's text, provides biographical information about pioneers.

Memoirs by Ewald (1888-1985) and Bragg (1890-1971), describe the first steps and subsequent developments in X-ray crystallography (Ewald, 1962; Bragg, 1975). Shafranovskii's history ends as X-rays are discovered. A full history of X-ray crystallography, a story of the 20th century, has yet to be written.

Here follows a biographical sketch of Shafranovskii, his table of contents, as well as a translation of the introduction to his two-volume opus, and a late chapter on Pierre Curie's Universal Principle of Symmetry.

2. Ilarion Ilarionovich Shafranovskii (1907-1994)

Ilarion Ilarionovich Shafranovskii (Anonymous, 1957, 1967, 1977, 1987, Figure 2), the son of a mathematician, was born in St. Petersberg. He first studied crystallography with Ansheles (1885 - 1957) at Leningrad University, graduating in 1931. In 1934, Shafranovskii began a professorship at the Leningrad Mining Institute, founded in 1907 by Fedorov (1853-1919), Ansheles' teacher. Shafranovskii received his doctoral degree in 1942 for studying diamond crystals with unusual morphologies. In 1946, he assumed the E. S. Fedorov Chair of Crystallography. Shafranovskii's name is frequently linked that of Federov. Shafranovskii wrote a biography of Fedorov (Shafranovskii, 1963), and in 1970 was awarded the E. S. Fedorov prize of the Academy of Sciences of the USSR for his work on the morphology of crystals and contributions to the history and popularization of crystallography.

Shafranovskii wrote some 500 articles and books. Among his major works are a textbook on crystallography with Popov, *Mineral Crystals* (1957), *Lectures on Crystallomorphology*, translated into English (1973), and *Outlines of Mineralogical Crystallography* (1974). In addition to the histories mentioned in the previous section, Shafranovskii published monographs on Koksharov (1818-1892) (Shafranovskii, 1964), Werner (Shafranovskii, 1968a), and Steno (Shafranovskii, 1972), among others, in addition to Fedorov, already mentioned. He wrote popular accounts of crystallography including *Diamonds* (1964) and *Symmetry in Nature* (1968b) that won the All-Union Knowledge Competition prize for the best popular science book.

In 1982, a mineral was named in Shafranovskii's honor, Shafranovskite, found the mountains of the Kola Peninsula, the eastward-jutting, thumb-shaped landmass atop Finland.

3. History of crystallography table of contents

4. A translation of the "Introduction" to *History of Crystallography*[1]

Goethe said, "The history of science is science itself" (Fink, 1991). Crystallography well illustrates his aphorism, at least as judged from its development in textbooks. Indeed, turning the pages of an elementary treatise in crystallography takes us from the simple to the complex following the chronological development of the science of crystals. For instance, the chronology of discoveries in geometrical crystallography mimics the order in which the associated concepts are presented in most textbooks. Pliny the Elder (AD 23 – 79) marveled at the extraordinarily flat faces of quartz crystals: "not even the most skillful lapidary could achieve such a finish" (Healy, 1999). A long time passed before the law of the constancy of interfacial angles was articulated in 17th and 18th centuries by Steno (1638-1686), Henkel (1678-1744), Lomonosov (1711-1765), and Romé de l'Lisle (1736-1790). Häuy (1743-1834) went further with law of rational indices, and the relationship between external shapes and internal structure. Weiss and Mohs deduced the zone law at the start of the 19th Century. Hessel, Bravais, and Gadolin (1828-1892) derived the finite symmetry classes, the 32 crystallographic point groups. Frankenheim (1801-1869), Bravias (1811-1863), and Sohncke (1842-1898) introduced the infinite symmetries of lattices. Fedorov and Schoenflies (1853-1928) carry us into the 20th Century and modern structural crystallography with derivations of the 230 space groups.

We could reconstruct the development of crystal physics likewise by tracing a path through discovery of double refraction in Iceland spar by Bartholinus (1669), to the correlation of optical and morphological symmetry by Brewster (1781-1868), to the correlation of all physical properties of crystals with symmetry by Neumann (1798-1895), and to the general symmetry principle of Curie (1859-1906) and modern solid state physics.

We thus might conclude that organizing a history of crystallography is a simple task. We need only enumerate in chronological order, and then elaborate on, all the achievements of crystallography. Of course, the situation is more complicated than it appears at first blush. The skeletal historical outlines above are idealized and purged of detours. Bewilderment, the lifeblood of the scientific enterprise, is nowhere in evidence. Such an accounting prejudicially selects only those developments that are organically incorporated into modern crystallography without disturbing the harmony of the imposing edifice. A faithful history of crystallography -- in all its fullness -- muddles the implicit history of the textbooks.

Foremost among the characteristics of crystals that have guided the development of crystallography is the problem presented by the stridently polyhedral shapes of crystals. "Crystals flash forth their symmetry"[2] according to Fedorov on the first page of his *Course in Crystallography* (1901). This fact had practical consequences: Agricola (1494-1555) instructed miners to identify minerals through their external "angular figures" (Agricola, 1556, 1950). Yet, Nature's well-facetted crystals presented a clearly defined problem to natural

---

[1] In order to provide more complete citations, we have added some sources that postdate Shafranovskii.
[2] The English rendering of this phrase was taken from Archard's translation of Shubnikov and Kopstik (1974).

philosophers that could not be solved without comprehensive geometrical analyses. Cardano (1501-1576) first proposed (1562) that the hexagonality of the rock crystal might arise from an internal structure consisting of densely packed spheres, anticipating Harriot (1560-1621) and Kepler, in part (Shafranovskii, 1975; Kahr, 2011). Ever since, the theoretical and empirical sciences of crystals developed in parallel. Albeit theory outpaced experiment until the 20th Century.

On the slight basis that crystals have geometrical shapes, are homogeneous, and anisotropic, theorists created a breathtaking mathematical crystallography. First articulated were laws that controlled the appearance of crystals of finite point symmetry. Like other mathematical disciplines, the development of theoretical crystallography was strictly logical, led to prediction, and guided subsequent experimental studies. The deduction of crystal classes (Hessel, 1830; Gadolin, 1867) was carried out before many of were illustrated by minerals; of the 32 crystal point groups, Gadolin found only 20 examples in nature. The laws governing crystal point symmetries were then extended to cover the symmetries of infinite crystal lattices. Indeed, at the end of the 19th Century, achievements in mathematical crystallography were so impressive that Fedorov proclaimed that its mathematical character rendered it "one of the most exact sciences" (Fedorov, 1901). Only now have advances in analysis matched those of theory, restoring balance to the science of crystals.

In the middle of 19th Century Frankenheim and Bravais developed the concept of the crystal lattice enumerating the 14 frameworks that form the basis of the modern structural crystallography. "Nature knelt before the hard theory, and the crystals positioned themselves in those classes where they should be according to the geometrical systems of points (space lattices)," expressively wrote Fedorov (1891). The 14 Bravais lattices and the 32 point groups were the constraints between which Fedorov, and independently Schoenflies (1853-1928), deduced in 1890-1891 the 230 possible space groups that restrict the mutual arrangement of building units (atoms, ions, molecules) inside crystals (1891). These far-seeing predictions were fully supported by experimental data subsequent to the discovery of X-ray diffraction by von Laue (1912), an achievement that is no less impressive than Mendeleev's expectations of undiscovered chemical elements on the basis of the periodic system. The derivation of the 230 space groups of Fedorov caps our history; it is the pinnacle in development of the classical science of crystallography.

Along the way, sharp conflicts between scientists were provoked. Romé de l'Lisle clashed with Häuy on the relationship between morphology and internal structure. The German physiographical school of Weiss (1780-1856), Mohs (1773-1839), and Naumann (1797-1873), conflicted with theoretical studies by Hessel (1796-1872) and Bravias. Mineralogists Koksharov and Eremeev (1830-1899) fiercely resisted the mathematical generalizations of the Fedorov.

In this history, chapters devoted to the development of important crystallography concepts alternate with chapters devoted to the lives, creative work, and struggles of the greatest crystallographers. Biographical details that inform certain advances are vital in that they color the local character or "microclimate" out of which those advances arose. Accounts of the fate of a discovery, involving the collective acceptance or negation of an idea by many scientists working in disparate countries over centuries, illustrate the global character of the history of crystallography. Experiment and theory drive one another while great currents sweep up individuals whose works and words broaden the stream.

The use of crystalline materials by various professionals, further confounds the author of a history of crystallography. Since ancient times minerals guided miners in search of raw materials. Subsequently, the growth of crystals became a part of problem solving in metallurgy, physics, chemistry, and pharmacology, connecting crystallography with many branches of pure and applied science. This prevented crystallography from coalescing as an independent science for a long time. Crystallography was variously considered as a part physics, chemistry, mathematics, or especially mineralogy. In the 19th Century, crystallography was "preparatory mineralogy". Young Fedorov called crystallography "geometrical mineralogy". Even after having placed the capstone on the science of classical crystallography with the derivation of the space groups, Fedorov wrote at the end of his life: "[Crystallography] plays an essential role at the heart of mineralogy and as part of mining science whose primary purpose is utilization of natural resources" (Fedorov, 1955). Only recently has the characterization of crystallography as a "servant of mineralogy" faded. Today even cell biologists, and biomedical researchers embrace crystallography although this aspect of the history of crystallography is not covered herein.

Metzger, it her doctoral dissertation *Genèse de la Science d'Cristaux* (1918), previously considered crystallography's emergence from other sciences. Nevertheless, there is backflow; advances in the aforementioned disciplines draw crystallography back in. For instance, according to Vernadsky, "Crystallography has not been separated from mineralogy. It embraced mineralogy in a new way, entered its foundations and changed it radically…Mineralogy does not need to free itself from the physical sciences. Rather we must build new relationships between crystallography and mineralogy so as to transform the latter" (Vernadsky, 1928). Similar things have been said about the relationship of crystallography to chemistry (Engels, 1954) and to pharmacy (Fabian, 1967).

The changing interrelations among the sciences and their sub-disciplines complicates a reconstruction of the history of crystallography. Important threads must be picked from the vast literature on mineralogy, mathematics, physics, chemistry, metallurgy, medicine, and biology among others disciplines. This extraction requires an enormous amount of time and effort. Obviously, the history of crystallography can be only conditionally likened to a continuous, smooth line. In reality, we face something like a dotted line diving in and out of the general tableaux of the development of science.

So, how shall we write a history of crystallography? We can follow Metzger and little by little separate crystallography from historically related sciences, stressing the increasing independence from other disciplines. Alternatively, we can consider the development of crystallography as a natural structure constrained by the symmetries of regular crystal packing that started with minerals and gradually subsumed a wider spectrum of objects from synthetic molecular crystals to semi-conductors to drugs to proteins. The development of crystallography validates both approaches. This happens due to dialectic process of the differentiation and synthesis of the sciences (Figurovsky, 1969). Indeed, specialization of the science of crystals results in great progress; narrow disciplines can probe ever more deeply. On the other hand, increasing contacts among a rising number of allied disciplines obscures the main themes that specifically delineate the development of crystallography.

These ideas fully correspond to the new conceptions of the development of sciences. It is interesting to note that Fedorov stands at the beginning of such a systems approach. In his philosophical treatise "Perfectionism" he wrote: "The scientist is perpetually faced with the generalization of proven laws. The higher the philosophical development of a scientist, the clearer he understands the need to generalize even further because the logic of philosophy requires complete reduction" (Fedorov, 1906). The same ideas expressed more emphatically can be found in his later papers: "Are there true boundaries between sciences? Maybe all the sciences constitute something united and indivisible. Maybe the boundaries of a science, as they are established, represent only artificial constructions adapted to current understanding" (Fedorov, 1917). Thus, we must follow the historically conditioned development of the science of crystals without becoming isolated behind "artificial partitions" established by other disciplines.

Crystallographic phenomenology is emblematic to scientific generalization. Now, scientists often invoke "isomorphic laws" in different fields of science. It is gratifying to witness symmetry laws, firstly discovered in crystals, transferred to other fields of science. The beautiful examples of "isomorphism" underscore the relationship of geometrical crystallography to chemistry; the Steno-Lomonosov-Romé de l'Lisle law of the constancy of crystal angles is "isomorphic" to the law of Proust (1754-1826) on the constancy of composition of "true chemical compounds". Lomonsov's mentor, Henkel, formulated the law of the constancy of crystal angles as follows: "Nature in the confusion of her varied combinations has chosen the structure and external appearance of substances according to their properties and corresponding to external conditions and circumstances. She does not deviate from this rule; she sets a compass and measures the angles establishing one substance for all time." (Marx, 1825). Of his eponymous law, Proust said: "A compound is a privileged product, that Nature has given a constant composition. Nature, even with the intercession of people, never produces a compound without balance in hand; everything is in accord with weight and measure" (Menshutkin, 1937). The similarity in the formulation of this statement with that of Henkel is startling.

The law of the constancy of angles combined with the observation of cleavage phenomena led Häuy to formulate the unique "polyhedral molecules" (crystal structures in modern parlance) for a given crystalline compound. In the 20th Century, Goldschmidt (1888-1947) interpreted this statement as "the primary basis of crystal chemistry" (Goldschmidt, 1937). The thesis of Häuy combined with Steno's law is the crystallographic analogue of the Proust's generalization in chemistry. The law of rational indices in crystals by Häuy is "isomorphous" to the basic law of chemistry, Dalton's (1766-1844) law of multiple proportions. Obviously, the older crystallographic laws played some role in establishing of latter ones. Thus, once again we see the impossible task of the historian keen to separate unadulterated crystallography from closely related disciplines of physics, chemistry, and mineralogy.

Periodization, the subdivision of a long history into stages of development, provides further practical problems for the historian. Lenin (1870-1924) provides a general guide: "From living contemplation to abstract thinking *and then to practice* – this is a dialectic way in perception of *truth*, perception of objective reality" (Lenin, 1967). These words agree well with a statement by Fedorov: "When the nearest practical consequences of a given theory become known, we acquire the power to control Nature...the task of any science is to obtain such a power. Therefore, everything that gives this power is scientifically true" (Fedorov, 1904).

According to Kedrov (1903-1985), there are three main stages in the development of any science: (1) empirical fact gathering, (2) theory and explanation, and (3) prognostication (Kedrov, 1971). In the history of crystallography, we can see all three periods. For example, previously, with Grigoriev, we divided the history of Russian mineralogy and crystallography into four stages: narrative-descriptive, exact-descriptive, theoretical, and synthetic (Grigoriev, Shafranovskii, 1949). To a certain extent this division agrees with Kedrov if the two descriptive stages are aligned with his empirical stage. While mindful of the dual theoretical and practical development of crystallography, we recognize that a strict division into stages is impossible. In fact, Kedrov admits the conditional character of his divisions. In Russian crystallography, these periods are intertwined, overlapped, and sometimes inverted. Sometimes all three Kedrov stages can be identified in the activity of one and the same scientist. Nevertheless, stages are evident when we take a course-grained, centuries-wise perspective of the most significant achievements that carried the science forward: rules of morphology by Steno (1669), formulation of descriptive and theoretical crystallography by Romé de l'Lisle and Häuy (1783-1784), the mathematical inventions of Fedorov (1881-1919). In the 20th Century we have to acknowledge two "great revolutions in crystallography" as they were called by academician Belov (1891-1982): the epochal discovery of X-ray diffraction by von Laue (1912) and revolutionary developments in the growth of technically important single crystals in the 1950s and1960s (Belov, 1972).

In this work, for operational purposes, we distinguish four periods in the history of crystallography:

1. Prehistory, from ancient times to Steno;
2. Emergence of crystallography as an independent science, from Steno to Romé de l'Lisle and Häuy;
3. Development of classical, geometrical, crystallography, from Häuy to Fedorov;
4. The modern period, from Fedorov and von Laue to the present day, with its powerful synergy of crystal physics, crystal chemistry, structural biology, and crystal growth technologies.

A finer grained division into stages requires accounting of the related scientific disciplines: geology (Tikhomirov & Khain, 1956; Gordeev, 1967; Batyushkova, 1973), mineralogy

(Povarennyh, 1962), physics (Dorfman, 1974), and chemistry (Figurovsky, 1969) among others.

5. Translation of "Universal Symmetry Principle – Curie"[3]

Pierre Curie (1859-1906, Figure 3) was crushed under the wheels of a horse drawn carriage on a Paris street, a great misfortune for the world science. One of the most splendid French scientists of all time died at the peak of his power. Curie's deep insights survive in just a few, unusually concise articles. For this reason, the impact of his ideas, especially those related to crystallography and the symmetry principle, were not fully realized for some time.

The life and scientific work of Curie is described in a modest book by his wife Marie Curie (1867-1934) (Curie, 1963). Her brief biography of her husband succeeded in fleshing-out some of Pierre's ideas on symmetry that were not found in his publications. Marie also conveyed a sense of her husband's simple character and his devotion to the abstract life of the mind. Marie wrote, "He could never accustom himself to a system of work which involved hasty publications, and was always happier in a domain in which but a few investigators were quietly working" (Curie, 1963).

Pierre Curie was born in Paris, the son and grandson of physicians. He was schooled at home, but began attending lectures at the Sorbonne at a comparatively early age. At 18 he obtained a licentiate in physics after which he worked as a laboratory assistant in charge of the practical operations of the *École municipale de physique et de chimie industrielles*. He served as an instructor in physics until his appointment as Professor at the Sorbonne in 1903.

Curie's first papers describing the discovery of piezoelectricity in tourmaline, quartz, and other crystals (1880-1882), were written with his brother Jacques. His doctoral dissertation (1895) was an investigation of magnetism and the distinctions among diamagnetic, paramagnetic, and ferromagnetic substances, especially their temperature dependences. Pierre was a collaborator in the studies of radioactivity initiated by his wife Marie Skłodowska Curie. This work led to their joint discovery of polonium and radium in 1898. In 1903 they were awarded the third Nobel Prize in physics, together with Henri Becquerel (1852-1908). However, less well known than Pierre's highly publicized and well recognized work on radioactivity, but arguably as important, were theoretical papers devoted to crystallography and symmetry.

---

[3] In his scientific work, Shafranovskii was driven to understand the well know fact that crystals frequently have lower morphological symmetry than that expressed by physical properties or by X-ray diffraction. He recognized that the dissymmetry of the medium was often responsible for "false" crystal morphologies. This relationship between dissymmetric cause and effect was understandable in terms of Pierre Curie's Universal Symmetry Principle. For this reason, the work of Curie was of special interest to Shafranovskii. And for this reason, we provide a translation of one of the last chapters of the second volume of the *History*: "University Symmetry Principle – Curie".

Figure 3. Pierre Curie.

Physics and crystallography, explained Marie in the foreword to Pierre's collected works, were "two sciences equally close to him and mutually complementary in spirit. For him, the symmetry of phenomena were intuitive." (Curie, 1908). Thus, he was perfectly positioned to fully apply symmetry to physical laws. Still, distractions of work on radioactivity, adverse health effects associated with handling radium, and the burdens of fame left him wanting of more time to devote to his first loves, symmetry and crystallography. In her biography, Marie wrote, "Pierre always wanted to resume his works on the symmetry of crystalline media...After he was named professor at the Sorbonne. Pierre Curie had to prepare a new course...He was left great freedom in the choice of the matter he would present. Taking advantage of this freedom he returned to a subject that was dear to him, and devoted part of his lectures to the laws of symmetry, the study of fields of vectors and tensors, and to the application of these ideas to the physics of crystals."

The crystallographic legacy of Pierre Curie consists of only 14 extremely brief articles, each a classic. Curie's earliest contributions to crystallography are devoted piezoelectricity. Then follow the papers on the Universal Symmetry Principle. Finally, there is a small article on the relationship of crystal form to surface energy (Curie, 1885). This is now known as the Gibbs-Curie-Wulff rule.

It is commonly stated that piezoelectricity of crystals was discovered by the Curie brothers in 1880. This assertion must be qualified. In 1817, Häuy published a communication "On the electricity obtained in minerals by pressure" (Haüy, 1817). Pierre and Jacques Curie rediscovered this lost and incompletely described phenomenon. For sphalerites, boracites, calamine, tourmaline, quartz, Rochelle salt and other compounds, the Curie brothers showed that piezoelectricity can be present only in hemihedral crystals with inclined faces – in other words in acentric crystals – and that electric dipole moments can arise only along polar directions. Thus, knowing the crystal symmetry it became possible to predict the orientation of electrical axes. "This was by no means a chance discovery. It was the result of much reflection on the symmetry of crystalline matter that enabled the brothers to foresee the possibilities of such polarization", wrote Marie (Curie, 1963).

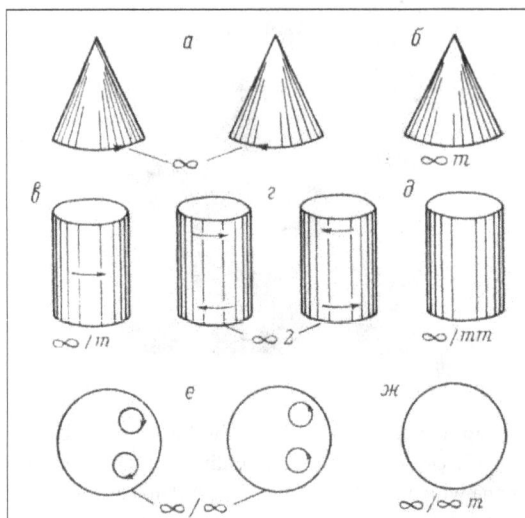

Figure 4. Seven infinite point groups of symmetry: rotating cone, cone at rest, rotating cylinder, twisted cylinder, cylinder at rest, rotating or chiral sphere, and sphere at rest.

Quartz crystals were studied in the most detail. The brothers Curie carried out a series of careful experiments that enabled them to establish general principles of piezoelectricity and define the magnitude of the quartz piezoelectric coefficient. The most complicated part of experimental work concerned the measurement of electrostriction, the deformation of piezoelectric crystals by applying an electric field (Curie, 1889). They proved the existence of this phenomenon, known as the inverse piezoelectric effect, first theoretically predicted by Lippmann (1845-1921). Finally, they invented and developed a series of devices for the study of piezoelectricy including a press with a manometer, a tool combining a lever and microscope for the measurement of electrostriction, and an extremely accurate electrometer in which metallized quartz surfaces were used to collect charges generated when pressure was applied to the quartz (Mouline & Boudia, 2009). Curies' works on piezoelectricity were inspirational to giants such as Röntgen (1845-1923), Kundt (1839-1894), Voigt (1850-1919), and Ioffe (1880-1960), among others. Langevin (1872-1946) utilized the piezoelectricity of quartz to produce ultrasound that is now used for measuring sea depth and detecting underwater objects.

At this same time, Curie worked out his theory of symmetry in a pair of papers (Curie, 1884, 1885b). Unlike Hessel, Bravais, and Fedorov, Curie's approach to symmetry fully integrated physics with mathematics. His lattices were made from physical objects, not geometrical points. The vectoral and tensorial physical properties of which he was so well aware through experimental work on magnetism and piezoelectricity were poorly accounted for by point lattices. "Significant difficulties arise", he said, "when points have associated properties related to direction in space. Such points should be represented by geometric figures embodying both magnitude and direction"(Curie, 1885). In searching for the proper figures, Curie was the first to establish the seven so-called "infinite point groups of symmetry" (Figure 4) with an infinite order axes (L_∞). Hessel identified only three: $L_\infty \infty P$ (∞m) (symmetry of the cone), $L_\infty \infty L_2 \infty PC$ (∞/mm) (symmetry of the bi-cone or cylinder), and $\infty L_\infty \infty PC$ ($\infty/\infty m$) (symmetry of the sphere). Curie completed this set by adding four additional infinite groups: L_∞ (∞) (symmetry of rotating cone), $L_\infty \infty PC$ (∞/m) (symmetry of rotating the cylinder), $L_\infty \infty L_2$ ($\infty 2$) (symmetry of the twisted cylinder), and ∞L_∞ (∞/∞) (symmetry of the sphere lacking mirror planes; all diameters of such a sphere are twisted to the right or left).

An illustration of seven infinite point groups after Shubnikov is given in Figure 4 (enantiomorphs are not shown). Curie illustrated these groups by examples from physics. The chiral sphere was associated with an optically active liquid. The $L_\infty \infty L_2$ case corresponded to two identical cylinders placed one onto another, filled with a liquid, and rotating with the same speed in opposite directions around their common axis L_∞. The symmetry of a cone ($L_\infty \infty P$) was compared with the symmetry of electric field, and the symmetry of a rotating cylinder ($L_\infty PC$) with the symmetry of the magnetic field (Curie, 1894). Infinite point groups are important because all other point groups are subgroups thereof.

Curie was the first to distinguish electric and magnetic dipoles. (Curie, 1894) Therefore, for example, in cubic crystals $m3m$ and 432 Curie considers the double number of axes compared to conventional notion: $6L_4$, $8L_3$, $12L_2$. Obviously, this approach was initiated by his studies of piezoelectricity in which it is essential to distinguish reversible and irreversible (polar) directions.

This profound approach to symmetry enabled Curie to discover a new symmetry element, the "periodically acting plane of symmetry." This symmetry element now corresponds to the improper rotation axis. Bravais, in his paper, *Note sur les polyèdres symétriques de la géométrie* (1849) "studied the symmetric polyhedra, but accounted only for proper rotation axes, centers of inversion, and mirror planes. He did not take into account periodically acting planes of symmetry," said Curie (1966). However, Curie did not know that this concept already had been proposed by Hessel in a different form, and by Gadolin in 1867 during his deduction of the 32 symmetry classes.

Almost simultaneously with Curie, Fedorov introduced mirror-rotation axes in his first book *Introduction to the Doctrine of Figures* (1855). Federov simultaneously discovered the mirror-rotation axes. In a letter to Schoenflies (1853-1926), Fedorov protested against calling the 32 crystal classes "Minnigerode groups". "In my opinion," he wrote, "this name is especially wrong, because in a paper by Curie as well as in my "Principles of doctrine on figures" (which, as I mentioned in my previous letter, was submitted for publication before Curie's paper) there were some new ideas, whereas the paper by Minnigerode (1837-1896) did not contain anything new" (Bokii & Shafranovskii, 1951). This question of priority lost its meaning when Sohncke (1842-1897) discovered that Hessel was in fact the first.

In 1885, Curie published a small but very important paper *Sur la formation des cristaux et sur les constants capillaires de leurs differrentes faces* (Curie, 1885a) in which he established that a crystal or an assemblage of crystals in equilibrium with a solution adopts a form that minimizes the surface energy. This result was obtained by Gibbs (1839-1903) in 1878, however, his work languished in the literature, unappreciated for a long time. In his classic paper "On the problem of growth and dissolution rates of crystal faces", Wulff (1863-1925) expressed this idea in terms that were easily applied (Wulff, 1952). The Wulff theorem states that "The minimum of the surface energy for a crystalline polyhedron of fixed volume is achieved, when the faces are spaced from the same point on distances that are proportional to the surface free energies" (Wulff, 1952). This theorem results in the important consequence that the growth rates of crystal faces are proportional to the specific surface energies of the faces. Wulff gave only an approximate proof of this theorem.

The theorem of Gibbs-Curie-Wulff was intensively debated. In 1915, Ehrenfest (1880-1933) emphasized that vicinal faces of real crystals have higher surface energies. This fact formed the basis of the objections to Curie's idea by the Dutch inorganic chemist, Van Arkel (1893-1976). But, this principle can be unconditionally applied only to the equilibrium shapes of the crystal.

In 1894, Curie published an especially important paper on symmetry: *Sur la symétrie dans les phénomènes physiques. Symétrie d'un champ électrique et d'un champ magnétique*. This paper begins with a following sentence: "I believe that it would be very interesting to introduce into the study of physical phenomena the property of symmetry, which is well known to crystallographers" (1894). This paper contains the most important ideas on the universal significance of symmetry. Reflections on these ideas can be found in the biographical sketch by Marie, *Pierre Curie, with the Autobiographical Notes of Marie Curie*: "It was in reflecting upon the relations between cause and effect that govern these phenomena that Pierre Curie was led to complete and extend the idea of symmetry, by considering it as a condition of space characteristic of the medium in which a given phenomenon occurs. To define this condition it is necessary to consider not only the constitution of the medium but also its condition of movement and the physical agents to which it is subordinated." And, "For this it is convenient to define the particular symmetry of each phenomenon and to introduce a classification which makes clear the principal groups of symmetry. Mass, electric charge, temperature, have the same symmetry, of a type called scalar, that of the sphere. A current of water and a rectilineal electric current have the symmetry of an arrow, of the type polar vector. The symmetry of an upright circular cylinder is of the type tensor" (Curie, 1963).

General statements found in the above paper are of great significance. "The characteristic symmetry of a given phenomenon is a maximal symmetry compatible with this phenomenon. The phenomenon can exist in the medium, which has a characteristic symmetry of this phenomenon or a symmetry of a subgroup of the characteristic symmetry. In the other words, some symmetry elements can coexist with some phenomena but they are not requisite. Some symmetry elements should be absent. That is, dissymmetry creates the phenomenon" (Curie, 1894).

Curie gave much broader interpretations to the concept "dissymmetry" than did Pasteur. He ascribed dissymmetry to the absence of symmetry elements that actuate some physical properties. For example, in the tourmaline crystal ($L_3 3P$ – $3m$) the absence of the perpendicular symmetry plane gives the polar character to the L_3 axis. This polarity makes pyroelectricity in tourmaline possible. For Curie, dissymmetry, the absence of symmetry, was as palpable as symmetry itself. He believed that the dissymmetric elements (e.g. a dissymmetry plane is any plane that is *not* a symmetry plane, a dissymmetry axis is any axis that is *not* a symmetry axis) could give a deeper insight into the physical meaning of phenomena. However, the infinite number of dissymmetry elements, unlike the very restricted number of symmetry elements, forces us to operate with the latter.

Shubnikov best characterized Curie's emphasis on dissymmetry: "symmetry must not be considered without its antipode – dissymmetry. Symmetry treats those phenomena at equilibrium, dissymmetry characterizes motion. The common conception of symmetry-dissymmetry is inexhaustible" (Shubnikov, 1946).

Curie formulated several important consequences to what is now called Curie's Universal Principle of Symmetry-Dissymmetry. "Superimposition of several phenomena in one and the same system results in addition of their dissymmetries. The remaining symmetry elements are only those that are characteristic of both phenomena considered separately. If some causes produce some effects, the symmetry elements of these causes should be present in the effects. If some effects reveal dissymmetry, this dissymmetry should be found in the causes" (Curie, 1894).

The statements cited above were illustrated by Curie with the infinite symmetry classes. He emphasized the special importance of class $L_∞ ∞P$: "Such a symmetry is associated with the axis of the circular cone. This is the symmetry of force, velocity, and the gravitational field, as well the symmetry of electric field. With respect to symmetry, all these phenomena may be depicted with an arrow" (Curie, 1894).

In fact, consequences of the association of symmetry $L_∞ ∞P$ with gravity are inexhaustible. For example, it explains evolution of the symmetry in organic life. The simplest organisms evolved in a medium of spherical symmetry ($∞L_∞ ∞PC$ ($∞/∞m$)) such as the protozoan suspended in a homogeneous fluid. Then the cone symmetry ($L_∞ ∞P$ ($∞m$)), that describes gravity begins to exert its influence pinning life to the ground. The plane symmetry $P(m)$ is actualized for moving organisms. Thus, the evolution of the organic life is controlled by the following sequence of desymmetrization of the medium: $∞/∞m > ∞m > m$ (Shafranovskii, 1968; Spaskii & Kravtsov, 1971).

Likewise, in mineralogy (Shafranovskii, 1974) detailed investigations of real, naturally occurring crystals requires a thorough knowledge on the medium in which the crystals were formed. Curie's principle does not allow us to consider the resulting crystal in the absence of its growth medium because the symmetry of the growth medium is superimposed on the symmetry of the growing crystal. The resulting form of the crystal can preserve only those symmetry elements that coincide with the symmetry elements of the growth medium. Of course, the internal symmetry, the crystal structure, does not change. The observed crystal morphology is a compromise resulting from the superimposition of two symmetries: internal symmetry of the crystal and the external symmetry of the medium. Thus, distorted crystal shapes, frequent in nature, are indicators of growth medium dissymmetry.

Curie's thoughts on symmetry have been only recently duly appreciated. Vernadsky was an advocate in his declining years. He wrote posthumously, "More than 40 year ago, in unfinished works interrupted first by the distraction of radium and then by death, Pierre Curie for the first time showed that the symmetry principle underlies all physical phenomena. Symmetry is as basic to physical phenomena as is the dimensionality of geometrical space because symmetry defines the physical state of the space – *état de l'éspace*. I have to stop here and emphasize the often forgotten importance of the force of personality. The premature depth of Curie at the peak of his powers stopped progress in this field for decades. Curie understood the significance of symmetry in physical phenomena before the causal relationship between symmetry and physical phenomena was not realized. He found the significance of this relationship previously overlooked" (Vernadsky, 1975).

Vernadsky writes: "The physically faithful definition [of symmetry], that we encounter throughout this book, was given by Curie...This is representation of a symmetry as a state of the earth, i.e. geological, natural space, or, more accurately as states of the space of natural bodies and phenomena of our planet Earth. Considering the symmetry as a state of the earth space it is necessary to emphasize the fact was expressed by Curie and recently stressed by A.V. Shubnikov, that the symmetry manifests itself not only in a structure but also in motions of natural bodies and phenomena" (Vernadsky, 1957).

Vernadsky knew Curie, whom he describes as "charming but lonely" (Vernadsky, 1965).

Detailed and very clear analyses of crystallographic ideas by Curie is presented in Shubnikov's paper "On the works of Pierre Curie in the field of symmetry" (Shubnikov, 1988): "P. Curie is known to broad audience of scientists as an author of influential works in the field of radioactivity. But he is almost unknown as the author of profound studies in the field of symmetry and its applications to physics. However, these studies, if they were continued by P. Curie, could have hardly less significance for development of natural science than his works on radioactivity for development of chemistry and physics."

Shubnikov noted that Curie's papers were "extremely concise", a style that did not lend itself to the general the acceptance of ideas that were before their time. He forecast that future generations would need to finalize Curie's ideas" (Shubnikov, 1988). At the same time, Shubnikov, with Koptsik argued that the Curie principle is part of a tradition, in that it is a generalization of the principles of his predecessors, Neumann and Minnegerode. This is true only in part. In fact, there is a vast difference between the scope of Curie's vision that expanded the significance of symmetry to all natural phenomena and the observations of Neumann and Minnegerode that were restricted to crystals. While, Curie is today rightly recognized as the forefather of the modern crystal physics, which is based entirely on symmetry laws, his ideas on symmetry in nature have penetrated into all branches of modern science.

5. Curtis Schuh, his *Biobibliography*, and Companion *History*

5.1. Curtis P. Schuh (1959-2007)

As we were working on the Shafranovskii translation, we became aware of three unfinished and unpublished documents on the website archives.org by Curtis P. Schuh (Figure 5): *Mineralogy & Crystallography: An Annotated Biobibliography of Books Published 1469 Through 1919, Volumes I & II* (Schuh, 2007a,b), as well as *Mineralogy & Crystallography: On the History of These Sciences From Beginnings Through 1919* (Schuh, 2007c). The *Biobibliography* has been incorporated into the Biographical Archive of the Mineralogical Record (2012). Schuh was an independent scholar working in Tucson, Arizona. He describes his 561 history based on the most complete bibliography of sources ever assembled (1562 pages) as a "derivative" study that no "true" historian would write. This is false. Though incomplete, it will have a lasting impact on future research in the history of crystallography for generations to come.

Curtis Schuh died prematurely in 2007. A sketch of his life was recorded in *The Mineralogical Record* by its editor and Schuh's friend, Wendell E. Wislon (2007, 2012). The following facts of Schuh's life were taken from Wilson's obituary, and also from an entry on the website Find a Grave by Bill Carr (2008).

Curtis Paul Schuh was born in Boulder, Colorado in 1959 and raised in the Denver area. After he graduated from high school, his father, a newly retired IRS agent, moved the family to Tucson. Schuh studied engineering and mathematics at the University of Arizona, earning three Bachelor of Science degrees. Subsequently, he worked in the field of computer support for a number of organizations in the Tucson area.

In both Colorado and Arizona, Schuh was fortunate to have found concerned and dedicated mentors in the mineralogy community who shared their love of minerals and books about minerals. The library of rare crystallography volumes belonging to Richard Bideaux, the owner of a local mineralogy shop in Tucson, inspired the preparation of a comprehensive bibliography of mineralogy and crystallography. The *Biobibliography* is dedicated to Bideaux who encouraged this decades-long undertaking. Schuh did not anticipate at the outset that he was embarking upon a lifelong project.

Schuh lived a quiet, solitary life of scholarship. Ill at age 48, Curtis Schuh ended his life in the Arizona desert. His abandoned car was found. He left a note claiming that "my body will never be found." It has not been.

We are grateful that before his death Schuh left behind electronic copies of his masterworks, freely available to anyone wishing to benefit from his labors (Schuh, 2007a,b,c).

There is no better way to appreciate the detail of Schuh than to download his documents (617 megabytes) and explore for one's self. Short of direct inspection, what can we say here?

Figure 5. Curtis P. Schuh (2005). Photograph courtesy of Wendell E. Wilson (2012).

5.2. Biobibliography

The *Biobibliography* has too many entries to count accurately. Figure 8 shows the first and last scientists illustrated, Abildgaard and Zittel. If an image of a significant survives, chances are very good that it can be found here.

Schuh's *Biobibliography* and *History* enable translation of Shafranovskii more than any other resource. For instance, Shafranovskii relies heavily the history of crystallography by C. M. Marx. What was this book? What can we learn about it short of locating a copy and reading it? Here is what Schuh says about this volume, the 3255rd entry of some 5170 likewise described:

> 3255. German, 1825. Geschichte Der Crystallkunde von Dr. C.M. Marx, Professor der Physik und Chemie in Braunschweig. [rule] Mit neun schwarzen Kupfertafeln und einer colorirten. [rule] Carlsruhe und Baden. D.R. Marx'sche Buchhandlung. [rule] 1825. Gedruckt bei Friedrich Bieweg und Sohn in Braunschweig. 8°: p7 1-198 206; 165?.; [i]-xiv, [2], [1]-313, [3] p., 10 plates (one folding and colored). Page size: 185 x 115 mm.
>
> Contents: [i-ii], Title page, verso blank.; [iii], Dedication to Count von Schmidt–Phiseldeck.; [iv], Blank.; [v-xii], Preface—signed Carl Michael Marx, 16 May 1825.; [xiii]-

xiv, "Uebersicht des Inhalts."; 1, "Geschichte der Crystallkunde."; [2], Quotation from Goethe concerning colors.; [3]-297, Text.; [298]-301, "Rückblick."; [302]- 309, "Zusätze."; [310]-313, "Namen–Verzeichniß."; [1 pg], "Berichtigungen."; [1 pg], "Abbildungen."; [1 pg], Blank.; [At end], 10 plates (one folding and hand-colored).

Figure 6. *Biobibliography* from Abildgaard to Zittel. Left: Peder Christian Abildgaard (1740-1801) founded the Veterinary School of Copenhagen but earns his place in Schuh for describing Cryolite from Greenland. Right: Karl Alfred von Zittel (1839-1904) served on the Geological Survey of Austria and rose to the Presidency of the Royal Bavarian Academy of Sciences.

Very rare. A highly respected work that develops an understanding of concepts in what was then modern crystallography through historical perspective. As a result, the book covers the history of crystallography from ancient times to 1824. The development is told by describing the contributions of the individuals in chronological order. The text is divided into six sections, each representing a specific time period. The first covers the ancient Greek and Roman researches. The others span (2) Albertus Magnus to Robert Boyle, (3) Nicolaus Steno to Johann Henckel, (4) Carl Linneaus to Jean Baptiste Louis Romé de l'Isle, (5) René Just Haüy to Henry James Brooke, and (6) Abraham Gotthelf Kästner to Friedrich Mohs. The name index lists about 300 researches [sic], whose contributions are described in the text. The plates illustrate various concepts brought forth in the discussion by reproducing recognizable figures from important crystallographic works.

Facsimile reprint, 1970: Geschichte...Wiesbaden Dr. Martin Sändig oHG. 8°: [i]-xiv, [2], [1]-313, [2], [1] blank p., 10 plates (one folding and colored). Photographic reprint of the original edition with a modified title page. ISBN 3500220002. References: BL: [726.c.34.].

Of direct relevance are the passages from Shafranovskii that Schuh has already translated. On Lomonsov's doctoral dissertation Shafranovskii worte, "His conceptions of the structure of crystals formulated in this dissertation are so significant that the year this dissertation was written might well be considered the origin of Russian scientific crystallography" (Grigorev & Shafranovskii, 1949). Regarding the doctoral dissertation of Vernadsky on crystallographic gliding, Shafronovkii says: "Here we find the richest synthesis of data relating to unique deformations of crystals, created as a result of gliding, that is the shifting of separate parts of a crystal along straight lines while preserving the volume, weight, and homogeneity of matter. Vernadsky revealed the connection between the planes of gliding, the crystalline facets and elements of symmetry. Here for the first time, he underlined the need to make several qualifications in our conceptions about the complete homogeneity of crystalline polyhedra in connection with changes in their physical features in their surface state. According to this idea, crystals are viewed not as abstract geometrical systems, but as real physical bodies (Shafranovskii, 1980)."

Perhaps you have wondered how many volumes comprised the *Materialy dlia Mineralogii Rossii* (1852) of Koksharov, another Shafranovskii favorite? Here is the answer which corrects a Shafranovskii pecadillo:

6 vols. plus atlas. [vol 1: 1852] 8°: [6], I-III, [1] blank, [1]-412, [4] p., illus. [vol 2: 1855] 8°: [4], [1]- 339, [1] blank, [4] p., illus. (Page numbers of the first signiature are reversed). [vol 3: 1858] 8°: [6], [1]-426, [4] p., illus. [vol 4: 1862] 8°: [4], 515, [5] p. [vol 5: 1866] 8°: [2], 373, [3] p., plates LXXV-LXXVII. [Atlas] 4°: 1-4 p., 74 plates (numbered I-LXXIV).

The bibliography of this Russian edition is difficult because of the rarity of the work. Contrary to what Sinkankas (1993) states this Russian edition did not exceed volume five as a separate publication, and contrary to what Grigoriev & Shafranovskii (1949) state volume six did not appear as a separate volume. Instead it made an appearance as an article in the *Gornoi Zhurnal*. In addition the plates are numbered I- LXXVII. In the copy examined, plates LXXV-LXXVII were bound in at the end of volume five and not included in the Atlas proper. Page size: 225 x 148 mm.

Schuh displays such an obsessive commitment getting the facts right that it is hard not to cheer him on in his solitary and unrewarded work.

Care to evaluate early editions of Giorgio Agricola's *De Re Metallica*, one of the most influential works of metallurgy? Now you can (Figure 7). And, is there a Polish edition, should you prefer it? Yes there is.

Figure 7. *De Re Metallica* by Giorgio Agricola. From upper left to lower right: Latin, 1530; Italian, 1550; German, 1557; German, 1580; Italian, 1563; Latin, 1657. See Schuh for many others editions and citations.

5.3. History

The *History* is labeled "(Rough Notes)". We would be grateful for the ability to produce "rough notes" mostly complete and so remarkably refined. Nevertheless, the *History* is incomplete. This is manifest as sections marked for insertion, sections taken verbatim from other sources, but always set-off with **"REWORK"** as a warning, and sections that were delivered directly from machine translators without refinement (In fact, Schuh was engaged in writing machine translating software, presumably to assist him in this work (Wilson, 2004, 2012)).

Schuh's *History* begins in pre-history, 25,000 years ago when humans first learned to distinguish quartz-rich flint rock from softer stones. He then discusses the ancients. Treatment of Islamic scholars is especially comprehensive. While Shafranovskii writes of the importance of al-Biruni's gemology, we learn from Schuh that this Persian Shiite scholar loathed Arabs, mined the emerald riches of the now lost Mount Muqattam, and made remarkably accurate measurements of specific gravity in the 11th century. He reviews the contributions and biographies of some three-dozen other Muslim mineralogists, emphasizing the curative properties of minerals purported in medieval texts as well as the use of minerals as poisons.

Chapter 5 covers physical crystallography. We read carefully the passages associated with Malus, Arago, Brewster, and Biot, pioneers in crystal optics whose work we have previously studied in detail (Kahr & McBride, 1991; Kahr & Claborn, 2008; Shtukenberg & Punin, 2007, Kahr & Arteaga, 2012). From these circumscribed aspects of the history of crystallography that we know best, we can declare that Schuh's understanding is accurate and deep, his comments nuanced and sophisticated. If we multiply this judgment by the thousands of episodes in the history of crystallography that he knows better than we do, it is hard to imagine how half a lifetime was enough for Schuh.

Certain subjects receive short shrift. For instance, section **8.5 Liquid Crystals**, says precisely this and no more. "Liquid crystals were discovered and studied in the 19th century and were studied primarily by Lehmann, Schenk and Vorlander. By 1908 a theoretical framework for liquid crystals was established and other theoretical studies by E. Bose, Max Born, F. Rhimpf, O. Lehmann, and G. Friedel were made. It was not until after World War II that practical applications for this class of substances were created. Today, every laptop computer, not to mention virtually every digital display utilizes liquid crystals as a display." We cannot know if he intended more for later – or whether this was enough for a subject somewhat tangential to Schuh's main love, mineralogy. We are fortunate to now have excellent liquid crystal histories including *Crystals that Flow* (Sluckin, Dunmur, & Stegemeyer, 2004) containing translations and reproductions of important papers with commentary, Schuh's principle resource for his brief remarks. See also the more accessible general history (Dunmur, Sluckin, 2010).

Section 11.0, "Regional Topographies", has "short histories outlining the development of mineralogy and crystallography in the countries of the world." He means, *all* the countries. He didn't make it through the >200 or so countries and territories, but there are 110 entries including those for Tasmania, the Faroe Islands, and Macedonia (Schuh is the Alexander of crystallography historians – he aspired to conquer the world).

In the chapter on "Mineral Representations", we learn of the first book illustration of a mineral crystal, gypsum from Meydenbach in 1491 (Figure 8, Pober, 1988,) and the fact that some minerals illustrated themselves – *Naturselbstdruckes* – by the direct transfer of mineral texture to paper with ink. Figure 9 shows striations printed from a meteor section (Schreibers, 1820).

Schuh includes chapters on nomenclature, journals, collectors and dealers, instruction, and instrumentation. The latter naturally contains a detailed discussion of the development of the goniometer, from the simplest protractors to the most artfully machined, multi-circle, reflecting instruments. More interesting, however, his discussion of how the goniometer was turned "inside-out", not for the purpose of indexing crystals but rather for constructing accurate plaster or wood models of crystal polyhedra. At first, apparatuses constructed by Fuess (Figure 10) for cutting precise sections from crystals were adopted to cut crystal models. Goldschmidt (Figure 10) published the first description of a device specifically designed to prepare models. His device was refined by Stöber (Figure 10).

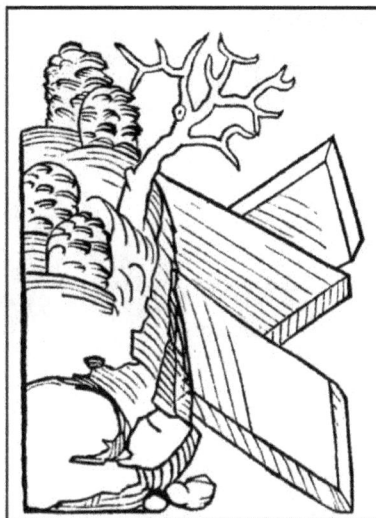

Figure 8. Left: Gypsum, Meydenbach, (1491).

Figure 9. *Naturselbstdruck.* Meteroite slice. Schreibers (1820).

Crystal drawing is surely a lost art. While it is unlikely to be recovered given crystal drawing software, Schuh allows us to appreciate it better than anyone else. Early representations of crystals from nature aimed at capturing the true symmetries, first appeared in the sixteenth century. See Bodt and Linnaeus, Figure 11a,b. Shading was used to capture three-dimensionality. It 1801, Haüy first introduced dashed lines to represent hidden faces (Figure 11c). This became standard. Twinning and concavities appeared in later plates, especially those of Dana in his *System of Mineralogy* (1877) (Figure 11i).

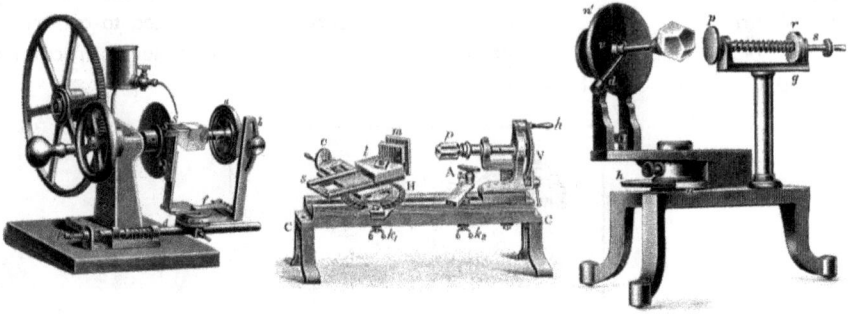

Figure 10. Crystal model making devices. From left to right: Fuess, 1889; Goldschmidt, 1908; Stöber, 1914.

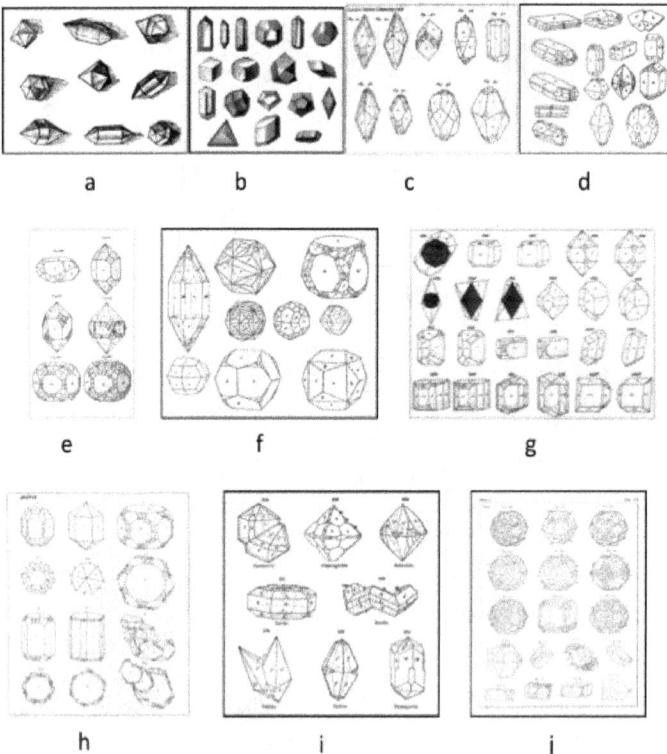

Figure 11. Crystal drawing from Schuh, 2007c. (a) Boodt, 1647; (b) Linneaus, 1768; (c) Haüy, 1801; (d) Dana, 1837; (e) Mohs, 1825; (f) Naumann, 1830; (g) Kopp, 1849; (h) Koksharov, 1853; (i) Dana, 1877; (j) Goldschmidt, 1913.

Section 15.3, "Minerals Illuminated in Colors", is the most luscious. We will indulge in a page of representations of in Figure 12 because we can in an on-line journal without consuming ink.

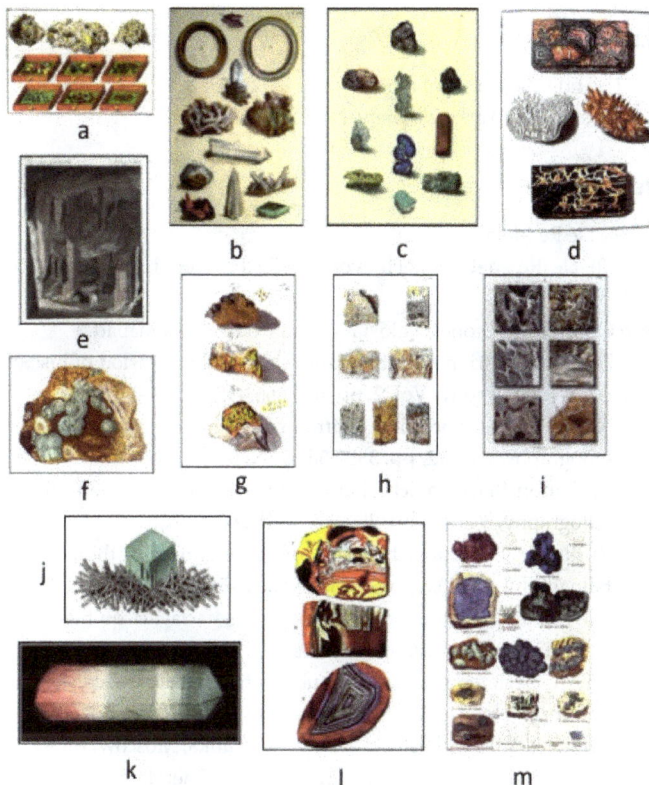

Figure 12. Color mineral illustrations from Schuh, 2007c. (a) Seba, 1734; (b) Knorr, 1754; (c). Rumphius, 1705; (d) Baumesiter, 1791; (e) Bertuch, 1798; (f) Wulfen, 1785; (g) Uibelaker, 1781; (h) Wirsing, 1775; (i) Sowerby, 1804; (j) Patrin, 1801; (k) Wilhelm, 1834; (l) Kurr, 1858; (m) Hamlin, 1873.

The *History* ends with a planned eighteenth chapter. Nothing was written but the chapter title: "18: STUDY OF CAVES". This is a foreboding final phrase. It represents all that remained unsaid by the author's premature death, and all that will remain hidden.

6. Conclusion

The range and detail of Schuh's *History,* supported by the *Biobibliography,* is unlikely to be surpassed for a very long time. It is an extraordinary achievement that deserves wider notice. It is the single narrative in English that we felt was lacking when we began the translation of Shafranovskii. The chasm is filled. The considerable effort of a full translation of Shafranovskii is not longer as urgent (if it ever was). We now terminate our translation project, having introduced English readers to the flavor of Shafranovskii's history, the most complete work of its kind until that of Curtis P. Schuh.

Author details

Bart Kahr and Alexander G. Shtukenberg
Department of Chemistry, New York University, New York City, USA

7. References

Agricola, G. (1556) De Re Metallica, Basil.

Agricola, G. (1950) De Re Metallica, Hoover, H. C.; Hoover, L. H. trans. Dover, ASIN: B0060T2202, New York.

Al-Biruni, (1989) On Precious Stones, Pakistani Hijra Council, Islamabad.

Anonymous (1957) Illarion Illarionovich Shafranovskii (on the occasion of his 50th birthday), Soviet Physics Crystallography, Vol. 2, pp. 665-666, ISSN: 0038-5638.

Anonymous (1967) Illarion Illarionovich Shafranovskii (on his sixtieth birthday), Soviet Physics Crystallography, Vol. 12, pp. 335-336, ISSN: 0038-5638.

Anonymous (1977) Illarion Illarionovich Shafranovskii (on his seventieth birthday), Soviet Physics Crystallography, Vol. 22, pp. 386-387, ISSN: 0038-5638.

Anonymous (1987) Illarion Illarionovich Shafranovskii: eightieth birthday greeting, Soviet Physics Crystallography, Vol. 32, pp. 771, ISSN: 0038-5638.

Bartholinus, E. (1669) Experimenta crystalli islandici disdiaclastici quibus mira & insolita refractio detegitur, Daniel Paulli, Copenhagen.

Bartholinus, E. (1959) Experiments with the double refracting Iceland crystal which led to the discovery of a marvelous and strange refraction, W. Brandt, trans. Westtown, Pa.

Batyushkova, I. V. (Ed.) (1973) The History of Geology, Nauka, Moscow, (in Russian).

Baumesiter, J. A. I. E. von (1763) Naturgeschichte des Mineralreich, Johannes Christian Dieterich, Gota.

Belov, N. V. (1972) Foreword. In: Koslova, O. G. Growth and Morphology of Crystals, Moscow, p. 5-6 (in Russian).

Bertuch, F. J. (1798) Bilderbuch für Kinder, Industre-Comptoir, Weimar.

Bokii, G. B., Shafranovskii, I. I. (1951) From correspondence between E. S. Fedorov and A. Schoenflies. In: Scientific Heritage, Vol. 2. Moscow, pp. 314-343 (in Russian).

Boodt, A. B. de (1647) Gemmarum et Lapidum Historia, Joannis Maire.

Bragg, W. L. (1993) The Development of X-ray Analysis, Dover, ISBN-13: 978-0486673165, New York.

Bravais, A. (1849) Note sur les polyèdres symétriques de la géométrie. Journal de Mathématiques pures et Appliquées, Vol. 14, pp. 137-140.

Bravais, A. (1866) Études crystallographiques, Gauthier-Villars, Paris.

Burke, J. G. (1966) Origins of the Science of Crystals, University of California Press, Berkeley.

Cardano, G. (1562) Somniorum Synesiorum...De Gemmis et Coloribus, Basil.

Carr, B. (2008) Curtis P. Schuh (1959 - 2007) - Find A Grave Memorial, www.findagrave.com/cgi-bin/fg.cgi?page=gr&GRid=25238101.

Curie, M. (1923) Pierre Curie, Macmillan, New York (http://etext.virginia.edu/toc/modeng/public/CurPier.html).

Curie, M. (1963) Pierre Curie, Kellogg, C. & Kellogg, V. trans. Dover, ISBN: 0486201996, New York.

Curie, P. (1884) Sur la Symétrie, Bulletin de la Société mineralogique de France, Vol. VII, pp. 418-454.

Curie, P. (1885a) Sur la formation des cristaux et sur les constants capillaires de leurs differrentes faces, Bulletin de la Société mineralogique de France, Vol. VIII, pp. 145-149.

Curie, P. (1885b) Sur les répétitions et la symétrie, Compte rendus de l'Academie des Sciences, III, pp. 1393-1396.

Curie, P. (1894) Sur la symétrie dans les phénomènes physiques, symétrie d'un champ électrique et d'un champ magnétique, Jouranl de physique, Vol. III, pp. 393-415.

Curie, P. (1889) Dilatation électrique de quartz (avec J. Curie), Journal de Physique, Vol. 8, pp. 149-168.

Curie, P. (1908) Oeuvres de Pierre Curie, Gauthier-Villars, Paris.

Dana, E. S. (1877) Textbook of Mineralogy, John Wiley & Sons, New York.

Dana, J. D. (1837) The System of Mineralogy, Durrie & Peck and Herrick & Noyes, New Haven.

Dorfman, Ya. G. (1974) Universal History of Physics, Moscow (in Russian).

Dunmur, D. A. & Sluckin, T. Soap, science, and flat-screen TVs: A history of liquid crystals, Oxford University Press, ISBN-13: 978-0199549405, Oxford.

Engels, F. (1954) Dialectics of Nature, Foreign Languages Publishing House, Moscow.

Ewald, P. P. ed. (1962) Fifty Years of X-ray Diffraction, International Union for Crystallography, Leiden.

Fabian, E. (1967) Historische Aspekte der Stellung der Kristallographie in den Naturwissenschaften, Freiberger Forschungshefte, Leipzig.

Fedorov, E. S. (1891) The symmetry of real systems of configurations. Zapiskii Mineralogicheskogo Obschestra, Vol. 28, pp. 1-146.

Fedorov, E. S. (1901) Course in Crystallography, Published by R. K. Rikker, Saint-Petersburg (in Russian).

Fedorov, E. S. (1904) From Results of 35th Anniversary, Moscow (in Russian, as cited in Shafranovskii).

Fedorov, E. S. (1906) Perfectionism, Izv. St.-Petersburg Biol. Lab. Vol. 8, N 1-2, 35-36 (in Russian).

Fedorov, E. S. (1917) Nature and Science. Priroda, N 4, 425-432 (in Russian).

Fedorov, E. S. (1955) Introduction to tables for the crystal chemical analysis "Kingdom of crystals". In: Crystallography, No. 3, pp. 5-80 (in Russian).

Figurovsky, N. A. (1969) Outlines of the General History of Chemistry, Nauka, Moscow (in Russian).

Fink, K. J. (1991) Goethe's History of Science, Cambridge University Press, Cambridge; p 163.

Frankenheim, M. L. (1842) System der Krystalle, Von Grass, Barth und Comp. Breslau.

Freudenthal, G. ed. (1990) Études Sur/Studies on Hélène Metzger, Brill, ISBN-13: 978-9004092105 Leiden.

Fuess, R. (1889) Über eine Orientirungsvorrichtung zum Schneiden und Schleifen von Mineralien nach bestimmten Richtungen, Neues Jahrbuch der Mineralogie, part 2, pp. 181-185. ISSN: 0077-7757.

Gadolin, A. V. (1867) Deduction of all crystallographic systems and their subdivisions by means of a single general principle, Ann. Imp. St. Peteresberg Mineralogical Society, Series 2, Vol. 4, pp. 112-200.

Gibbs, J. W. (1878) On the equilibrium of heterogeneous substances, Transactions of the Connecticut Academy of Sciences, Vol. 3, pp. 108-248 & 353-524.

Goldschmidt, V. (1908) Krystall-Modellierapparat 2. Zeitschrift für Krystallographie, Vol. 45, pp. 573-576. ISSN: 0044-2968.

Goldschmidt, V. (1913) Atlas der Krystallformen, Carl Winters Universitätsbuchhandlung, Heidelberg.

Goldschmidt, V. M. (1937) Crystal Chemistry, Khimtheoretisdat, Leningrad (in Russian).

Gordeev, D. I. (1967) The History of Geological Sciences, Vol. 1. Moscow State University Press, Moscow (in Russian).

Grigoriev, D. P. & Shafranovskii, I. I. (1949) Outstanding Russian Mineralogists, Academy of Sciences Press, USSR, Moscow, Leningrad (in Russian).

Groth, P. v. (1906-1919) Chemische Kristallographie, Vol. I-V, Wilhelm Engelmann, Leipzig,.

Groth, P. v. (1926) Entwicklungsgeschichte der mineralogischen Wissenschaften, Julius Springer, Berlin.

Hamlin, A. C. (1873) The Tourmaline, James R. Osgood & Co. London.

Haüy, R. J. (1801) Traité de Minéralogie, Chez Louis, Paris.

Haüy, R. J. (1817) Sur l'électricité produite dans les minéraux à l'aide de la pression, Mémoires de le Museum D'histoire naturelle, Vol. III, pp. 223-228.

Healy, J. F. (1999) Pliny the Elder on Science and Technology, Oxford University Press, ISBN:0-I9-8I4687-6, Oxford, p 187.

Hessel, J. F. C. (1831) Krystallometrie oder Krystallonomie und Krystallographie, E. B. Schwickert, Leipzig.

Kahr, B. (2011) Murder charges against close-pack, Crystal Growth Design, Vol. 11, pp. 4-11.

Kahr, B.; Arteaga, O. (2012) Arago's best paper, ChemPhysChem, Vol. 13, No. 1, pp. 79-88; ISSN: 1439-7641.

Kahr, B.; Claborn, K. (2008) The lives of Malus and his bicentennial law, ChemPhysChem, Vol. 9, No. 1, 43-58, ISSN: 1439-7641.

Kahr, B.; McBride, J. M. (2003) Optically anomalous crystals, Angewandte Chemie International Edition in English, Vol. 31, No. 1, pp. 1-26, ISSN: 1521-3773.

Kedrov, B. M. (1971) The History of Science and Principles of its Research, Voprosy Filosofii, No. 9, 78-89 (in Russian).

Lima-de-Faria, J. (1990) Historical Atlas of Crystallography, Springer, ISBN-13: 978-0792306498, Dordrecht.

Knorr, G. W. (1754) Deliciae Natureae Selectae, Georg Wolfgang Knorr, Nürnberg.

Kobell, F. v. (1856) Die Urzeit der Erde: ein Gedicht, Literarisch-artistische Anstalt, München.

Kobell, F. v. (1864) Geschichte der Mineralogie von 1650-1860, Historische Commission bei der Königl. Academie der Wissenschaften, München.

Koksharov, I. V. (1852) Materialy Dlia Mineralogii Rossii, Saint Petersberg, I. I. Glazunova.

Kopp, H. F. M. (1849) Einleitung in die Krystallographie, Friedrich Vieweg und Sohn, Brauschweig.

Kurr, J. G. von (1858) Das Mineralreich in Bildern, Schreiber und Schill, Stuttgart und Esslingen.

Lemmlein, G. G. (1940) Lomonosov's thoughts on crystals, Works of the Academy of Science History Commission. Linneaus, C. (1768) Systema Naturae, Impensis Direct. Laurentii Salvii, Holmiae.

Lomonosov, Academy of Sciences Press, USSR, Moscow, Leningrad (in Russian).

Lenin, V. I. (1967) Philosophical notebooks, in Collected Works, Political Literature Press, Moscow, pp. 152-153 (in Russian).

Marx, C. M. (1825) Geschichte der Cristallkunde, Friedrich Bieweg und Sohn, Karlsruhe und Baden.

Menshutkin, B. N. (1937) Chemistry and Ways of its Development, Academy of Sciences Press, USSR, Moscow, Leningrad (in Russian).

Metzger, H. (1918) La genèse de la Science des Cristaux, Félix Alcan, Paris.

Meydenbach, J. (1491) Hortus Sanitatis, Mainz.

Mohs, F. (1822) Grundriss der Mineralogie, Arnoldischen Buchhandlung, Dresden.

Mineralogical Record (2012), Bibliographic Archive, http://www.minrec.org/index.asp.

Molinié, P.; Boudia, S. (2009) Mastering picocoulombs in the 1890s: The Curies' quartz-electrometer instrumentaion, and how it shaped early radioactivity history, J. Electrostatics, Vol. 67, pp. 524-530, ISSN: 0304-3886.

Naumann, C. F. (1830) Lehrbuch der reinen und angewandten Krystallographie, F. A. Brockhaus, Leipzig.

Patrin, M. L. (1801) Histoire Naturelle Des Minéraux, L'Imprimerie De Crapelet, Paris.

Pober, S. E. (1988) The first book illustration of a mineral crystal. Matrix, Vol. 1, No. 3, p. 44.

Povarennyh, A. S. (1962) On the question of periodization in the history of mineralogy. In: Outlines of the History of Geological Knowledge, Vol. 10, Academy of Sciences Press, USSR, Moscow (in Russian).

Rumf, G. E. (1705) D'Amboinische Rariteitskamer, François Halma, Amsterdam.

Schoenflies, A. (1891) Kristallsysteme und Kristallstruktur, Teubner Verlag, Leipzig.

Schreibers, K. A. F. R. v. (1820) Beiträge zur Geschichte und Kenntniß meteorischer Stein und Metall Massen und der Erscheinungen welche deren Niederfallen zu begleiten pflegen, J. G. Heubner, Wien.

Schuh, C. P. (2007a) Mineralogy & Crystallography: On the History of These Sciences From Beginnings Through 1919, archives.org: http://www.archive.org/details/History_Mineralogy_2007.

Schuh, C. P. (2007b) Mineralogy & Crystallography: An Annotated Biobibliography of Books Published 1469 through 1919, Volume I, archives.org: http://www.archive.org/details/BioBib_Mineralogy_2007_Vol_1.

Schuh, C. P. (2007c) Mineralogy & Crystallography: An Annotated Biobibliography of Books Published 1469 through 1919, Volume II, archives.org: http://www.archive.org/details/BioBib_Mineralogy_2007_Vol_2.

Seba, A. (1734) Locupletissimi Rerum Naturalium Thesauri, Apud Janssonio-Waesbergios, & J. Wetstenium, & Gul. Smith, Amsterdam.

Shafranovskii, I. I. (1952) Mineral Crystals, Leningrad State Publishing House, Leningrad.

Shafranovskii, I. I. (1957) Pierre Curie – Crystallographer. Papers of the Institute of History of Natural and Engineering Sciences, Moscow, Vol. 19, pp. 84-94.

Shafranovskii, I. I. (1964) Diamonds, Nauka, Moscow, Leningrad (in Russian).

Shafranovskii, I. I. (1962) History of Crystallography in Russia, Leningrad State Publishing House, Leningrad (in Russian).

Shafranovskii, I. I. (1963) Evgraf Stepanovich Fedorov, Academy of Science Press, USSR, Moscow, Leningrad (in Russian).

Shafranovskii, I. I. (1964) Nikolai Ivanovich Koksharov, Nauka, Moscow, Leningrad (in Russian).

Shafranovskii, I. I. (1968a) A. G. Werner: A Famous Mineralogist and Geologist (1749-1817), Nauka, Leningrad (in Russian).

Shafranovskii, I. I. (1968b) Symmetry in Nature, Nedra, Leningrad (in Russian).

Shafronoviskii, I. I. (1972) Nicolaus Steno: A Crystallographer, Geologist, Paleontologist, and Anatomist, Nauka, Leningrad (in Russian).

Shafranovskii, I. I. (1973) Lectures on Crystallomorphology, National Bureau of Standards, Washington D.C.

Shafranovskii, I. I. (1974) Outlines of Mineralogical Crystallography, Nedra, Leningrad (in Russian).

Shafranovskii, I. I. (1975) Kepler's crystallographic ideas and his tract "The six-cornered snowflake", Vistas in Astronomy, Vol. 18, pp. 861-876, ISSN: 0083-6656.

Shafranovskii, I. I. (1978) The History of Crystallography. Vol. I. From Ancient Times to the Beginning of the Nineteenth Century, Nauka, Leningrad (in Russian).

Shafranovskii, I. I. (1980) The History of Crystallography. Vol. II. The Nineteenth Century, Nauka, Leningrad (in Russian).

Shafranovskii, I. I. & Belov, N. V. (1962) In Memoriam: E.S. Fedorov, Fifty years of X-ray diffraction, pp. 341-350, Ewald, P.P. ed. International Union of Crystallographers, ISBN-13: 978- 9027790293, Utrecht.

Shtukenberg, A. G. & Punin, Yu. Optically Anomalous Crystals, (Kahr, B. ed.) Springer, ISBN-13: 978-1-4020-5287-3, Dordrecht.

Shubnikov, A. V. (1946) Dissymmetry. In: Questions in Mineralogy, Geochemistry, and Petrography, Academy of Science Press, USSR, Moscow, Leningrad, pp. 158-163 (in Russian).

Shubnikov, A. V. (1988) On the Works of Pierre Curie on Symmetry, pp. 357-364, in Hargittai, I.; Vainshtein, B. K. eds. Crystal Symmetries: Shubnikov Centennial Papers, Pergamon Press, ISBN: 0080370144 9780080370149 , Oxford.

Shubnikov, A. V. & Kopstik, V. A. (1974) Symmetry in Art and Science, Archard, G. D. trans. Plenum Press, ISBN: 0-306-30759-6, New York.

Sinkankas, J. (1973) Gemology. An Annotated Bibliography, The Scarecrow Press, ISBN-13: 978-0810826526, Metuchen N.J.

Sluckin, T.; Dunmur, D. A. & Stegemeyer, H. eds. (2004) Crystals That Flow: Classic Papers from the History of Liquid Crystals. CRC Press, ISBN-13: 978-0691037448, New York.

Sowerby, J. (1804) British Mineralogy: or, Coloured Figures Intended to Elucidate the Mineralogy of Great Britain, R. Taylor & Co. London.

Spaskii, N. Ya., & Kravtsov, A. G. (1971) Evolution of animals and the symmetry. In: Symmetry in Nature, Leningrad, pp. 367-369 (in Russian).

Stöber, P. (1914), Über einige neue krystallographische Apparate, Zeitschrift für Krystallographie, Vol. 54, pp. 273-288.

Tertsch, H. (1947) Das Geheimnis der Kristallwelt. Roman einer Wissenschaft, Gerlach & Wiedling, Wien.

Tikhomirov, V. V. & Khain, V. B. (1956) Brief Outlines of the History of Geology, Gosgeoltekhisdat, Moscow, (in Russian).

Uibelaker, P. F. (1781) System des Karlsbader Sinters unter Vorstellung schöner und seltener Stücke, Kosten Wolfgang Walthers, Erlangen.

Vernadsky, V. I. (1904) Fundamentals of Crystallography, Moskva Gosuniversitet, Moscow

Vernadsky, V. I. (1928) Tasks of mineralogy in our country (1917-1927), Priroda, No. 1, pp. 21-40 (in Russian).

Vernadsky, V. I. (1957) Problems of biogeochemistry. III. On the state of space in geological phenomena of Earth in relation to scientific growth in the XXth century. Moscow branch of the archive of the Academy of Science, USSR, Vol. 3, as cited in Shafranovskii (1957).

Vernadsky, V. I. (1965) Chemical Structure of the Biosphere of Earth and its Surrounding, Nauka, Moscow (in Russian).

Vernadsky, V. I. (1975) Thoughts of Naturalist. Space and Time in Inanimate and Animate Nature, Nauka, Moscow (in Russian).

Wilhelm, G. T. (1834) Unterhaltungen aus der Naturgeschichte. Des Mineralreichs, J. A. Schloßers Buch und Kunsthandlung, Augsburg.

Wilson, W. E. (2007) The Mineralogical Record, Vol. 38, No. 6, pp. 426-427. ISSN: 0026-4628.

Wislon, W. E. (2012) Obituary, Curtis P. Schuh, Mineralogical Record Bibliographical Archive, at www.mineralogicalrecord.com.

Wirsing, A. L. (1775) Marmora et adfines Aliquos Lapides, Kupferstecher und Kunsthandler, Nürnberg.

Wulfen, F. X. von (1785) Abhandlung vom kärnthnerischen Bleyspate, Johann Paul Kraußischen, Wien.

Wulff, Yu. (G.) V. (1952) Selected Papers in Crystal Physics and Crystallography, Gostekhisdat, Moscow, Leningrad (in Russian).

Experimental Techniques

X-Ray N-Beam Takagi-Taupin Dynamical Theory and N-Beam Pinhole Topographs Experimentally Obtained and Computer-Simulated

Kouhei Okitsu, Yasuhiko Imai and Yoshitaka Yoda

Additional information is available at the end of the chapter

1. Introduction

In the field of X-ray crystal structure analysis, while the absolute values of structure factors are directly observed, phase information is lost in general. However, this problem (phase problem) has been overcome mainly by the direct method developed by Hauptmann and Karle except for protein crystals. In the case of protein crystal structure analysis, the isomorphous replacement method and/or anomalous dispersion method are mainly used to solve the phase problem. Phasing the structure factors is sometimes the most difficult process in protein crystallography.

On the other hand, It has been recognized for many years since the suggestion by Lipscomb in 1949 [12] that the phase information can be physically extracted, at least in principle, from X-ray diffraction profiles of three-beam cases in which transmitted and two reflected beams are simultaneously strong in the crystal. This suggestion was verified by Colella [3] that stimulated many authors [2, 4, 5, 21, 22] and let them investigate the multiple-beam (n-beam) method to solve the phase problem in protein crystallography.

The most primitive n-beam diffraction is the cases $n = 3$. The shape of three-beam rocking curve simply depends on the triplet phase invariant. In the case of protein crystallography, however, it is extremely difficult to realize such three-beam cases that transmitted and only two reflected beams are strong in the crystal, which is due to the extremely high density of reciprocal lattice nodes owing to the large size of unit cell of the crystal. Therefore, X-ray n-beam dynamical diffraction theory is necessary to solve the phase problem in protein crystallography. The Ewald-Laue (E-L) dynamical diffraction theory [7, 11] was extended to the three-beam cases in the late 1960's [8–10]. The numerical method to solve the n-beam ($n \geq 3$) E-L theory was given by Colella [3]. Colella's method [3] to solve the n-beam E-L theory is applicable only to the case of crystals with planar surfaces.

On the other hand, Okitsu and his coauthors [13, 14, 16, 17] extended the Takagi-Taupin (T-T) dynamical diffraction theory [18–20] to n-beam cases ($n \in \{3, 4, 5, 6, 8, 12\}$) and presented a numerical method to solve the theory. They showed six-beam pinhole topographs experimentally obtained and computer-simulated based on the n-beam T-T equation, between which excellent agreements were found. In reference [16], it was shown that the n-beam T-T equation can deal with X-ray wave field in an arbitrary-shaped crystal.

In the n-beam method to solve the phase problem in the protein crystal structure analysis, one of the difficulty is the shape of the crystal which is complex in general. Then, the above advantage of n-beam T-T equation over the E-L dynamical theory is important. The present authors have derived an n-beam T-T equation applicable for arbitrary number of n, which will be published elsewhere.

The n-beam T-T equation was derived in references [13, 17] from Takagi's fundamental equation of the dynamical theory [19]. In section 2 of the present chapter, however, the n-beam E-L theory is described at first. The n-beam T-T equation is derived by Fourier-transforming the n-beam E-L dynamical theory. Then, it is also described that the E-L theory can be derived from the T-T equation. This reveals a simple relation between the E-L and T-T formulations of X-ray dynamical diffraction theory. This equivalence between the E-L and T-T formulations has been implicitly recognized for many years but is explicitly described for the first time. In section 5, experimentally obtained and computer-simulated pinhole topographs are shown for $n \in \{3, 4, 5, 6, 8, 12\}$, which verifies the theory and the computer algorithm to solve it.

2. Derivation of the n-beam Takagi-Taupin equation

2.1. Description of the n-beam Ewald-Laue dynamical diffraction theory

The fundamental equation with the E-L formulation is given by [1, 2]

$$\frac{k_i^2 - K^2}{k_i^2} \mathcal{D}_i = \sum \chi_{h_i - h_j} \left[\mathcal{D}_j \right]_{\perp \mathbf{k}_i}. \tag{1}$$

Here, k_i is wavenumber of the ith numbered Bloch wave whose wave vector is $\mathbf{k}_0 + \mathbf{h}_i$ where $\mathbf{k}_0$ is the wave vector of the forward-diffracted wave in the crystal, $K(= 1/\lambda)$ is the wavenumber of X-rays in vacuum, $\mathcal{D}_i$ and $\mathcal{D}_j$ are complex amplitude vectors of the ith and jth numbered Bloch waves, $\sum$ is an infinite summation for all combinations of i and j, $\chi_{h_i - h_j}$ is Fourier coefficient of electric susceptibility and $[\mathcal{D}_j]_{\perp \mathbf{k}_i}$ is component vector of $\mathcal{D}_j$ perpendicular to $\mathbf{k}_i$, respectively.

By applying an approximation that $k_i + K \approx 2k_i$ to (1), the following equation is obtained,

$$\xi_i \mathcal{D}_i = \frac{K}{2} \sum \chi_{h_i - h_j} \left[\mathcal{D}_j \right]_{\perp \mathbf{k}_i}, \tag{2}$$
$$\text{where } \xi_i = k_i - K.$$

Let the electric displacement vector $\mathcal{D}_i$ be represented by a linear combination of scalar amplitudes as follows:

$$\mathcal{D}_i = \mathcal{D}_i^{(0)} \mathbf{e}_i^{(0)} + \mathcal{D}_i^{(1)} \mathbf{e}_i^{(1)}.$$

Here, $e_i^{(0)}$ and $e_i^{(1)}$ are unit vectors perpendicular to s_i, where s_i is a unit vector parallel to k_i. s_i, $e_i^{(0)}$ and $e_i^{(1)}$ construct a right-handed orthogonal system in this order. (2) can be described as follows:

$$\xi \cos \Theta_B \mathcal{D}_i^{(l)} = -K \left(S_{i,0}^{(0)} \beta^{(0)} + S_{i,0}^{(1)} \beta^{(1)} \right) \mathcal{D}_i^{(l)}$$

$$+ \frac{K}{2} \sum_{j=0}^{n-1} \sum_{m=0}^{1} C_{i,j}^{(l,m)} \chi_{h_i - h_j} \mathcal{D}_j^{(m)}, \tag{3}$$

$$\text{where,} \quad i,j \in \{0, 1, \cdots, n-1\}, n \in \{3, 4, 5, 6, 8, 12\},$$

$$l, m \in \{0, 1\}.$$

Here, S and C are polarization factors defined by

$$e_j^{(m)} = S_{i,j}^{(m)} s_i + C_{i,j}^{(0,m)} e_i^{(0)} + C_{i,j}^{(1,m)} e_i^{(1)}, \tag{4}$$

where i and j are ordinal numbers of waves ($i, j \in \{0, 1, 2, \cdots, n-1\}$) and l and m are ordinal numbers of polarization state ($l, m \in \{0, 1\}$). When deriving (3) from (2), all reciprocal lattice nodes lying on the surface of Ewald sphere are assumed to be on a circle in reciprocal space. Then number of waves n are limited to be $n \in \{3, 4, 5, 6, 8, 12\}$ even in the case of cubic crystals with the highest symmetry. Θ_B is the angle spanned by $\overrightarrow{PQ}$ and k_i which is an identical value for every i ($i \in \{0, 1, 2, \cdots, n-1\}$), where P and Q are centers of the Ewald sphere and the circle on which the reciprocal lattice nodes lie, respectively. ξ is such a value that

$$\overrightarrow{P_1 P_1'} = -\xi \overrightarrow{PQ} / \left| \overrightarrow{PQ} \right|, \tag{5}$$

where P_1' is the common initial point of k_i [whose terminal points are H_i ($i \in \{0, 1, \cdots, n-1\}$)] and P_1 is a point on the sphere whose distance from the origin O of reciprocal space is K. Hereafter, this surface of sphere is approximated as a plane whose distance from O is K in the vicinity of the Laue point La whose distance from H_i ($i \in \{0, 1, \cdots, n-1\}$) is the identical value K. For description in the next section, it is described here that $\overrightarrow{P_1 P_1'}$ is represented by a linear combination of s_i, $e_i^{(0)}$ and $e_i^{(1)}$ as follows:

$$\overrightarrow{P_1 P_1'} = -\xi \left(\cos \Theta_B s_i + \eta_i^{(0)} \sin \Theta_B e_i^{(0)} + \eta_i^{(1)} \sin \Theta_B e_i^{(1)} \right).$$

P_1 is such a point that

$$\overrightarrow{P_1 La} = K \left(\beta^{(0)} e_0^{(0)} + \beta^{(1)} e_0^{(1)} \right). \tag{6}$$

(2) and (3) can also be represented using matrices and vector as follows:

$$\xi \cos \Theta_B E \mathcal{D} = A \mathcal{D}. \tag{7}$$

Here, $\mathbf{E}$ is a unit matrix of size $2n$, $\mathcal{D}$ is a amplitude column vector of size $2n$ whose qth element is $D_j^{(m)}$ $(q = 2j + m + 1)$ and $\mathbf{A}$ is a square matrix of size $2n$ whose element $a_{p,q}$ is given by

$$a_{p,q} = \frac{K}{2} \chi_{h_i - h_j} C_{i,j}^{(l,m)} - \delta_{p,q} K \left(S_{i,0}^{(0)} \beta^{(0)} + S_{i,0}^{(1)} \beta^{(1)} \right).$$

Here, $p = 2i + l + 1$ and $\delta_{p,q}$ is Kronecker delta. $2n$ couples of ξ and $\mathcal{D}$ can be obtained by solving eigenvalue-eigenvector problem of (7). This problem was solved by Colella [3] for the first time. Dispersion surfaces on which the initial point P_1' of wave vectors of Bloch waves should be, is given as $2n$ sets of eigenvalues for (7).

2.2. Derivation of the n-beam Takagi-Taupin equation from the Ewald-Laue theory

In this section, the n-beam theory of T-T formulation is derived by Fourier-transforming the n-beam E-L theory described by (3).

A general solution of dynamical diffraction theory is considered to be coherent superposition of Bloch plane-wave system when X-ray wave field $\tilde{\mathbf{D}}(\mathbf{r})$ is given as follows:

$$\tilde{\mathbf{D}}(\mathbf{r}) = \sum_{i=0}^{n-1} \sum_{l=0}^{1} \mathbf{e}_i^{(l)} D_i^{(l)}(\mathbf{r}) \exp\left(-i2\pi \overrightarrow{LaH_i} \cdot \mathbf{r} \right), \tag{8}$$

where $\mathbf{r}$ is the location vector. For the following description, $\mathbf{r}$ is described by a linear combination of $\mathbf{s}_i$, $\mathbf{e}_i^{(0)}$ and $\mathbf{e}_i^{(1)}$ as follows,

$$\mathbf{r} = s_i \mathbf{s}_i + e_i^{(0)} \mathbf{e}_i^{(0)} + e_i^{(1)} \mathbf{e}_i^{(1)}. \tag{9}$$

The amplitude of the ith component wave whose polarization state is l is described as,

$$\mathcal{D}_i^{(l)}(\Delta \mathbf{k}) \exp\left(-i2\pi \overrightarrow{P_1'H_i} \cdot \mathbf{r} \right) = \mathcal{D}_i^{(l)}(\Delta \mathbf{k}) \exp\left(-i2\pi \Delta \mathbf{k} \cdot \mathbf{r} \right) \exp\left(-i2\pi \overrightarrow{LaH_i} \cdot \mathbf{r} \right),$$

$$\text{where } \Delta \mathbf{k} = \overrightarrow{P_1'La}.$$

In this section, the amplitude of plane wave whose wave vector is $\mathbf{k}_i$ and polarization state is l is denoted by $\mathcal{D}_i^{(l)}(\Delta \mathbf{k})$ in place of $\mathcal{D}_i^{(l)}$ in order to clarify this value depends on $\Delta \mathbf{k}$. $D_i^{(l)}(\mathbf{r})$ in (8) is represented by superposing coherently $\mathcal{D}_i^{(l)}(\Delta \mathbf{k})$ as follows:

$$D_i^{(l)}(\mathbf{r}) = \int_{\Delta \mathbf{k}}^{D.S.} \mathcal{D}_i^{(l)}(\Delta \mathbf{k}) \exp\left(-i2\pi \Delta \mathbf{k} \cdot \mathbf{r} \right) dS_k. \tag{10}$$

Substituting (4) with $j = 0$, (5), (6) and (9) into (10),

$$D_i^{(l)}(\mathbf{r}) = \int_{\Delta \mathbf{k}}^{D.S.} \mathcal{D}_i^{(l)}(\Delta \mathbf{k})$$
$$\times \exp\left\{ -i2\pi \left[\left(\xi \cos \Theta_B + K\beta^{(0)} S_{i,0}^{(0)} + K\beta^{(1)} S_{i,0}^{(1)} \right) s_i + T_i(\xi, e_i^{(0)}, e_i^{(1)}) \right] \right\} dS_k. \tag{11}$$

X-Ray N-Beam Takagi-Taupin Dynamical Theory and N-Beam Pinhole Topographs Experimentally Obtained and Computer-Simulated

43

Here, $\int_{\Delta k}^{D.S.} dS_k$ means an integration over the dispersion surfaces in reciprocal space and $T_i(\xi, e_i^{(0)}, e_i^{(1)})$ is a term that does not depend on s_i. $\partial D_i^{(l)}(\mathbf{r}) / \partial s_i$ can be calculated as follows:

$$\frac{\partial}{\partial s_i} D_i^{(l)}(\mathbf{r}) = \frac{\partial}{\partial s_i} \int_{\Delta k}^{D.S.} \mathcal{D}_i^{(l)}(\Delta k) \exp\left(-i2\pi \Delta \mathbf{k} \cdot \mathbf{r}\right) dS_k$$

$$= \int_{\Delta k}^{D.S.} \frac{\partial}{\partial s_i} \left[\mathcal{D}_i^{(l)}(\Delta k) \exp\left(-i2\pi \Delta \mathbf{k} \cdot \mathbf{r}\right) \right] dS_k$$

$$= -i2\pi \int_{\Delta k}^{D.S.} \left[\xi \cos \Theta_B + K \left(S_{i,0}^{(0)} \beta^{(0)} + S_{i,0}^{(1)} \beta^{(1)} \right) \right] \mathcal{D}_i^{(l)}(\Delta k) \exp\left(-i2\pi \Delta \mathbf{k} \cdot \mathbf{r}\right) dS_k.$$

$$(12)$$

Substituting (3) into (12),

$$\frac{\partial}{\partial s_i} D_i^{(l)}(\mathbf{r}) = -i\pi K \int_{\Delta k}^{D.S.} \sum_{j=0}^{n-1} \sum_{m=0}^{1} C_{i,j}^{(l,m)} \chi_{h_i - h_j} \mathcal{D}_j^{(m)}(\Delta k) \exp\left(-i2\pi \Delta \mathbf{k} \cdot \mathbf{r}\right) dS_k$$

$$= -i\pi K \sum_{j=0}^{n-1} \sum_{m=0}^{1} C_{i,j}^{(l,m)} \chi_{h_i - h_j} \int_{\Delta k}^{D.S.} \mathcal{D}_j^{(m)}(\Delta k) \exp\left(-i2\pi \Delta \mathbf{k} \cdot \mathbf{r}\right) dS_k. \quad (13)$$

Incidentally, when the crystal is perfect, the electric susceptibility $\chi(\mathbf{r})$ is represented by Fourier series as $\chi(\mathbf{r}) = \sum_{h_i} \chi_{h_i} \exp[-i2\pi \mathbf{h}_i \cdot \mathbf{r}]$. However, when the crystal has a lattice displacement field of $\mathbf{u}(\mathbf{r})$, the electric susceptibility is approximately given by $\chi[\mathbf{r} - \mathbf{u}(\mathbf{r})]$ and represented by Fourier series as follows,

$$\chi[\mathbf{r} - \mathbf{u}(\mathbf{r})] = \sum_{h_i} \chi_{h_i} \exp[i2\pi \mathbf{h}_i \cdot \mathbf{u}(\mathbf{r})] \exp(-i2\pi \mathbf{h}_i \cdot \mathbf{r}).$$

Then, in the case of crystal with a lattice displacement field of $\mathbf{u}(\mathbf{r})$, $\chi_{h_i - h_j}$ can be replaced by $\chi_{h_i - h_j} \exp[i2\pi (\mathbf{h}_i - \mathbf{h}_j) \cdot \mathbf{u}(\mathbf{r})]$. Therefore, the following equation is obtained from (13),

$$\frac{\partial}{\partial s_i} D_i^{(l)}(\mathbf{r}) = -i\pi K \sum_{j=0}^{n-1} \sum_{m=0}^{1} C_{i,j}^{(l,m)} \chi_{h_i - h_j} \exp\left[i2\pi (\mathbf{h}_i - \mathbf{h}_j) \cdot \mathbf{u}(\mathbf{r})\right] D_j^{(m)}(\mathbf{r}), \quad (14)$$

where, $\quad i, j \in \{0, 1, \cdots, n-1\}, n \in \{3, 4, 5, 6, 8, 12\},$

$\quad l, m \in \{0, 1\}.$

The above equation is nothing but the n-beam T-T equation that appeared as eq. (4) in reference [17].

2.3. Derivation of the n-beam E-L dynamical theory from the T-T equation

In this section, it is described that the n-beam E-L theory given by (3) can be derived from the n-beam T-T equation (14).

When plane-wave X-rays are incident on the crystal to excite $2n$ tie points on the dispersion surfaces, each Bloch plane-wave system is described by

$$\check{D} = \sum_{i=0}^{n-1} \sum_{l=0}^{1} \mathbf{e}_i^{(l)} \mathcal{D}_i^{(l)} \exp\left(-i2\pi\Delta\mathbf{k}\cdot\mathbf{r}\right) \exp\left(-i2\pi\overrightarrow{LaH_i}\cdot\mathbf{r}\right).$$

Even when $D_i^{(l)}(\mathbf{r}) = \mathcal{D}_i^{(l)} \exp(-i2\pi\mathbf{k}_i^{(l)}\cdot\mathbf{r})$, $D_i^{(l)}(\mathbf{r})$ should satisfy (14) with $\mathbf{u}(\mathbf{r}) = 0$,

$$\frac{\partial}{\partial s_i}\left[\mathcal{D}_i^{(l)}\exp\left(-i2\pi\Delta\mathbf{k}\cdot\mathbf{r}\right)\right] = -i\pi K \sum_{j=0}^{n-1} \sum_{m=0}^{1} C_{i,j}^{(l,m)} \chi_{h_i-h_j}\left[\mathcal{D}_j^{(m)}\exp\left(-i2\pi\Delta\mathbf{k}\cdot\mathbf{r}\right)\right]. \quad (15)$$

Applying the same procedure as used when deriving (11),

$$\frac{\partial}{\partial s_i}\left[\mathcal{D}_i^{(l)}\exp\left(-i2\pi\Delta\mathbf{k}\cdot\mathbf{r}\right)\right]$$

$$= \mathcal{D}_i^{(l)}\frac{\partial}{\partial s_i}\exp\left\{-i2\pi\left[\left(\xi\cos\Theta_B + K\beta^{(0)}S_{i,0}^{(0)} + K\beta^{(1)}S_{i,0}^{(1)}\right)s_i + T_i(\xi, e_i^{(0)}, e_i^{(1)})\right]\right\}$$

$$= -i2\pi\left(\xi\cos\Theta_B + K\beta^{(0)}S_{i,0}^{(0)} + K\beta^{(1)}S_{i,0}^{(1)}\right)\left[\mathcal{D}_i^{(l)}\exp\left(-i2\pi\Delta\mathbf{k}\cdot\mathbf{r}\right)\right]. \quad (16)$$

Comparing (15) and (16), the same equation as (3) is obtained. The equivalence between the n-beam E-L and T-T X-ray dynamical diffraction theories ($n \in \{3, 4, 5, 6, 8, 12\}$) described by a Fourier transform as defined by (10) is verified. As far as the present authors know, description on this equivalence between the E-L and T-T dynamical diffraction theories for two-beam case is found just in section 11.3 of Authier's book [1].

3. Algorithm to solve the theory

Figure 1(a) and 1(b) are schematic drawings for explanation of the algorithm to solve the n-beam T-T equation (14) for a six-beam case whose computer-simulated and experimentally obtained results are shown in Figure 9 of the present chapter. Vectors $\overrightarrow{R_i^{(0)}R^{(1)}}$ in Figure 1(a) are parallel to $\mathbf{s}_i$. When the length of $\overrightarrow{R_i^{(0)}R^{(1)}}$ is sufficiently small compared with the extinction length $-1/(\chi_0 K)$ of the forward diffraction, The T-T equation (14) can be approximated by

$$\frac{D_i^{(l)}(R^{(1)}) - D_i^{(l)}(R_i^{(0)})}{\left|\overrightarrow{R_i^{(0)}R^{(1)}}\right|}$$

$$= -i\pi K \sum_{j=0}^{n-1} \sum_{m=0}^{1}\left\{\chi_{h_i-h_j}\exp\left[i2\pi\left(\mathbf{h}_i - \mathbf{h}_j\right)\cdot\mathbf{u}(Rm_i)\right]\right.$$

$$\times C_{i,j}^{(l,m)}\left[D_j^{(m)}(R_i^{(0)}) + D_j^{(m)}(R^{(1)})\right]/2\right\}. \quad (17)$$

X-Ray N-Beam Takagi-Taupin Dynamical Theory and N-Beam Pinhole Topographs Experimentally
Obtained and Computer-Simulated

45

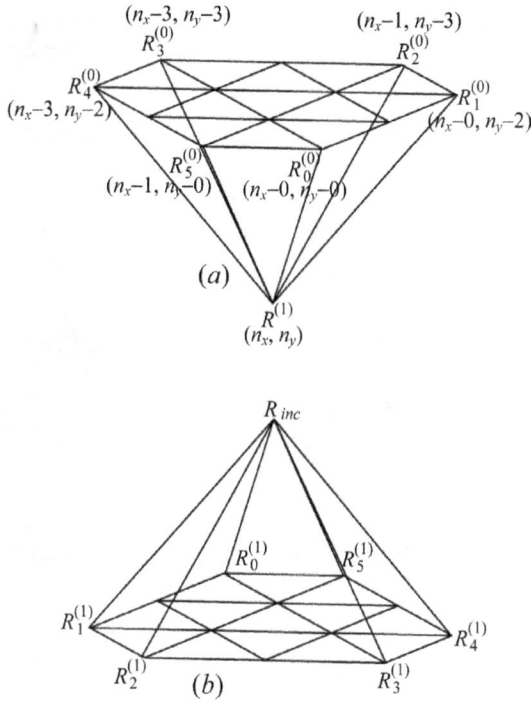

Figure 1. This Figure shows small hexagonal pyramids used when solving the T-T equation (14) in a six-beam case whose results are shown in Figure 9.

The above equation (17) can be described using matrix and vectors as follows:

$$\mathbf{A} = \mathbf{BD} \tag{18}$$

where, $a_p = -\mathrm{i}\dfrac{1}{2}\pi K \displaystyle\sum_{j=0}^{n-1}\sum_{m=0}^{1}\chi_{h_i-h_j}\exp\left[\mathrm{i}2\pi(\mathbf{h}_i-\mathbf{h}_j)\cdot\mathbf{u}(Rm_i)\right]C_{i,j}^{(l,m)}D_j^{(m)}(R_i^{(0)})$

$\qquad\qquad + \dfrac{D_i^{(l)}(R_i^{(0)})}{\left|\overrightarrow{R_i^{(0)}R^{(1)}}\right|},$

$\quad b_{p,q} = \mathrm{i}\dfrac{1}{2}\pi K\chi_{h_i-h_j}\exp\left[\mathrm{i}2\pi(\mathbf{h}_i-\mathbf{h}_j)\cdot\mathbf{u}(Rm_i)\right]C_{i,j}^{(l,m)} + \dfrac{\delta_{p,q}}{\left|\overrightarrow{R_i^{(0)}R^{(1)}}\right|},$

$\quad d_q = D_j^{(m)}(R^{(1)}),$

$\quad p = 2i+l+1,$

$\quad q = 2j+m+1.$

Here, $\mathbf{A}$ and $\mathbf{D}$ are column vectors of size $2n$ whose pth and qth elements are a_p and d_q, respectively, and $\mathbf{B}$ is a square matrix of size $2n$ whose element of the pth column and the qth raw is $b_{p,q}$.

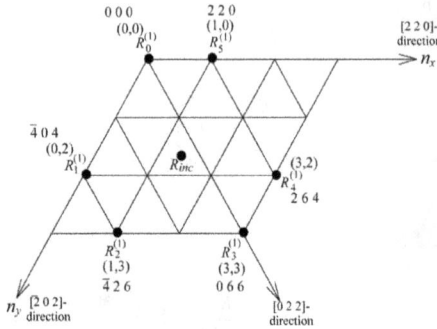

Figure 2. This Figure shows a top view of Figure 1(b).

Figure 2 is a top view of Figure 1(b). The X-ray amplitudes $D_j^{(m)}(R_i^{(1)})$ were calculated from the X-ray amplitudes at the incidence point $D_0^{(l)}(R_{inc})$ of the crystal surface. In this case, 0 0 0-forward-diffracted and $\bar{4}$ 0 4-, $\bar{4}$ 2 6-, 0 6 6-, 2 6 4- and 2 2 0-reflected X-rays are simultaneously strong. The angle spanned by n_x- and n_y-axes is 120°. $\overrightarrow{R_{inc}R_i^{(1)}}$ in Figure 1(b) are parallel to the wave vectors of 0 0 0-forward-diffracted and $\bar{4}$ 0 4-, $\bar{4}$ 2 6-, 0 6 6-, 2 6 4- and 2 2 0-reflected X-rays. As a boundary condition on the crystal surface, amplitude array $D_{even}(i,l,n_x,n_y)$ has nonzero value (unity) when $(i,l,n_x,n_y)=(0,0,0,0)$ or $(i,l,n_x,n_y)=(0,1,0,0)$. On the first layer, nonzero X-ray amplitudes $D_{odd}(j,m,n_x,n_y)$ are calculated when $(n_x,n_y) = [n_x'(i),n_y'(i)]$ $(i \in \{0,1,\cdots,n-1\})$. Here, $[n_x'(i),n_y'(i)] = (0,0),(0,2),(1,3),(3,3),(3,2)$ and $(1,0)$ for $i = 0, 1, 2, 3, 4, 5$, respectively. In general, $D_{even}(i,l,n_x,n_y)$ [or $D_{odd}(i,l,n_x,n_y)$] is calculated as $D_i^{(l)}(R^{(1)})$ by substituting $D_{odd}(j,m,n_x - n_x'(i),n_y - n_y'(i))$ [or $D_{even}(j,m,n_x - n_x'(i),n_y - n_y'(i))$] into $D_j^{(m)}(R_i^{(0)})$ in (17). The calculation was performed layer by layer scanning n_x and n_y in a range of $N Min[n_x'(i)] \leq n_x \leq N Max[n_x'(i)]$ and $N Min[n_y'(i)] \leq n_y \leq N Max[n_y'(i)]$, where N is the ordinal number of layer. The values of $\chi_{h_i - h_j}$ were calculated by using XOP version 2.3 [6].

4. Experimental

4.1. Phase-retarder system

When taking four-, five-, six- and eight-beam pinhole topographs shown in section 5, the horizontally polarized synchrotron X-rays monochromated to be 18.245 keV with a water-cooled diamond monochromator system at BL09XU of SPring-8 were incident on the 'rotating four-quadrant phase retarder system' [15, 17].

Figure 3 shows (a) a schematic drawing of the phase retarder system and (b) a photograph of it. [1 0 0]-oriented four diamond crystals PR_n ($n \in \{1,2,3,4\}$) with thicknesses of 1.545,

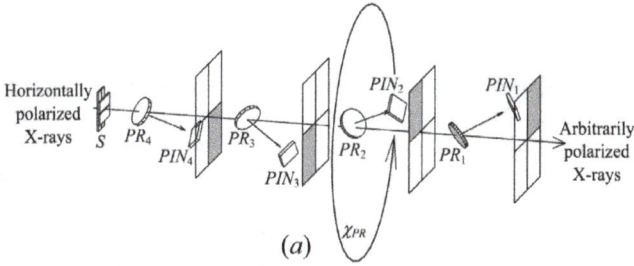

(a)

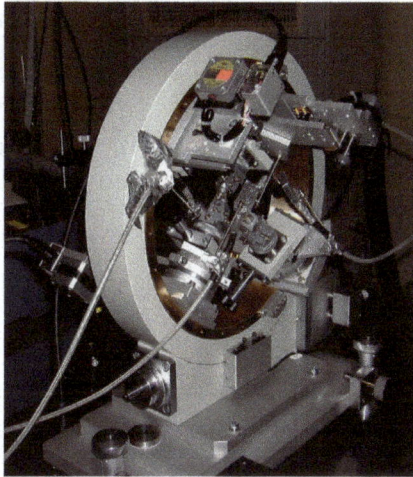

(b)

Figure 3. (a) is a schematic drawing of the 'rotating four-quadrant phase retarder system' (reproduction of Figure 3 in reference [17]). (b) is a photograph of it.

2.198, 1.565 and 2.633 mm were mounted on tangential-bar type goniometers such that the deviation angles $\Delta\theta_{PR_n}$ from the exact Bragg condition of 1 1 1 reflection in an asymmetric Laue geometry can be controlled. See Figure 4 in reference [17] for more detail. The four tangential-bar type goniometers were mounted in a χ-circle goniometer [see Figure 3 (b)] such that the whole system of the phase retarders can be rotated around the beam axis of transmitted X-rays. The rotation angle of the χ-circle χ_{PR} and $\Delta\theta_{PR_n}$ were controlled as summarized in Table 3 in reference [17] such that horizontal-linearly (LH), vertical-linearly (LV), right-screwed circularly (CR), left-screwed circularly (CL), $-45°$-inclined-linearly (L-45) and $+45°$-inclined-linearly (L$+45$) polarized X-rays were generated to be incident on the sample crystal.

In the cases of three- and twelve-beam pinhole topographs, horizontally polarized synchrotron X-rays monochromated to be 18.245 keV and to be 22.0 keV, respectively, but not transmitted through the phase retarder system were incident on the sample crystals.

4.2. Sample crystal

Figure 4 is a reproduction of Fig 7 in reference [17] showing the experimental setup when the six-beam pinhole topographs shown in reference [17] were taken. Also in the case of n-beam pinhole topographs ($n \in \{3,4,5,6,8,12\}$) the $[1\,\bar{1}\,1]$-oriented floating-zone silicon crystal with thickness 9.6 mm (for three-, four-, five-, six- and eight-beam topographs) and 10.0 mm (for twelve-beam topographs) were also mounted on the four-axis goniometer whose χ-, ϕ-, ω- and θ-axes can be rotated. Transmitted X-rays through the sample and two reflected X-rays were searched by three PIN photo diodes as shown in Figure 4. The positions of the two PIN photo diodes for detecting the reflected X-rays were determined using a laser beam guide reflected by a mirror. The mirror was mounted at the sample position on the goniometer whose angular positions were calculated such that the mirror reflects the laser beam to the direction of X-rays to be reflected by the sample crystal.

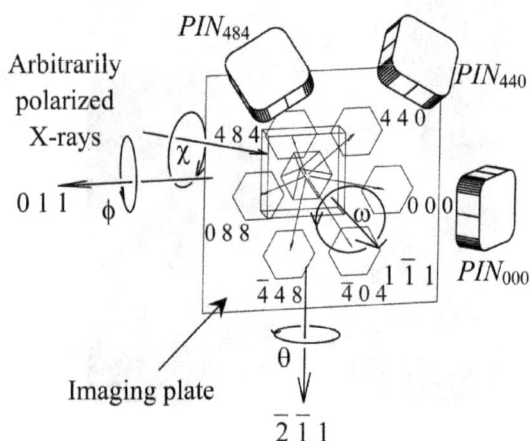

Figure 4. A schematic drawing of the goniometer on which the sample crystal was mount (reproduction of Figure 7 in reference [17]).

After adjusting the angular position of the goniometer such that the n-beam simultaneous reflection condition was satisfied, the size of slit S in Figure 3 (a) was limited to be 25×25 μm.

N images of n-beam pinhole topographs were simultaneously recorded on an imaging plate placed behind the sample crystal.

5. Results and discussion

5.1. Three-beam case

Three-beam case is the most primitive case of X-ray multiple reflection. Figures 5[$E(a)$] and 5[$S(a)$] are 0 0 0-forward-diffracted and $\bar{4}$ 0 4- and 0 4 4-reflected X-ray pinhole topograph images. Figures 5[$E(b)$] and 5[$S(b)$] are enlargements of 0 4 4-reflected images from Figures 5[$E(a)$] and 5[$S(a)$], respectively. Fine-fringe regions ♯1 and ♯2 ([$FFR(1)$] and [$FFR(2)$]) and

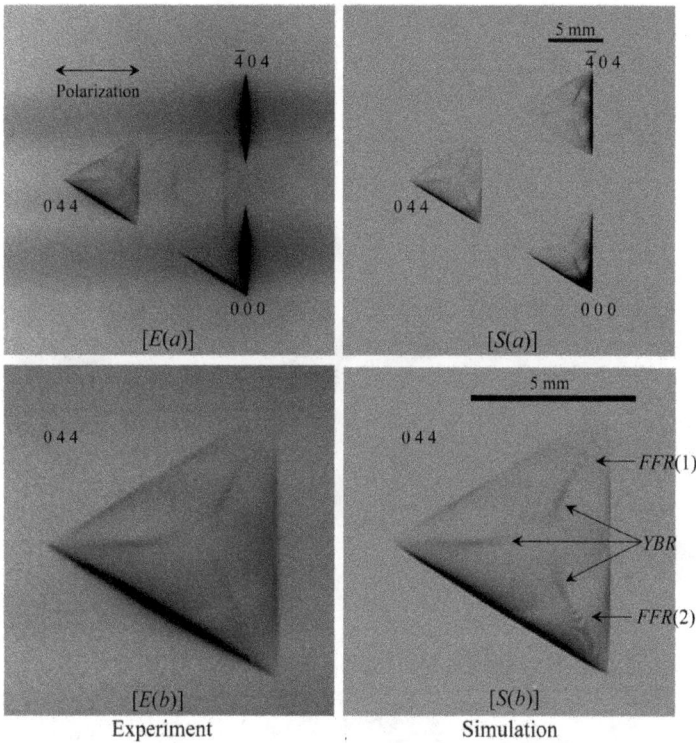

Figure 5. $[E(a)]$ and $[S(a)]$ $(x \in \{a,b\})$ are experimentally obtained and computer-simulated three-beam X-ray pinhole topographs with an incidence of horizontal-linearly polarized X-rays whose photon energy was 18.245 keV. $[Y(b)]$ $(Y \in \{E,S\})$ are enlargements of 0 4 4-reflected X-ray images in $[Y(a)]$. The exposure time for $[E(x)]$ was 600 s.

'Y-shaped' bright region (YBR) indicated by arrows in Figure 5$[S(b)]$, are found also in Figure 5$[E(b)]$.

5.2. Four-beam case

Figures 6$[E(x)]$ and 6$[S(x)]$ $(x \in \{a,b,c\})$ show experimentally obtained and computer-simulated pinhole topographs of 0 0 0-forward-diffracted, and 0 6 6-, $\overline{6}$ 2 8- and $\overline{6}$ $\overline{2}$ 4-reflected X-ray images, respectively. $[Y(a)]$, $[Y(b)]$ and $[Y(c)]$ $(Y \in \{E,S\})$ were obtained with an incidence of $+45°$-inclined-linearly, $-45°$-inclined-linearly and right-screwed-circularly polarized X-rays, respectively, generated with the phase retarder system or assumed in the simulation.

Figures 7$[E(x)]$ and 7$[S(x)]$ $(x \in \{a,b,c\})$ are enlargements of $\overline{6}$ 2 8-reflected X-ray images from Figures 6$[E(x)]$ and 6$[S(x)]$. In Figure 7$[S(a)]$, fine-fringe region ♯1 $[FFR(1)]$, fine-fringe region ♯2 $[FFR(2)]$ and knife-edge line (KEL) are indicated by arrows. These characteristic patterns are also observed in Figure 7$[E(a)]$. In Figures 7$[E(b)]$ and 7$[S(b)]$, while $FFR(2)$ is

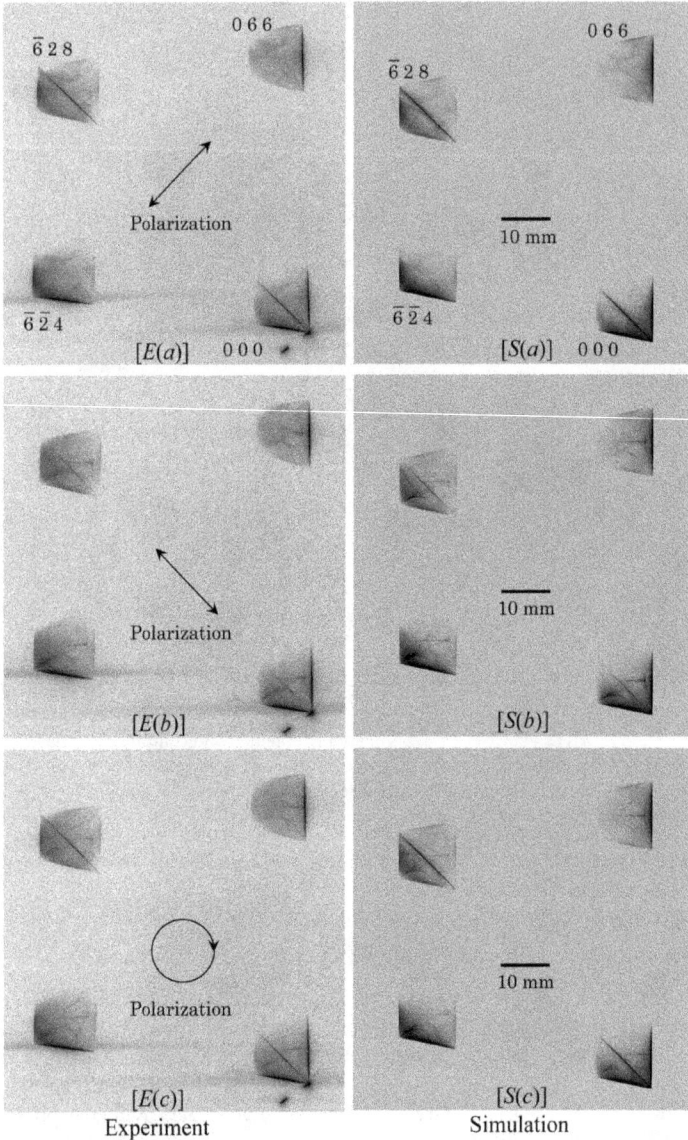

Figure 6. $[E(x)]$ and $[S(x)]$ $(x \in \{a,b,c\})$ are experimentally obtained and computer-simulated four-beam X-ray pinhole topographs with an incidence of $+45°$-inclined-linearly, $-45°$-inclined-linearly and right-screwed-circularly polarized X-rays whose photon energy was 18.245 keV. The exposure time for $[E(x)]$ was 1800 s.

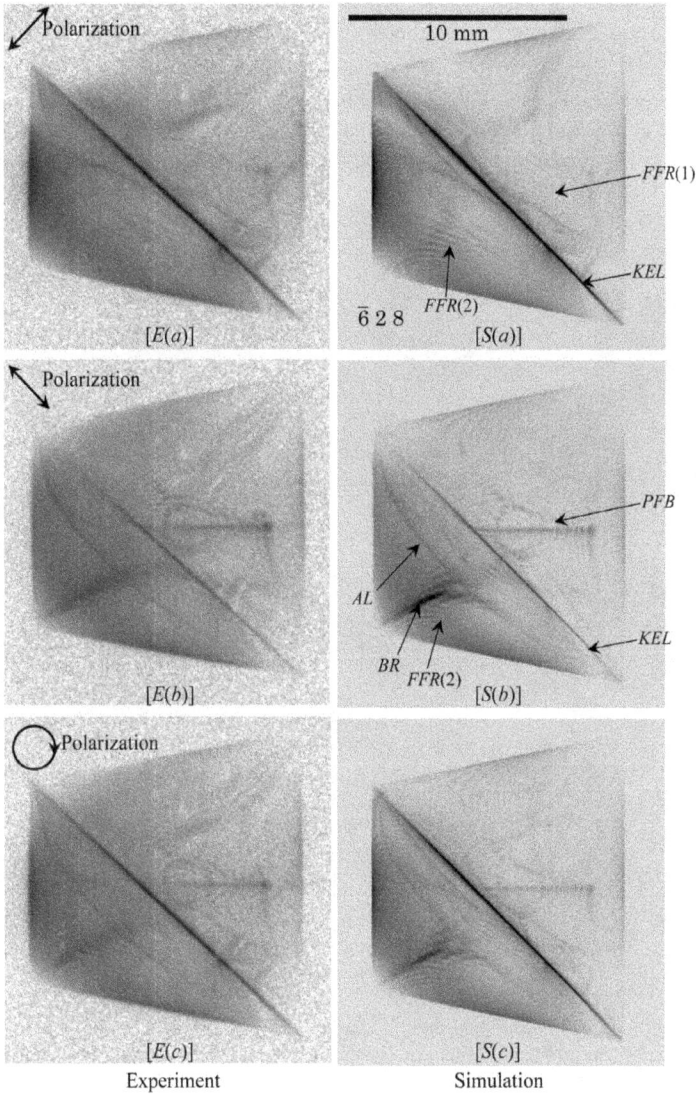

Figure 7. $[E(x)]$ and $[S(x)]$ $(x \in \{a, b, c\})$ are enlargements of $\bar{6}$ 2 8-reflected X-ray images in Figures 6 $[E(x)]$ and 6 $[S(x)]$.

observed at the same position, a pattern like a fish born (PFB) is observed in place of [$FFR(1)$]. KEL in Figures 7[$E(b)$] and 7[$S(b)$] are fainter. Furthermore, an arched line (AL) and a bright region (BR) not observed in Figures 7[$E(a)$] and 7[$S(a)$] are observed in Figure 7[$E(b)$] and 7[$S(b)$]. In Figures 7[$E(c)$] and 7[$S(c)$], almost all the characteristic patterns above-mentioned are observed.

Between the horizontal and vertical components of incident X-rays, there is difference not in amplitude but in phase among Figures [$Y(a)$], [$Y(b)$] and [$Y(c)$] ($Y \in \{E, S\}$), which reveals that the wave fields excited by horizontal- and vertical-linearly polarized components of the incident X-rays interfere with each other.

5.3. Five-beam case

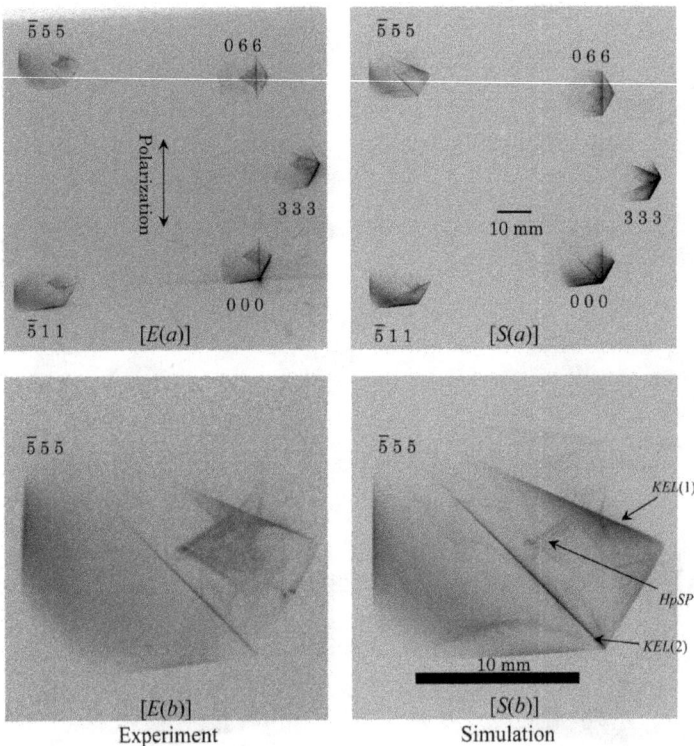

Figure 8. [$E(x)$] and [$S(x)$] ($x \in \{a, b\}$) are experimentally obtained and computer-simulated five-beam X-ray pinhole topographs with an incidence of vertical-linearly polarized X-rays whose photon energy was 18.245 keV. [$Y(b)$] ($Y \in \{E, S\}$) are enlargements of $\bar{5}\,5\,5$-reflected X-ray images in [$Y(a)$]. The exposure time for [$E(x)$] was 1800 s.

In the case of cubic crystals, five reciprocal lattice nodes (including the origin of reciprocal space) can ride on a circle in reciprocal space. For understanding such a situation, refer to Figure 1 of reference [17].

Figures $8[E(a)]$ and $8[S(a)]$ are experimentally obtained and computer-simulated five-beam pinhole topographs. Figures $8[E(b)]$ and $8[S(b)]$ are enlargements of $\bar{5}$ 5 5-reflected X-ray images from Figures $8[E(a)]$ and $8[S(a)]$. Knife-edge patterns $\sharp 1$ and $\sharp 2$ $[KEL(1)$ and $KEL(2)]$ and 'harp-shaped' pattern $(HpSP)$ indicated by arrows in Figure $8[S(b)]$ are observed also in Figure $8[E(b)]$.

Remarking on the directions of $KEL(1)$ and $KEL(2)$, these knife-edge patterns are directed to 0 0 0-forward-diffracted and 3 3 3-reflected X-ray images, respectively. Then, $KEL(1)$ and $KEL(2)$ are considered to suggest the strong energy exchange mechanism between 0 0 0-forward-diffracted and $\bar{5}$ 5 5-reflected X-ray wave fields and between 3 3 3- and $\bar{5}$ 5 5-reflected X-ray wave fields. Such knife-edge patterns are found also in three-, four-, six- and eight-beam pinhole topograph images shown in the present chapter.

5.4. Six-beam case

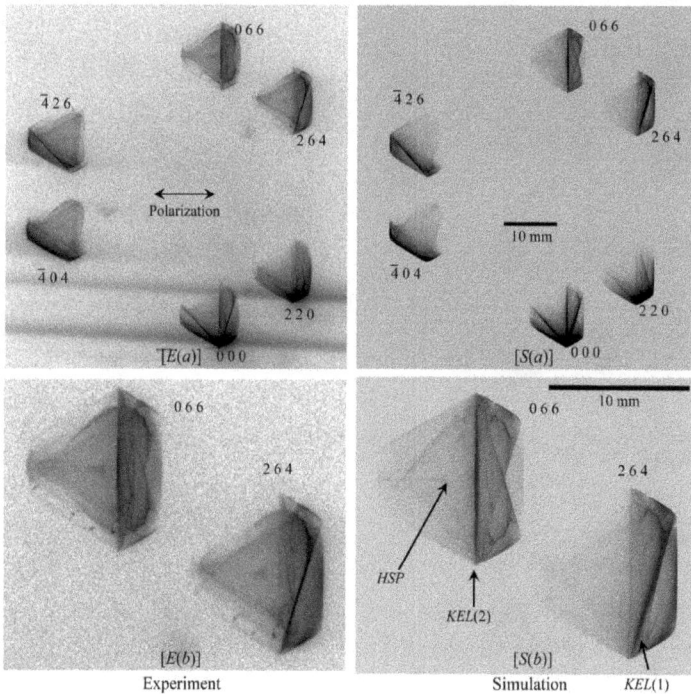

Figure 9. $[E(a)]$ and $[S(a)]$ are experimentally obtained and computer-simulated six-beam X-ray pinhole topographs with an incidence of horizontal-linearly polarized X-rays with a photon energy of 18.245 keV. $[E(b)]$ and $[S(b)]$ are enlargements of 2 6 4- and 0 6 6-reflected X-ray images in $[E(a)]$ and $[S(a)]$. The exposure time for $[E(a)]$ and $[E(b)]$ was 300 s.

While experimental and computer-simulated six-beam pinhole topograph images whose shapes are regular hexagons have been reported in reference [14, 16, 17], shown in this section are six-beam pinhole topographs whose Borrmann pyramid is not a regular hexagonal pyramid.

Such six-beam pinhole topographs experimentally obtained and computer-simulated are shown in Figure 9. Figures 9[$E(b)$] and 9[$S(b)$] are enlargements of 2 6 4- and 0 6 6-reflected X-ray images from Figures 9[$E(a)$] and 9[$S(a)$]. Knife-edge patterns [$KEL(1)$ and $KEL(2)$] indicated by arrows in Figure 9[$S(b)$] are found also in Figure 9[$E(b)$]. Circular patterns that were found in the central part of the six-beam pinhole topographs [14, 16, 17] cannot be found in the present case. A 'heart-shaped' pattern (HSP) is found also in Figure 9[$E(b)$].

5.5. Eight-beam case

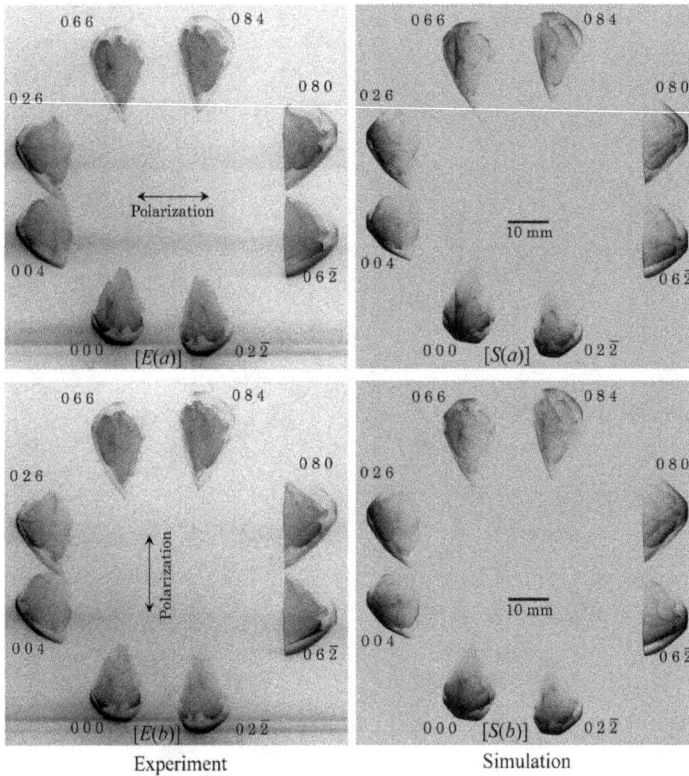

Experiment   Simulation

Figure 10. [$E(x)$] and [$S(x)$] ($x \in \{a, b\}$) are experimentally obtained and computer-simulated eight-beam X-ray pinhole topographs with an incidence of (a) horizontal-linearly and (b) vertical-linearly polarized X-rays whose photon energy was 18.245 keV. The exposure time for [$E(x)$] was 240 s.

Figure 10[$E(a)$] and 10[$S(a)$] are eight-beam X-ray pinhole topographs experimentally obtained and computer-simulated, respectively, with an incidence of horizontal-linearly

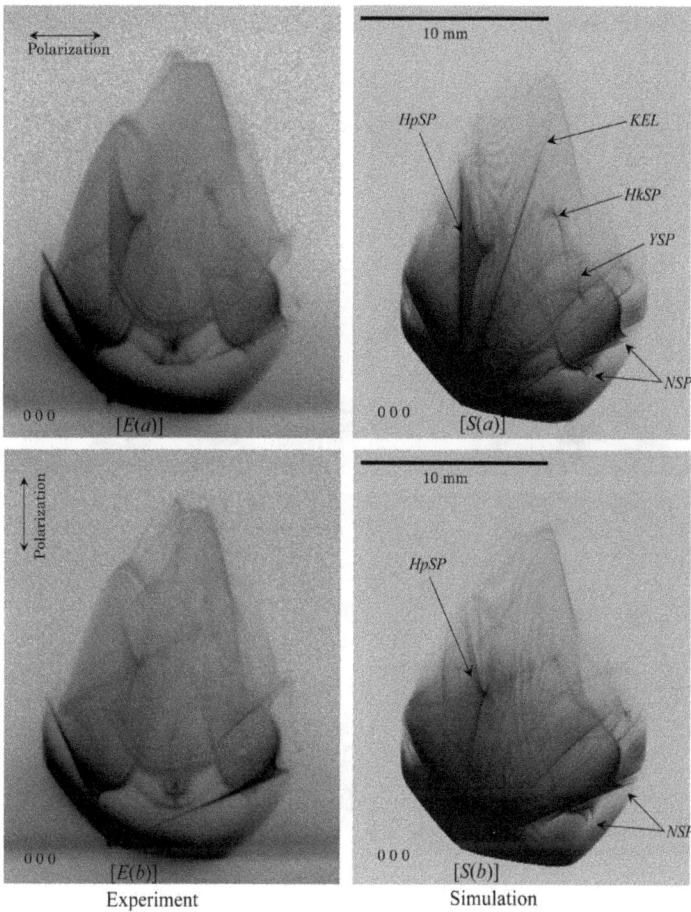

Experiment   Simulation

Figure 11. $[E(x)]$ and $[S(x)]$ ($x \in \{a, b\}$) are enlargements of 0 0 0-forward-diffracted X-ray images in Figures 10 $[E(x)]$ and 10 $[S(x)]$.

polarized X-rays. Figure 10$[E(b)]$ and 10$[S(b)]$ were obtained with an incidence of vertical-linearly polarized X-rays. Figure 11$[E(x)]$ and 11$[S(x)]$ ($x \in \{a, b\}$) are enlargements of 0 0 0-forward-diffracted X-ray images from Figures 10$[E(x)]$ and 10$[S(x)]$, respectively.

In Figure 11$[S(a)]$, A 'harp-shaped' pattern ($HpSP$), knife-edge line (KEL), 'hook-shaped' pattern ($HkSP$), 'Y-shaped' pattern (YSP) and 'nail-shaped' patterns are indicated by arrows. All these characteristic patterns are observed also in Figure 11$[E(a)]$. NSP is also observed in Figures 11$[E(b)]$ and 11$[S(b)]$. However, $HpSP$ in Figures 11$[E(b)]$ and 11$[S(b)]$ are rather fainter compared with Figures 11$[E(a)]$ and 11$[S(a)]$.

5.6. Twelve-beam case

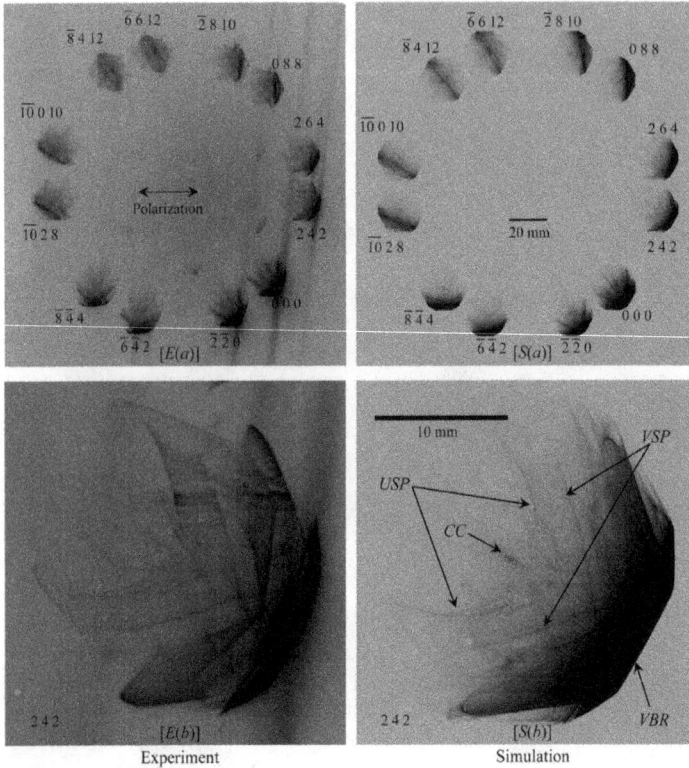

Figure 12. $[E(a)]$ and $[S(a)]$ are experimentally obtained and computer-simulated twelve-beam X-ray pinhole topographs with an incidence of horizontal-linearly polarized X-rays whose photon energy was 22.0 keV. $[E(b)]$ and $[S(b)]$ are enlargements of 2 4 2-reflected X-ray images in $[E(a)]$ and $[S(a)]$. The exposure time for $[E(a)]$ and $[E(b)]$ was 300 s.

Twelve is the largest number of n for the n-beam T-T equation (14) that restricts a condition that n reciprocal lattice nodes should ride on a circle in reciprocal space. Figures $12[E(a)]$ and $12[S(a)]$ are experimentally obtained and computer-simulated twelve-beam pinhole topographs. Figures $12[E(b)]$ and $12[S(b)]$ are enlargements of 2 4 2-reflected X-ray images from Figures $12[E(a)]$ and $12[S(a)]$.

A very bright region (VBR), 'V-shaped' pattern (VSP), central circle (CC) and 'U-shaped' pattern indicated by arrows in Figure $12[S(b)]$ are found also in Figure $12[E(b)]$.

6. Summary

The n-beam ($n \in \{3,4,5,6,8,12\}$) Takagi-Taupin equation and computer algorithm to solve it were verified from excellent agreements between experimentally obtained and computer-simulated three-, four-, five-, six-, eight- and twelve-beam pinhole topographs.

The equivalence between the E-L and T-T formulations of the n-beam X-ray dynamical diffraction theory, which has been implicitly recognized for two-beam case theory, was explicitly described in the present chapter. Whereas the former theory can be calculated by solving an eigenvalue-eigenvector problem, the latter can be calculated by solving a partial differential equation. This equivalence is very similar to that between the Heisenberg and Schrödinger pictures of quantum mechanics and is very important.

Whereas this chapter has been described with focusing on the n-beam case that $n \in \{3,4,5,6,8,12\}$, the n-beam X-ray dynamical diffraction theory applicable to the case of arbitrary number of n, which is effective and important for solving the phase problem in protein crystal structure analysis, will be described elsewhere. In the case of protein crystallography, the situation that arbitrary number of reciprocal lattice nodes are very close to the surface of the Ewald sphere, cannot be avoided. In protein crystallography, the n-beam X-ray dynamical diffraction theory for arbitrary number of n is necessary.

Acknowledgements

The part of theoretical study and computer simulation of the present work was conducted in Research Hub for Advanced Nano Characterization, The University of Tokyo, supported by the Ministry of Education, Culture, Sports, Science and Technology (MEXT), Japan.

HITACHI SR-11000 and SGI Altix ICE 8400EX super computer systems of the Institute for Solid State Physics of The University of Tokyo were used for the computer simulations.

The preliminary experiments were performed at AR-BL3A of the Photon Factory AR under the approval of the Photon Factory Program Advisory Committee (Proposals No. 2003G202 and No. 2003G203). The main experiments were performed at BL09XU of SPring-8 under the approval of Japan Synchrotron Radiation Research Institute (JASRI) (Proposals No. 2005B0714 and No. 2009B1384).

The present work is one of the activities of Active Nano-Characterization and Technology Project financially supported by Special Coordination Fund of the Ministry of Education, Culture, Sports, Science and Technology of the Japan Government.

The authors are indebted to Dr. Y. Ueji Dr. X.-W. Zhang and Dr. G. Ishiwata for their technical support in the present experiments and also to Professor Emeritus S. Kikuta for his encouragements and fruitful discussions for the present work.

Author details

Kouhei Okitsu

Nano-Engineering Research Center, Institute of Engineering Innovation, Graduate School of Engineering, The University of Tokyo, 2-11-16 Yayoi, Bunkyo-ku, Tokyo 113-8656, Japan

Yasuhiko Imai and Yoshitaka Yoda
Japan Synchrotron Radiation Research Institute, SPring-8, 1-1-1 Kouto, Mikazuki-cho, Sayo-gun, Hyogo 679-5198, Japan

7. References

[1] Authier, A. [2005]. *Dynamical Theory of X-Ray Diffraction, Reprinted with Revisions 2004, 2005.*, Oxford University Press.
[2] Chang, S.-L. [2004]. *X-Ray Multiple-Wave Diffraction, Theory and Application*, SPringer.
[3] Colella, R. [1974]. *Acta Cryst.* A30: 413–423.
[4] Colella, R. [1995a]. *Comments Cond. Mat. Phys.* 17: 175–198.
[5] Colella, R. [1995b]. *Comments Cond. Mat. Phys.* 17: 199–215.
[6] del Rio, M. S. & Dejus, R. J. [1998]. *Proc. SPIE* 3448: 340–345.
[7] Ewald, P. P. [1917]. *Ann. Phys. 4. Folge* 54: 519–597.
[8] Ewald, P. P. & Héno, Y. [1968]. *Acta Cryst.* A24: 5–15.
[9] Héno, Y. & Ewald, P. P. [1968]. *Acta Cryst.* A24: 16–42.
[10] Hildebrandt, G. [1967]. *Phys. Stat. Sol.* 24: 245–261.
[11] Laue, M. v. [1931]. *Ergeb. Exakten Naturwiss* 10: 133–158.
[12] Lipscomb, W. N. [1949]. *Acta Cryst.* 2: 193–194.
[13] Okitsu, K. [2003]. *Acta Cryst.* A59: 235–244.
[14] Okitsu, K., Imai, Y., Ueji, Y. & Yoda, Y. [2003]. *Acta Cryst.* A59: 311–316.
[15] Okitsu, K., Ueji, Y., Sato, K. & Amemiya, Y. [2002]. *Acta Cryst.* A58: 146–154.
[16] Okitsu, K., Yoda, Y., Imai, Y. & Ueji, Y. [2011]. *Acta Cryst.* A67: 550–556.
[17] Okitsu, K., Yoda, Y., Imai, Y., Ueji, Y., Urano, Y. & Zhang, X.-W. [2006]. *Acta Cryst.* A62: 237–247.
[18] Takagi, S. [1962]. *Acta Cryst.* 15: 1311–1312.
[19] Takagi, S. [1969]. *J. Phys. Soc. Jpn.* 26: 1239–1253.
[20] Taupin, D. [1964]. *Bull. Soc. Fr. Minéral. Cristallogr.* 87: 469–511.
[21] Weckert, E. & Hümmer, K. [1997]. *Acta Cryst.* A53: 108–143.
[22] Weckert, E. & Hümmer, K. [1998]. *Cryst. Res. Technol.* 33: 653–678.

Chapter 3

X-CHIP: An Integrated Platform for High-Throughput Protein Crystallography

Nickolay Y. Chirgadze, Gera Kisselman, Wei Qiu, Vladimir Romanov, Christine M. Thompson, Robert Lam, Kevin P. Battaile and Emil F. Pai

Additional information is available at the end of the chapter

1. Introduction

High-throughput protein crystallography can be a time consuming and resource intensive endeavor. Although recent years have seen many advances in the field, screening for suitable crystallization conditions using common commercially available platforms still requires considerable amounts of protein and reagents. Furthermore, diffraction quality testing and data collection typically involve physical extraction and cryogenic freezing of the crystal samples, which may have a significant impact on the integrity of the crystal (Garman, 1999). To acquire high-quality diffraction data, both the crystallization conditions and the cryoprotectants must be further optimized. These steps can be time consuming and are often restricted to experienced users (Alcorn & Juers, 2010). In response to these concerns, the last decade has seen a significant surge of developments in crystallography-aimed microtechnology, specifically the use of crystallization chips. So far, the field is dominated by a range of microfluidic devices (Li & Ismagilov, 2010), with one of the most significant differences between them being the type of crystallization technique they employ. Several devices have been developed, and even commercialized (Topaz® crystallizer, Fluidigm Corp., CA, USA; The Crystal Former, Microlytic Inc., MA, USA) that utilize free interface diffusion (FID) (Hansen et al., 2002). Other chips employ nanochannels to create counter-diffusion crystallization (Hasegawa et al., 2007, Ng et al., 2008, Dhouib et al., 2009) or nanodroplets that simulate batch crystallization (Zheng et al., 2003). There are two clear, parallel implications in all these devices. They are all striving to increase efficiency of the hit identification process, and are offering the possibility of *in situ* X-ray analysis and, in favorable cases, diffraction data collection for structure determination (Zheng et al., 2004, Hansen et al., 2006, Ng et al., 2008, May et al, 2008, Dhouib et al., 2009).

The X-CHIP (Chirgadze, 2009) addresses the same challenges of high-throughput crystallography with an alternative approach, and has a number of unique additional advantages. In contrast to microfluidic chips, the crystallization process takes place on the chip surface, in droplet arrays of aqueous protein and crystallization reagents mixtures under a layer of oil. These microbatch arrays are made possible by altering the chip surface with a unique coating, creating defined areas of varying hydrophobicity. This paper presents the design of the device and accompanying tools for setting up crystallization trials and mounting the chip for data collection, as well as the important benefits, limitations and implications that are inherent to this platform. It also describes proof-of-concept experiments in which this technology was utilized for crystal growth, visual inspection, X-ray diffraction data collection and structure determination of two native and one selenomethionine-labeled protein targets. The presented results show that large, well-diffracting crystals can be grown and high-quality data sets sufficient for structure determination can be collected on a home as well as a synchrotron X-ray source.

2. X-CHIP design and application

The principal idea behind the X-CHIP was to create a platform that presents an alternative to the conventional crystallographic pipeline by consolidating the processes of crystallization condition screening, crystal inspection and data collection onto one device, streamlining the entire process and eliminating crystal handling and arduous cryogenic techniques (Fig. 1).

Figure 1. The X-CHIP was designed and developed as a miniaturized and integrated alternative to conventional methods of crystallization and data collection.

The chip is made from a material chosen for its visual light transparency and relatively low absorption of X-ray radiation. An X-CHIP with a thickness of 0.375 mm absorbed approximately thirty percent of the X-ray intensity of the primary synchrotron beam that was attenuated by 1,800 - 2,000 times to avoid excessive radiation damage to crystals during data acquisition. Designed to be compatible with most standard goniometers, the device inserts into a chip-base (possessing a machined slit and locking screw) for support and simple mounting (Fig. 1). A plastic receptacle holds multiple chips mounted on bases, providing rigidity for set up, storage and visual inspection of the crystallization drops and can be covered with a special lid to prevent dust contamination. The chip, along with supporting tools, is shown in Fig. 2.

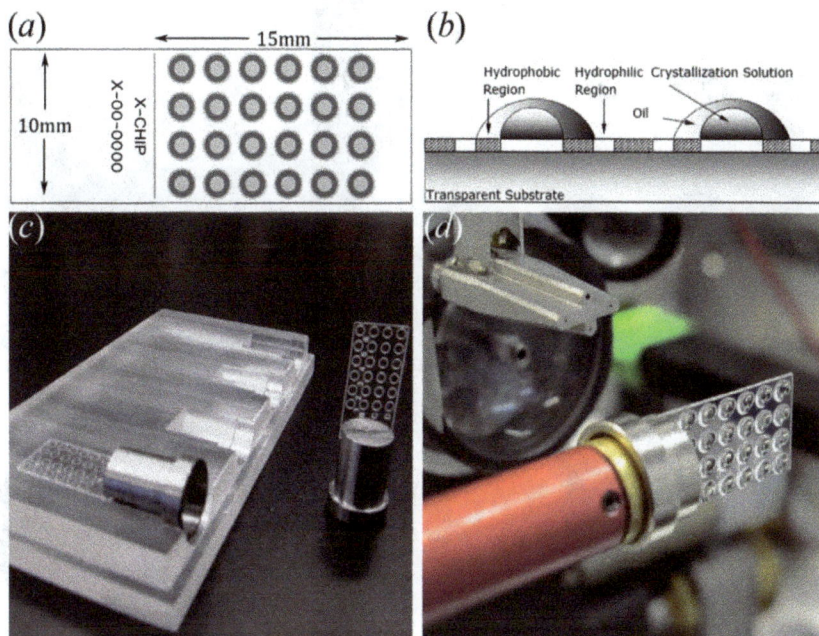

Figure 2. Schematics and images of the X-CHIP (a) Top-view schematic of the 24-drop format chip, hydrophilic and hydrophobic areas are shown in light and dark grey, respectively (other formats include 6-drops; not shown) (b) Cross-section of the chip (c) X-CHIP on a base and a 4-chip receptacle device (d) X-CHIP with 24 crystallization drops, mounted on a goniometer.

The described system applies principles of the microbatch crystallization method, the high effectiveness and unique benefits of which have been described elsewhere (D'Arcy et al., 1996, Chayen, 1998, D'Arcy et al., 2003). On the surface of the chip, circular hydrophilic areas are inscribed in hydrophobic annuli in ordered arrays (Fig. 2a, 2b). Nanoliter volumes of aqueous protein and precipitant solutions are mixed onto the hydrophilic circle by sequential addition and then covered by an mineral oil layer, which is dispensed on top of the drop and is stabilized on the surrounding hydrophobic ring. The interactions between the aqueous phase, oil layer and coated surface create highly defined droplets of predictable volume and thickness and prevent drops not only from drying but from fusing with each other with each other during crystallization set up and data acquisition. The design of the chip currently uses 1 x 6 and 4 x 6 formats and its size permits visual inspection of the entire chip in one image (Fig. 3a).

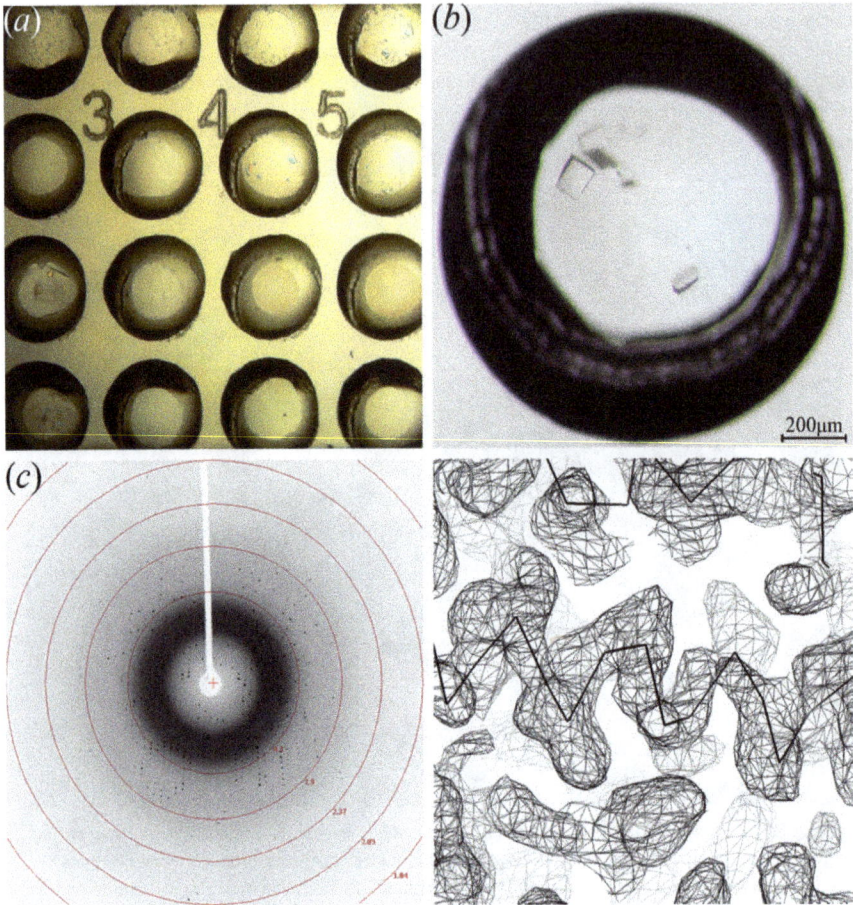

Figure 3. Experimental results: (*a*) section of a 4x6 X-CHIP with a two-dimensional optimization of two crystallization conditions for native PA0269, taken two weeks after initial set up (*b*) crystals of EphA3 grown overnight, crystal size approximately 250μm in length (*c*) On-the-chip diffraction image for an EphA3 crystal, collected on a Rigaku FR-E rotating anode with R-AXIS HTC detector (*d*) Part of an experimental electron density map generated using the SAD PA0269SM data set collected directly from the crystal grown on the X-CHIP, superimposed with the protein C$_\alpha$-trace, shown as a black solid line.

3. Materials and methods

Previously investigated targets, the protein kinase domain of human Ephrin Receptor Tyrosine Kinase A3 (EphA3) (Davis *et al.*, 2008) and the *Pseudomonas aeruginosa* alkylhydroperoxidase D protein (PA0269) (McGrath *et al.*, 2007) were selected as model proteins to demonstrate the feasibility of crystal growth and *in situ* data collection using the X-CHIP. Following several rounds of on-chip optimization, both projects were crystallized at conditions similar to those in the cited literature. EphA3, at 15 mg ml^{-1}, and native PA0268, at 10 mg ml^{-1}, were crystallized using a 1:1 (vol/vol) with 0.2M $(NH_4)_2SO_4$, 0.1M Hepes pH 7.5, 25% PEG3350 and 0.8M $(NH_4)_2SO_4$, 0.1M sodium citrate pH 4.0 respectively. For crystallization drop set up, a protein sample volume of 200-250nl was mixed with an equal volume of precipitant solution and then covered by an mineral oil volume of 0.75-1.25µl; all solutions were dispensed with a Gilson P2 pipette. Prior to set up, the chip was inserted into the holder, which was afterward covered with a lid to prevent contamination. Chips were stored for 1-4 weeks at room temperature without significant evaporation and transported by road to the synchrotron beamline. Crystals both from stored chips as well as from chips set up at the synchrotron were used to collect diffraction data sets.

In-house data sets were collected on a *Rigaku FR-E* Superbright rotating anode X-ray source equipped with a *Rigaku R-AXIS HTC* image plate as a detector (Rigaku, The Woodlands, TX, USA). The synchrotron data sets were collected on the *IMCA-CAT ID-17* beamline at the Advanced Photon Source (APS) facility with appropriate beam attenuation, using the *Pilatus 6M* detector (Dectris Ltd., Baden, Switzerland). The X-CHIP was manually mounted onto the goniometers as shown in Fig. 2d. Individual samples were initially optically centered, and then centered using diffraction to refine crystal position. In both cases the cryo-stream was blocked and data collection was performed at room temperature (~298K).

4. Results

Two important aspects of the described system were investigated throughout this study; the capacity of the chip to produce diffraction quality crystals and the feasibility of diffraction data acquisition (*in situ*) of sufficient quality for *de novo* structure determination. To assess the first task, the reproducibility of previous crystal hits obtained by sitting drop vapor diffusion technique was tested. For both EphA3 and PA0269 projects, vapor diffusion crystallization conditions resulted in high-quality crystals on the X-CHIP (Fig. 3a, 3b). For PA0269, on-chip optimizations further improved the crystal size and quality and decreased the number of crystals per drop (Fig. 3a). These results demonstrated that the X-CHIP can be successfully used to obtain and optimize crystallization hits and grow single crystals that are large enough for straightforward data collection.

Proof-of-concept experiments for on-the-chip data collection were carried out on the rotating anode source and the synchrotron beamline. The initial data collection trials on the in-house X-ray source led to the acquisition of a complete EphA3 data set. While the experiment was conducted at room temperature, diffraction data could still be obtained with sufficient completeness, even for crystals of such low symmetry space group as P2$_1$ (Table 1). At the synchrotron beamline, data sets for EphA3, PA0269 and a PA0269 selenomethionine derivative (SAD) were collected. The high sensitivity and ultrafast readout time of the *Pilatus 6M* detector allowed complete data sets to be collected quickly at room temperature without significant degradation of the sample and with excellent processing statistics. Owing to the finely focused beam (50 x 50µm), it was possible to collect data from multiple small crystals grown within the same drop, without any obvious impact on the diffraction quality of neighboring crystals. A particularly interesting result can be observed by comparing the mosaic spread between the X-CHIP and the benchmark data (*i.e.* cryo loop) in Table 1. It is evident that the mosaic spread was consistently lower for data set experiments collected using the chip, and in the case of PA0268 was as low as 0.046°. Furthermore, based on resolution range alone, EphA3 crystals only started showing radiation damage after as much as ten minutes of continuous X-ray exposure, more than twice the time needed for obtaining a full data set (data not shown).

	EphA3	*EphA3*	EphA3	*PA0269 SM**	PA0269	PA0269S M *
Crystal mount method	X-CHIP	*Cryo Loop*	X-CHIP	*Cryo Loop*	X-CHIP	X-CHIP
Sample Temperature (K)	295	*100*	295	*100*	295	295
X-ray Source	Rigaku FR-E	*BM-17 APS*	ID-17 APS	*ID-17 APS*	ID-17 APS	ID-17 APS
λ (Å)	1.54	*1.00*	1.00	*0.97934*	1.00	0.97938
Detector	R-AXIS HTC	*MarCCD M300*	Pilatus 6M	*ADSC Q210*	Pilatus 6M	Pilatus 6M
Space group	$P2_1$	*$P2_1$*	$P2_1$	*$P6_322$*	$P6_322$	$P6_322$
Resolution (Å)	2.00	*1.93*	1.95	*1.75*	1.95	1.95
High resolution shell (Å)	(2.10-2.00)	*(2.03-1.93)*	(2.05-1.95)	*(1.84-1.75)*	(2.05-1.95)	(2.05-1.95)
Data Collection Time (min)	100	*29*	5.3	*74*	3.0	3.3
$\Delta\phi_{total}$ (°)	100†	*190*	160	*185*	90	100
Mosaic spread (°)	0.100	*1.048*	0.360	*0.364*	0.046	0.160
Completeness (%)	85.3†	*98.4*	96.2	*99.8*	100	100
Multiplicity	2.4	*3.6*	2.7	*10.6*	9.5	10.3
$<I/\sigma(I)>$	4.5 (2.2)	*12.9 (5.0)*	7.5 (2.7)	*19.4 (3.5)*	12.1 (2.7)	20.8 (3.9)
R_{merge} (%)	8.8 (34.6)	*4.9 (9.7)*	10.9 (34.6)	*7.5 (49.9)*	8.2 (47.5)	6.9 (50.2)

Benchmark data is in *italic font*.
Values in parentheses refer to the highest resolution shell.
*Single-wavelength anomalous dispersion (SAD) data collection, using anomalous signal from seleno-methionine.
† Completeness of 99% was achievable from the same set of crystals with a total oscillation angle of 140 degrees.

Table 1. Summary of selected data sets.

For crystallization of EphA3 and PA0269, paraffin oil was used to coat the crystallization drops after protein and precipitant solution had been dispensed. Other oils have been explored, such as Hampton's Al's Oil (50/50 paraffin/silicon oil mixture), silicon oil and a 50/50 mix of paratone/paraffin oils. Higher viscosity oils (paraffin, paratone/paraffin) performed better on the X-CHIP by being highly restricted to the hydrophobic ring boundaries. The thinner silicon oil was found to flow outside of these boundaries causing drop merging. Al's oil required more careful application compared to higher viscosity oils, but proved to stay within the hydrophobic boundaries. Crystallization conditions containing ethanol, 2-methyl-2,4-pentanediol (MPD) and detergents were also tested on the X-CHIP. Ethanol tolerance was tested with a 5-30% gradient using paraffin oil as a cover. The phase separation within the crystallization drops remained intact for the entire gradient range. Crystallization drops containing MPD in combination with different oils tolerated up to 8% before they began to disperse beyond the hydrophobic area the hydrophobic area. While this can exclude some MPD based conditions from being used on the X-CHIP, the impact on the overall versatility is low since most commercially available initial screens from Hampton Research and Emerald Biosystems (e.g. Wizard I&II, Index, Crystal) only have an average of 5-6% total conditions containing MPD. Detergent tolerance was tested with n-dodecyl-N,N-dimethylamine-N-oxide (DDAO) and n-octyl-β-D-glucoside in combination with a paraffin oil covering. Separation between the phases remained intact with the 0.05% n-octyl-b-D-glucoside condition but, DDAO was not tolerated even at very low concentrations.

5. Discussion

The series of initial experiments on the X-CHIP crystallization platform described above demonstrated the chip's applicability for high-throughput protein crystallography and provided insight into the benefits and limitations of this system. Crystallization using the microbatch method on the chip was shown to be suitable for crystal growth and also offered additional benefits. Oil covered drops evaporate very slowly (days to weeks), simplifying both manual and automated set up. Furthermore, changing the composition of the top oil layer with various oil mixtures makes it possible to vary the rate of water evaporation over a wide range, adding another favorable dimension to crystallization screening (D'Arcy et al., 2004). Inherently, the system is economical since crystallization hit determination and optimization trials require up to five times less volume of protein sample and five hundred times less reagent solution than standard vapor diffusion methods. Theoretically, the volumes can be decreased even further by incorporating robotic liquid handling systems, but are currently limited by the accuracy of manual dispensing. In addition, the simplicity of the device results in low manufacturing costs and the platform design eliminates the time and expenses associated with cryogenic techniques. The small size of the chip offers more convenient and faster visual inspection, as all the crystallization drops can be viewed under a microscope simultaneously. Furthermore, the system design provides a non-invasive means of diffraction testing and screening, as the developed device can be mounted on most in-house and synchrotron beamline data acquisition systems without any modification of the chip or adjustments to that system. These capabilities of the X-CHIP make it a potentially useful platform for high-throughput initiatives such as fragment-based screening by co-crystallization.

The X-CHIP system has the potential to completely remove the "user factor" between crystal growth and X-ray diffraction data collection, eliminating crystal manipulation. The feasibility of *in situ* data collection has significant implications. Firstly, data collection at room temperature eliminates the need for the tedious and often limiting step of cryo-condition optimization with the added advantage that crystal structures determined at room temperature are more representative of the physiological state. Additionally, experimentally obtained SAD data displayed excellent processing statistics, clearly of sufficient quality for *de novo* structure determination. Interestingly, in at least one of the cases investigated, undisturbed crystals have shown significantly lower mosaic spread than that of cryogenically frozen samples, suggesting the potential application of this system to samples of high sensitive or ones with a large unit cell (Table 1). Once mounted on the goniometer, navigation along the chip and alignment of any crystal in the drops is quite straightforward, presenting the potential for data collection in a high-throughput mode. This approach eliminates the necessity for mounting of individual loops as in conventional robotics systems and may save hours of valuable synchrotron beam time. Finally, the elimination of manual crystal handling opens the opportunity for full automation of the crystallization to data acquisition pipeline allowing streamlining of the entire process.

Current developments on the project are aimed at scaling down the drop volumes of the X-CHIP system. Attempting to do so using manual set up has proven to be challenging, but application of a liquid handling robotics system can address this issue. The *Mosquito* crystallization robotic system (Molecular Dimensions Ltd., Suffolk, UK) has already been used to successfully set up crystallization experiments with total drop volumes as low as 200nl. The X-CHIP is also being applied to the crystallization trials of additional protein targets. As a point of interest, experiments with highly sensitive and/or small crystal samples could greatly benefit from the use of this system, as the non-invasive data collection approach would likely resolve many problems that arise from crystal handling. We are also exploring the application of the chip for projects in which low mosaic spread is essential for successful outcome.

6. Conclusions

From the initial studies of the device it is evident that not only does the X-CHIP have the potential to increase efficiency and offer on-the-chip *in situ* data collection for *de nevo* structure determination, but it also has a range of additional benefits including the opportunity for full automation. Even though the recent growth of microchip crystallization technology has seen the development of several useful devices, the X-CHIP platform offers previously unprecedented simplicity with comparable or in some cases even better performance. As seen in the recent growth of commercially available and published microchip crystallization devices, there is continuing interest in improving the efficiency and effectiveness of the diffraction pipeline. With its intuitive design, minimalistic support platform and compatibility with most existing beamline setups, the X-CHIP platform offers a new integrated tool for protein crystallization and X-ray diffraction data collection.

Author details

Nickolay Y. Chirgadze*
Canada Campbell Family Cancer Research Institute, Ontario Cancer Institute, Princess Margaret Hospital, University Health Network, Toronto, Ontario, Canada
Department of Pharmacology and Toxicology, University of Toronto, Toronto, Ontario, Canada

Gera Kisselman, Wei Qiu, Vladimir Romanov, Christine M. Thompson and Robert Lam
Canada Campbell Family Cancer Research Institute, Ontario Cancer Institute, Princess Margaret Hospital, University Health Network, Toronto, Ontario, Canada

Kevin P. Battaile
Hauptman-Woodward Medical Research Institute, IMCA-CAT, Advanced Photon Source, Argonne National Laboratory, Argonne, Illinois, USA

Emil F. Pai
Canada Campbell Family Cancer Research Institute, Ontario Cancer Institute, Princess Margaret Hospital, University Health Network, Toronto, Ontario, Canada
Departments of Biochemistry, Molecular Genetics, and Medical Biophysics, University of Toronto, Toronto, Ontario, Canada

Acknowledgement

For providing technical support and access to the in-house and synchrotron X-ray sources we are thankful to Aiping Dong of the Toronto Structural Genomics Consortium and the IMCA-CAT staff at the APS, respectively. We also extend our gratitude to Dr. Tara Davis for providing bacterial cell cultures for purification of EphA3, Kathy Jones for her help with initial crystallization experiments, Joe Miller for help with business development and Jason C. Ellis for machining accessory items. Use of the IMCA-CAT beamline 17-ID at the Advanced Photon Source was supported by the companies of the Industrial Macromolecular Crystallography Association through a contract with the Hauptman-Woodward Medical Research Institute. Use of the Advanced Photon Source was supported by the U.S. Department of Energy, Office of Science, Office of Basic Energy Sciences, contract No. DE-AC02-06CH11357.

7. References

Alcorn, T., Juers, D. H. (2010) Acta Crystallogr D Biol Crystallogr D66, 366-373

---

* Corresponding Author

Chayen, N. E. (1998). Acta Crystallogr D Biol Crystallogr 54, 8-15.

Chirgadze, N. Y. *et. al.* (2009). Patent Application Number WO 2009/073972 A1

D'Arcy, A., Elmore, C., Stihle, M. & Johnston, J. E. (1996). Journal of Crystal Growth 168, 175-180.

D'Arcy, A., Mac Sweeney, A., Stihle, M. & Haber, A. (2003). Acta Crystallogr D Biol Crystallogr 59, 396-399.

D'Arcy, A., Sweeney, A. M. & Haber, A. (2004). Methods 34, 323-328.

Davis, T. L., Walker, J. R., Loppnau, P., Butler-Cole, C., Allali-Hassani, A. & Dhe-Paganon, S. (2008). Structure (London, England : 1993) 16, 873-884.

Dhouib, K., Khan Malek, C., Pfleging, W., Gauthier-Manuel, B., Duffait, R., Thuillier, G., Ferrigno, R., Jacquamet, L., Ohana, J., Ferrer, J. L., Theobald-Dietrich, A., Giege, R., Lorber, B. & Sauter, C. (2009). Lab Chip 9, 1412-1421.

Garman, E. (1999). Acta Crystallogr D Biol Crystallogr 55, 1641-1653.

Hansen, C. L., Classen, S., Berger, J. M. & Quake, S. R. (2006). J Am Chem Soc 128, 3142-3143.

Hansen, C. L., Skordalakes, E., Berger, J. M. & Quake, S. R. (2002). Proc Natl Acad Sci U S A 99, 16531-16536.

Hasegawa, T., Hamada, K., Sato, M., Motohara, M., Sano, S., Kobayashi, T., Tanaka, T., Katsube, Y. (2007). Presented at 24th European Crystallographic Meeting, Marrakech, Morocco.

McGrath, T.E., Battaile, K., Kisselman, G., Romanov, V., Wu-Brown, J., Virag, C., Ng, I., Kimber, M., Edwards, A.M., Pai, E.F., Chirgadze, N.Y. (2007) 2O4D Protein Data Bank

Li, L. & Ismagilov, R. F. (2010). Annu Rev Biophys 39, 139-158.

May, A., Fowler B., Frankel, K. A., Meigs, G., Holton J. M. (2008). Acta Cryst A64, C133-134

Ng, J. D., Clark, P. J., Stevens, R. C. & Kuhn, P. (2008). Acta Crystallogr D Biol Crystallogr 64, 189-197.

Zheng, B., Roach, L. S. & Ismagilov, R. F. (2003). J Am Chem Soc 125, 11170-11171.

Zheng, B., Tice, J. D., Roach, L. S. & Ismagilov, R. F. (2004). Angew Chem Int Ed Engl 43, 2508-2511.

Chapter 4

Crystallography Under Extreme Conditions: State of the Art and Perspectives

Legrand Vincent

Additional information is available at the end of the chapter

1. Introduction

Sir William Henry Bragg and his son William Lawrence Bragg were the pioneers of crystallography (Bragg, W. H., 1912, 1913a, 1915a, 1915b; Bragg, W. L., 1920). In 1913, they published several articles, notably *The reflection of X-rays by crystals* (Bragg, W. H., 1913b) and *The structure of diamond* (Bragg, W. H. & Bragg, W. L., 1913) were they wrote: "We have applied the new methods of investigation involving the use of X-rays to the case of the diamond, and have arrived at a result which seems of considerable interest. The structure is extremely simple". Two years later, they were jointly awarded the Nobel Prize in Physics for their works in the analysis of crystal structure by means of X-rays.

A century after the first crystallographic experiment, new computing facilities, modern technologies and new diffraction sources (synchrotron, neutron sources…) offer a large range of possibilities and opportunities for crystallographers to probe matter. Crystallography appears nowadays as a new science.

Performing structural analyses at ambient conditions or at low temperature (i.e. above 100 K using nitrogen jet-stream) is very common and popular in laboratories to obtain the structure of powder and single-crystal materials. To understand the mechanisms governing the behaviour of materials it is essential to well know the close relations between the structure and the physical and chemical properties. First, X-ray diffraction is a unique tool to obtain routinely a detailed description of atomic structure and thermal vibrations by analysing the diffracted intensities I_{hkl} of the crystallographic hkl reflections:

$$I_{hkl} = S.C_{hkl}.\left|F_{hkl}\right|^2 \qquad (1)$$

$$F_{hkl} = \sum_{j=1}^{n} f_j.T_j.\exp(2i\pi(h.x_j + k.y_j + l.z_j)) \qquad (2)$$

where S is a scale factor, C_{hkl} is an experimental corrections term (including absorption, extinction, Lorentz-polarization correction…), F_{hkl} is the structure factor of a *hkl* reflection which depends of f_j the form factor of the atom j with (x_j, y_j, z_j) coordinates in the cell and of T_j the Debye-Waller factor given in the case of isotropic harmonic vibrations by:

$$T_j = \exp\left(-B_j \frac{\sin^2\theta}{\lambda^2}\right) = \exp\left(-8\pi^2 \langle U_j^2 \rangle \frac{\sin^2\theta}{\lambda^2}\right) \tag{3}$$

where $\langle U_j^2 \rangle$ is the mean square atomic displacement of the atom j.

If one wants to increase the data quality, the best choice, without consideration about crystal quality or particular experimental conditions, is to perform a low temperature measurement, usually at 110 K, to reduce the atomic displacement parameters which affect the value of the atomic structure factor. In those experimental conditions, if high resolution X-ray diffraction measurement is performed up to large momentum transfers ($\sin\theta/\lambda >$ 1 Å^{-1}) the electron density distribution of a molecule can be determined thanks to the Hansen-Coppens pseudo-atomic multipolar expansion (Hansen & Coppens, 1978):

$$\rho(\vec{r}) = \rho_{core}(r) + P_{val}\kappa^3\rho_{val}(\kappa r) + \sum_l \kappa'^3 R_{nl}(\kappa' r) \sum_{m=0}^{l} P_{lm\pm} y_{lm\pm}(\theta,\varphi) \tag{4}$$

where $\rho_{core}(r)$ and $\rho_{val}(r)$ are spherically averaged core and valence electron densities calculated from Clementi Hartree-Fock wave functions for ground state isolated atoms (Clementi & Roetti, 1974). κ and κ' are contraction-expansion parameters and P_{val} is the atomic valence shell electron population. The deformation of the valence electron shell is projected on real spherical harmonics $y_{lm\pm}(\theta,\phi)$ times Slater-type radial functions $R_{nl}(r)$. $P_{lm\pm}$ are the multipole population parameters.

In the later case, the accuracy of the results is highly dependant of the experimental conditions (Destro et al., 2004; Zhurov et al., 2008) and particularly of the crystal quality and the measurement temperature which must be chosen to reduce at the maximum the thermal vibrations of the atoms. These *usual experiments* at static low temperature give basic structural properties, and their variations as a function of temperature can reveal particular behaviour of matter as phase transition for example with changes in structural, electronic, optical and/or magnetic properties. But, in a general consideration, performing crystallographic measurements under external perturbations is of prime importance. Nowadays, experiments at different temperatures or hydrostatic pressures can be done almost routinely and exotic sample environments are more and more used to explore materials properties. If we just consider for the moment experiments involving temperature or pressure variations, a Web of Science search for "high pressure", "low temperature" and "high temperature" in the field of X-ray and neutron diffraction between the years 2005 and 2012 gave about 3500 experiments performed in 2005 but nearly 5000 in 2011 (figure 1). Two comments can be done observing figure 1. First, a large amount of the extreme conditions (temperature and pressure) experiments are carried out using X-ray diffraction compared to neutron diffraction. This difference is of course due to the large number of X-ray diffractometers available in laboratories, which are more and more outstanding in terms of source power and detector efficacy, combined with numerous sample environments especially designed for laboratory equipments. But this is also due to the building of several 3rd generation of synchrotron sources which offer the possibility to conduct very quick measurements on a very small quantity of matter, allowing the measurement of materials under non ambient conditions, particularly in the domains of chemistry and biology sciences (neutron diffraction experiment demands more matter quantity, in general at least some mm³). The second observed aspect on figure 1 concerns the evolution of the cumulative number (neutron and X-ray) of publications during the considered period: about 300 supplementary published papers per year appear between 2005 and 2009, but in the years 2009-2011, a distinct decrease of this number is noticed. This effect is directly related to the number of synchrotron radiation facilities over the world. One can count about 69 particle accelerators and accelerator laboratories in 2005 and about 76 in 2012 (including about 40 synchrotrons in 2011), with most of the new facilities operative in 2006-2008.

As said before, crystallography under extreme conditions is now more and more used and plays a key role as it offers helpful understanding of the physical, chemical and mechanical properties of the solid state. The term *extreme conditions* was first used to define non ambient thermodynamical conditions of pressure or temperature. It is also now employed when out-of-equilibrium conditions are applied such as light irradiation, external magnetic or electric fields, specific chemical environments (e.g. under liquid or gas flux…) or applied strain. In situ measurements can also be considered as extreme conditions, for examples in the cases of time-resolved experiments (picosecond diffraction…) or chemical kinetic reactions (dehydrogenation reaction, diffusion process, decomposition pathway…). The present challenge is to combine two or more extreme conditions to explore new states of matter and new material properties, taking advantage of last generations of high brilliance sources (synchrotron and neutron sources) (figure 2).

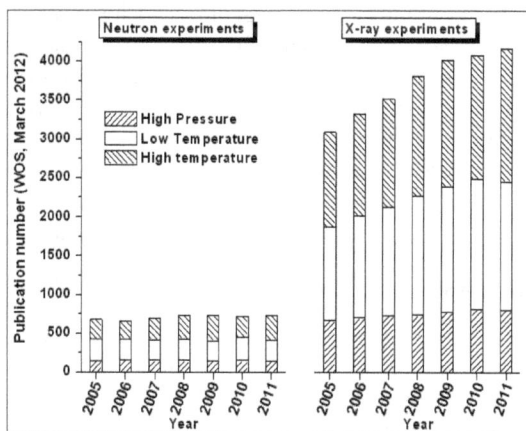

Figure 1. Publication number related to neutron or X-ray diffraction experiments under extreme conditions of temperature and pressure during the period 2005-2011 (Web of Science, March 2012).

When extreme conditions are applied to a material, various changes occur. They may involve sample state variations (gaseous, liquid and solid phases), phase transitions (magnetic and structural), electronic structure modifications (the chemical bonds can change from covalent to ionic or metallic) and atomic bond lengths variations (which induce variations of the atomic vibrations, of the coordination numbers, of the diffusion processes...). A large number of the material properties are modified under extreme conditions, leading sometimes to metastable states, which open new investigation opportunities for the crystallographer, but also for all scientist who wants to go deeper in the understanding of the structure-properties relationships in order to design future materials.

Figure 2. The neutron source of the Institut Laue-Langevin (ILL) and the European Synchrotron Radiation Facility (ESRF) in Grenoble, France (credit: V. Legrand).

This chapter is reviewing different aspects of the use of crystallography under extreme conditions to investigate the nature, the mechanisms and the dynamics of out-of-equilibrium phase transitions as well as transformations driven by external applied conditions. The part 2. is focus on the advantages of using Large scale facilities around the world and what are now the extreme sample environments available to explore unknown properties of materials. Finally, an overview of perspectives and future developments in diffraction instruments and sample environments are presented in part 3. showing that modern crystallography is in perpetual evolution and that the present century is certainly the one of *crystallography under extreme conditions*.

2. Large scale facilities and extreme sample environments

2.1. State-of-the-art sample environments at Laboratories

As noticed before, there are more than 5000 publications per year dealing with extreme conditions of temperature and pressure. Nevertheless, those conditions illustrate only 2 of the numberless non ambient thermodynamical and exotic conditions which can be used to probe the close links between structure and properties of materials. Changing the pressure or the temperature in a crystallographic experiment is performed quite routinely in laboratories. But if one wants for examples to apply electric or magnetic fields, to explore superconductivity properties below 1 K, to perform very fast measurements of transient species or quick chemical reactions, or to understand phase transitions under combined sample environments, the diffractometer performances become overriding and in particular the maximum power of the source, the quality of the beam collimators and the efficacy of the detector. This is very important to obtain exploitable data as the sample environment in those later examples is very bulky and could dramatically affect the beam intensity. Moreover, it is very rare for a laboratory to have sample equipments to perform sub-kelvin, time-resolved or high-magnetic field measurements as these sample environments are very expensive, maintenance demanding and hard to exploit on conventional diffractometers. For a long time, extreme conditions measurements were the privilege of expert laboratories in cryogenic, high fields or high pressures with ability and technical resources to build and to use such devices. Those techniques are nowadays distributed in a larger number of laboratories and, for some of them, take place to semi-industrial devices merchandised by high technology industries. Since ten or so years, apparatus able to be used on conventional diffractometers were punctually developed to respond to the requirement of probing new properties of materials. We can cite some of them. First, to probe the structural and the electronic properties of materials under light irradiation at low temperature, *photocrystallographic* experiments (Coppens et al., 1998; Fomitchev et al., 2000) were developed first to study transient species of transition metal nitrosyl complexes. Coppens et al. (1998) wrote that "the study of photo-induced processes in crystals is a frontier area of crystallographic research, which requires development of novel experimental and computational methods". This is particularly true as the live-time of transient species is in general not related to the time of the experiment required to collect accurate data for structural refinement or experimental electron density modelling. The success of kind experiments consists in in situ laser irradiation of a cryogenically cooled crystal, with a N_2 or He gas-flow system for less restricted access than closed cryostat (White et al., 1994), mounted on a diffractometer equipped with a CCD detector for fast data collection (Graafsma et al., 1997; Muchmore, 1999). More recently, Legrand (2005) has developed a crystallographic experimental methodology approach (Legrand et al., 2005; 2007a; 2007b) and shown that it is possible to measure with accuracy photo-induced metastable states and to refine its experimental electron density below 35 K using conventional X-ray sources (Legrand et al., 2006; Pillet et al., 2008).

We can also mention an other development to do pressure measurements using particular pressure cell, up to 3.0 GPa and for T > 9 K (Guionneau et al., 2004) (figure 3). This last device was developed to probe phase transitions and structure-properties relationships in molecular conductor (Chasseau et al., 1997) and spin transition complexes (Guionneau et al., 2001). Doing in situ laboratory pressure investigations is still difficult as the pressure cell induce several perturbations affecting the quality of the data collection. The originality of this high pressure cell is that the X-ray beam goes through the gasket and not through the diamonds, offering a wide diffraction angle of 342° rotation. In this case, Guionneau et al. (2004) indicate that using a CCD area detector to collect Bragg peaks improves data and reduces the measurement time.

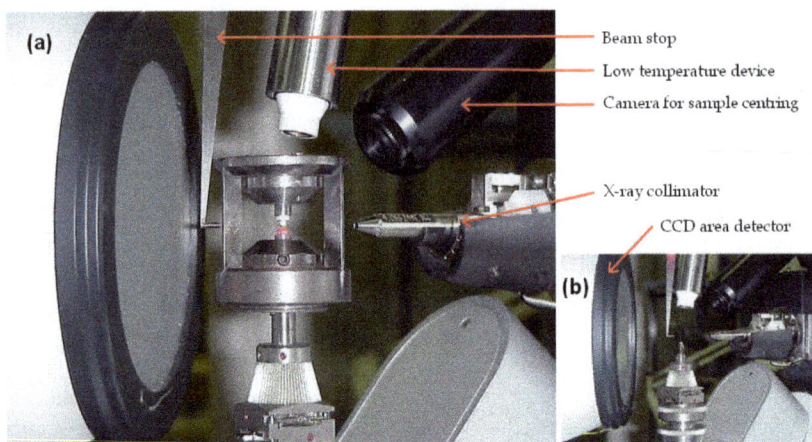

Figure 3. The high pressure cell developed by Guionneau et al. (2004): (a) view of the high pressure cell mounted on a 4-circle diffractometer equipped with low temperature device and CCD area detector; (b) comparison of the spatial occupied volume when a conventional goniometer head is used instead of the pressure cell. (Reprinted by permission from IOP Publishing Ltd).

To finish this non-exhaustive description, we can briefly mentioned two device developments in the field of very low temperature measurements. A mini-goniometer for X-ray diffraction studies down to 4 K on a four-circle diffractometer equipped with a CCD area detector (Fertey et al., 2007). The authors used a helium-bath orange cryostat (Brochier, 1977) and added a remarkable evolution: a magnetically coupled two-rotation-axis mini-goniometer permanently installed in the sample chamber of the helium-bath cryostat. This original apparatus was tested with success on organic charge-transfer compounds (Garcia et al., 2005; Garcia et al., 2007). An other low temperature system was developed to analyse Jahn-Teller distortion of $TmVO_4$ using conventional X-ray diffractometer between 0.22 and 3.5 K (Suzuki et al., 2002). The authors used a diffractometer equipped with a counter-monochromator and a scintillation counter, and an X-ray generator with rotating Cu Anode. To perform the ultra-low temperature measurements, a modified ^3He-^4He dilution refrigerator was mounted on the two-fold axis goniometer. In this later case, the study was also a success and the Jahn-Teller effect was fully investigated down to 0.22 K. In their paper, Suzuki et al. mentioned what was noticed above in this section concerning the use of bulky devices: "the x-ray beam generated by a Cu target, which passed through 4 walls of Be 2 mm thick and 4 aluminized mylar walls was reduced to approximately 1/100". This beam attenuation is generally inevitable when doing extreme conditions experiments and to guaranty good data collection, it must be used first class equipments on the diffractometer. Even if a lot of progresses concerning the performances of new diffractometers were obtained last years, extreme conditions measurements are sometimes impossible to perform at laboratories. Then, the best to do is to use Large scale facilities which benefits are now more and more highlighted. The term *Large scale facilities* stands for synchrotron research institutes and neutron facilities (spallation sources and reactors).

2.2. Why using large scale facilities in extreme crystallography?

The synchrotron working principle was shown during the Second World War by Oliphant (1943, University of Birmingham, UK). Three years later, it was shown for the first time (Goward & Barnes, 1946; Elder et al., 1947) that it is possible to built synchrotrons providing a satisfactory means of producing high energy electrons and X-rays. Nearly 70 years later, several synchrotrons of 3rd generation are operational throughout the world and are of first necessity for all the scientists communities (Bilderback et al., 2005). The X-ray beam, produces by a synchrotron, is a light beam (electromagnetic wave) which interacts with electrons surrounding the nuclei of heavy elements (Born, 1933, 1935; Born & Schrödinger, 1935). The actual phenomenal success of the synchrotron for performing extreme conditions crystallographic measurements comes from numerous undeniable advantages and particularly it's very intense (brilliance) and focussing beam and it's time structure allowing the measurement of small or ultra-dilute samples at different time scales. Thus, it is possible to study transient species and fast chemical reactions or to perform biological and geological studies using bulky sample environment allowing temperature and pressure changes without affecting the data.

Neutron diffraction also plays a key role in actual crystallographic studies under extreme conditions. The first nuclear reactor was built in 1942 by Enrico Fermi and Leó Szilárd (University of Chicago, USA) (Fermi, 1947). In October 2011, the International Atomic Energy Agency (IAEA) database indexed 241 operational research reactors in the world. Generally, experiments using neutrons to explore the properties of a material are based not only on the intensity of the beam (to exploit this criteria, synchrotron radiation would be preferable) but on the very nature of neutrons. It is a particle beam, allowing interactions with the nuclei and the magnetic moment of unpaired electrons in the sample, scattered by all elements also the light ones like hydrogen and its isotopes, with a deep penetration depth enables bulk studies of materials (Fermi, 1930, 1934, 1940; Szilárd, 1935). As neutrons can penetrate deeper in matter (unless exceptions like cadmium, samarium, europium or gadolinium elements for examples), sample environments requiring to put the sample inside them (e. g. pressure cells, furnaces, gas handling inside a cryostat…) can be used even if further walls, generally in aluminium, must be crossed by neutrons before to hit the sample. This neutron property is very appreciable and neutron facilities now include very well equipped services and highly qualified staff to carry out invaluable experiments as the Services for Advanced Neutron Environment (SANE) at the Institut Laue-Langevin in France, where the Orange Cryostat (Brochier, 1977) and the Cryopad (Tasset, 1989), among others equipments, were invented and built.

Synchrotron and neutron facilities are complementary techniques which allow to analyse structural details of materials at the atomic scale. Microscopic properties are then known and the challenge is to derive the macroscopic ones. To achieve that goal, investigations must concern both surface and bulk phenomena of materials in both dynamic and steady states. Large scale facilities can give fine details on the nuclear structure of materials, on the physical and chemical properties, on the electronic and magnetic structures, they can give crucial information about surface and interface states. All materials are concerned: metals, alloys, polymers, composites, colloids, glass, cement, biological macromolecules (viruses, enzymes, proteins…), minerals, food products, superconductors… The world-wide scientific community can observe since about ten years state of the art measurements in the above mentioned fields.

One field where X-ray and neutron measurements are really complementary is the domain of spin-transition molecular magnetic materials which is highly studied under *exotic environments* since 20 years (Gütlich & Goodwin, 2004; Bousseksou et al., 2011). These materials are especially interesting owing to their bistability properties which give them promising applications as data storage elements, thermal switches or display devices. These materials consist of a d^4 to d^7 transition metal coordinated to specific containing aromatic ligands whose ligand field is intermediate between weak and strong. Correlatively, the electron configuration of the central metal ion can be driven by several kinds of external perturbations such as pressure, intense magnetic field or light excitation. In the case of iron(II) complexes, the spin conversion is related to the electronic configuration of the ion changing from $t_{2g}^6 e_g^0$ in the low spin (LS) state to $t_{2g}^4 e_g^2$ in the high spin (HS) state. In addition to this electron redistribution, spin transition complexes present drastic structural variations between both phases, principally observed in the iron coordination sphere: an increase of the iron octahedron distortion and of the Fe-ligand bond length, by typically 0.2 Å, are observed for the LS → HS transition, generating a decrease of the molecular volume by several $Å^3$. Thus, owing to their high electronic and structural contrast, spin transition materials are good candidates for crystallographic studies under extreme conditions especially using Large scale facilities. Pressure dependence of the lattice parameters of several iron(II) complexes were derived from neutron powder diffraction (Legrand et al., 2008) and single-crystal X-ray synchrotron (Legrand, private communication) measurements at ambient and low temperatures. These studies highlight very promising results in the field of spin crossover phenomena under constraint, where crystallographic investigations under pressure are still very rare, especially because experimental conditions are difficult and sometimes not favourable. Concerning investigations under high magnetic field, Goujon et al. (2003) reported the photo-induced magnetization density of the iron(II) spin crossover compound $Fe(ptz)_6(BF_4)_2$ using an experimental setup allowing combined application of light irradiation (λ = 473 nm) at low temperature (2 K) and high magnetic field (5 T) (figure 4a). Thanks to this polarized neutron diffraction measurement, the temperature dependence of the magnetization was well described and the thermal relaxation of the photo-induced state well observed.

Moreover, electronic and structural properties of iron(II) spin transition complexes were studied under light irradiation and low temperature (< 15 K) using neutron Laue (Goujon et al., 2006) and X-ray synchrotron (Legrand et al., 2007; Pillet et al., 2008) diffraction (figure 4b). These latter authors notably refined for the first time the experimental electron density of a photo-induced metastable state in a spin transition compound. Finally, probing photo-induced phase transitions in molecular materials can also be performed using ultra-fast measurements at a picosecond time scale (Coppens et al., 2004; Collet et al. 2012a; 2012b).

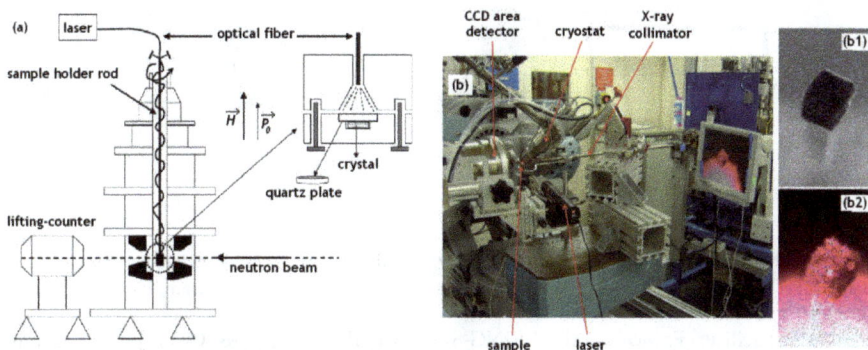

Figure 4. (a) Schematic experimental setup of the polarized neutron diffractometer; in the insert, the sample holder allowing light irradiation (Reprinted figure with permission from [Goujon et al., *Phys. Rev. B* 67, 220401(R) 2003]. Copyright 2012 by the American Physical Society). (b) The photocrystallographic experiment used by Legrand et al. (2007) showing the 6-circle diffractometer (BM01 beamline, ESRF - France) equipped with a large CCD area detector, an open flow He cryosystem and a He-Ne laser; inserts show the measured single-crystal (b1) without and (b2) with laser irradiation.

Real time investigation of temporally varying molecular structures during chemical reactions and phase transitions is a great challenge due to their ultrashort time scales. Collet et al. shown that it is possible to track complex reactions by time-resolved X-ray diffraction to temporal and spatial resolutions of 100 ps and 0.01 Å. Kind measurements give access to *molecular movies* during the transformation of matter induced by light irradiation. The authors succeed to describe for the first time the different steps of the dynamical process of several molecular materials and give their thought for the future: "these results pave the way for structural studies away from equilibrium and represent a first step toward femtosecond crystallography".

Other examples using the advantages of synchrotron radiations are numerous in structural biology, geology and materials sciences. The atomic structure of the bluetongue virus core (700 Å in diameter), containing about 1000 protein components self-assembled, was determined at a resolution of 3.5 Å (Grimes et al., 1998). The atomic structure of large ribosomal subunits were refined (Ban et al., 2000; Wimberly et al., 2000) providing a wealth of information about RNA and protein structure, protein-RNA interactions and ribosome assembly. The high brilliance of synchrotron sources in combination with fast detectors was used to perform time-resolved radiography of metal foam formation (Banhardt et al., 2001) (figure 5a). The possibility to have a micro-focus beam was applied to the study of the structure of spider silk (Riekel et al., 1999; Riekel & Vollrath, 2001) (figure 5b). The beam coherence properties enable new imaging techniques to do quantitative phase imaging and holotomography to probe materials with a micrometer resolution (Cloetens et al., 1999).

Nowadays, highlight articles are still numerous and new instrumental and device developments allow unthinkable experiments even ten years ago. Using highly brilliant X-ray beams at the APS (Advanced Photon Source, USA), ESRF (European Synchrotron

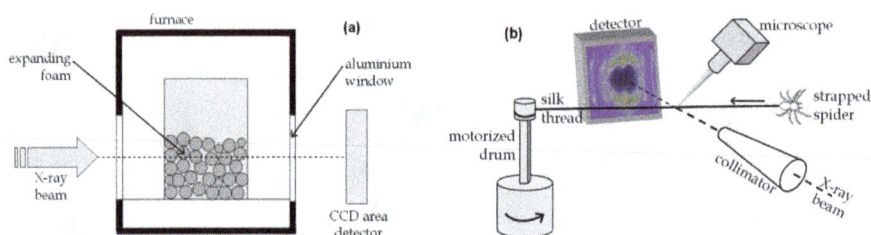

Figure 5. Schematic views of the experimental setup for real-time studies (a) of metal foam formation (Banhardt et al., 2001) and (b) of structural properties of spider silk (Riekel & Vollrath, 2001).

Radiation Facility, France) and SPRING-8 (Super Photon ring-8 GeV, Japan), teams uncovered subtle details of how cyan fluorescent proteins store incoming energy and retransmit it as fluorescent light (Goedhart et al., 2012), or investigated structure-properties relationships of biological mesocrystal in the adult sea urchin spin (Seto et al., 2012). Combination of in situ synchrotron X-ray absorption techniques and molecular dynamics simulations were used to determine the density range of primitive lunar melts at pressures equivalent to those in the lunar interior (4.5 GPa and 1800 K) (Van Kan Parker et al., 2012): these extreme conditions were generated with tiny samples heated thanks to high electric current while squashing them in a press (figure 6); knowing the attenuation of the synchrotron X-ray beam through both the solid and molten parts of the sample, the density at high pressure and high temperature could be measured. High pressure in situ X-ray diffraction and specific volume measurements on isotactic poly(4-methyl-1-pentene) melt have uncovered abrupt changes in the pressure dependence of microscopic structure as well as that of macroscopic density (Chiba et al., 2012). It was proved another time that pressure has an essential role in the production and control of superconductivity (Sun et al., 2012): it is reported that in the superconducting iron chalcogenides, a second superconducting phase suddenly re-emerges at a critical temperature Tc reaching 48.0 K and above 11.5 GPa, after the Tc drops from the first maximum of 32 K at 1 GPa. Metallic liquid silicon at 1787 K was investigated using X-ray scattering (Okada et al., 2012): the results show persistence of covalent bonding in liquid silicon and provide support for the occurrence of theoretically predicted liquid-liquid phase transition in supercooled states. To finish, it can be pointed out the crystallographic structure refinement of a protein, PthXo1 (which is a transcription activator–like, TAL, effector), encoded by an important group of harmful plant pathogens (Mak et al., 2012): understanding DNA recognition by TAL effectors may facilitate rational design of DNA-binding proteins with biotechnological applications.

Using neutron diffraction on reactors and spallation sources, researchers also push back the limits of measurements as for examples in the observation of a roton-like excitation in a monolayer of liquid ^{3}He, a Fermi liquid (Godfrin et al., 2012). A new class of magnetic ionic liquid surfactants showing remarkable effects on surface and interfacial tension and allowing access to magneto-responsive emulsions and new methods of separation, recovery, catalysis, and potential magnetophoretic applications were analyses by small-angle neutron scattering (SANS) (Brown et al., 2012). Using neutron diffraction studies, along with

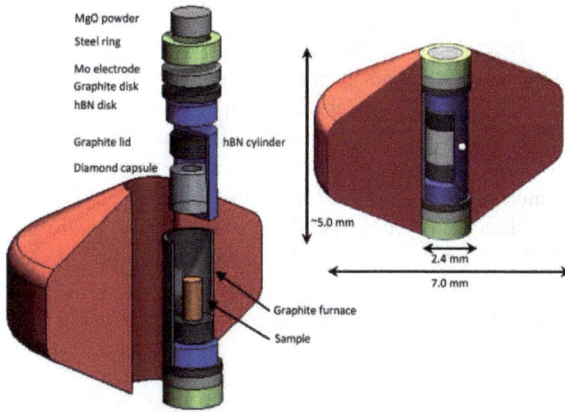

Figure 6. Drawing of the high-pressure cell assembly for the synchrotron X-ray experiments used to study the density range of primitive lunar melts (Van Kan Parker et al., 2012). The artificial moon rock samples (orange) were placed inside the ring-shaped, natural diamond sample holder (grey) which in turn was surrounded by a large, disk-shaped container (red). (Reprinted by permission from Macmillan Publishers Ltd: *Nature Geoscience*, copyright 2012)

computational chemistry, the mystery of lead oxide, the battery anode material of lead-acid batteries, was solved (Scanlon et al., 2011): oxygen vacancies give metallic behaviour by freeing up electrons to carry electrical current, neutron experiment put in evidence that commercial lead oxide powder is oxygen deficient by showing that oxygen sites were 1.6% vacant. Using high resolution neutron diffraction under high temperature and high magnetic field, it was shown that the thermal evolution of the helimagnetic state in CoMnSi is accompanied by a change in interatomic distances of up to 2%, the largest ever found in a metallic magnet (Barcza et al., 2010); this important result could lead to more efficient fridges and cooling systems.

The above overview has to remain incomplete with a number of fields left out. However, one can realize the wealth of extreme condition experiments which can be performed nowadays at laboratories and, particularly, using Large scale facilities which benefit of the last innovations in terms of sample environments.

2.3. Crystallographic experiments: advances in sample environments

Sample environment equipments like pressure cells, cryostats, etc, are essential adjuncts to perform with success the analysis of materials into a state or a phase with special behaviours and properties. This section is a brief overview of the most employed stately sample environments available mainly on Large scale facilities. The technical points of the discussed devices are not detailed here and the reader could find further information in the references given through section 2.3 and in specific reviews as written by Bailey (2003) or Mignot (2008). Specific Information is also related in reviews from the Workshop "sample environments in neutron and X-ray experiments" held at the Institut Laue-Langevin (France) in 1984 (Vettier et al., 1984). There are various ways of tuning materials properties. In this section, the most employed sample environments are described, but the possibilities are miscellaneous and it is not possible to make an exhaustive list. *Exotic* sample environments like, among others, electric fields (Guillot et al., 2002; Hansen et al., 2004) or in situ reactions using gas handling (Walker et al., 2009; Price et al., 2010), as well as experiments performed with time resolved methods (Gembicky & Coppens, 2006; Nozawa et al., 2007) are not discussed. The investigations using these increasingly sophisticated experimental methods contribute enormously to the understanding of structural properties and progress in solid state physics and chemistry, biology and in geosciences. Advances in sample environments help to perform crystallographic experiments in extreme conditions and extend current areas of investigations.

2.3.1. Low temperature

Cryogenic conditions are by far the most used sample environments principally to reduce vibrations and thermal motions of atoms, thus the structural refinement is improved. Moreover, phenomena like phase transitions are also generally induced at low temperature. Different devices could be used for low temperature experiments depending the desired temperature range: liquid helium bath cryostat (Orange Cryostat (Brochier, 1977)) for 300 K to 1.5 K; ^{3}He sorption systems for 3 K to 300 mK; ^{3}He/^{4}He dilution refrigerator systems below 1 K. Of course, there exist many others systems allowing to reach low and very low temperatures as N_2 or He gas-flow systems, helium closed cycle refrigerators (Displex), etc.

The Orange cryostat is now regarded as the international standard for low temperature experiments performed above 1.5 K. Some variations exist differing in the sample stick geometry (diameters and length). Cryofurnace and Cryomagnet versions were also developed. Nowadays, more and more experiments are based on combining sample environments. Thus, special versions of the Orange cryostat or Displex were built to meet the demand of users. Among the list of special purposes, it can be noted versions allowing the use of a high pressure cell (even Paris-Edinburgh cell) (figure 7), a gas handling for in situ chemical reactions analysis, an omega rotating for single crystals measurements, a sample changer. Because these cryostats are very common at Large scale facilities and are user friendly, recently improvements were made to have the possibility to do ultra-low temperature measurements using new sample stick inserts including high performance dilution refrigerator with a base temperature of less than 15 mK (Neumaier et al., 1984).

2.3.2. High temperature

Contrary to cryogenic devices, there is no standard high temperature apparatus. The designs depend of the working temperature range, of the type of experiment, of the diffraction beam (X-ray or neutron). However, a typical furnace configuration places a sample into the centre of a thin cylindrical metal foil element, of about 50 μm thick, surrounded by heat shields. According to the chosen metal, the achieved temperature could

Figure 7. (a) Details of the high-intensity 2-axis diffractometer D20 (Hansen et al., 2008) at the Institut Laue-Langevin (France) with a special version of the helium closed cycle refrigerator (Displex) allowing the use of (c) Paris-Edinburgh high pressure cell. Cryostats are now fully self-governed (b) with automated cold valve, pressure transducer which monitors the exhausting helium gas pressure and temperature controller which adjust the liquid helium flow to the set temperature value.

be typically up to 1400-2000 K. Closed-shell furnaces (Kuhs et al., 1993) and mirror furnaces (Lorenz et al., 1993) were developed for powder and single-crystal analysis at high temperature. This last furnace type has the advantage to allow air-operated experiments with no heat shield, thus removing apparatus diffraction. As for cryogenic, variations are developed to combine for example high temperature and high pressure (Falconi et al., 2005). Finally, the last developments concern the design and built of furnace using aerodynamic levitation for reaching ultra-high temperatures up to 3000 K. This technique is well described by Hennet et al. (2007): "A spherical sample (3mm in diameter) is placed on a levitator that contains a convergent-divergent nozzle enabling the diffusion of a regulated gas flow onto the sample from below. This enables the sphere to remain in a stable position without any contact with the nozzle. When the sample levitates, it is heated by two CO_2 laser from the bottom through the hole in the nozzle. The temperature is measured with one or two optical pyrometers". Aerodynamic levitation apparatus were optimized for neutron (Hennet et al., 2006) and X-ray (Hennet et al., 2002; Sakai et al., 2005) scattering measurements. They were applied with success for examples to study the liquid states and the solidification of $CaAl_2O_4$ (Hennet et al., 2008) and Y_2O_3 (Hennet et al., 2003; Cristiglio et al., 2007).

2.3.3. High magnetic field

Studying materials under high magnetic field could be done ether by X-ray diffraction (Katsumata, 2005) or by neutron scattering (Brown et al., 2002; Givord et al., 2004).

However, one of the advantages of neutrons is their possible interaction with the unpaired electron magnetic field. Thus, magnetic studies deal with a wide proportion of the total neutron diffraction experiments. As magnetic properties variations are generally observed at low temperature, superconductor cryomagnets are used for kind studies, with applied vertical or horizontal fields. The available magnetic field range is wide and nowadays fields around 10 T are commonly obtained, some devices also produce fields between 14.5 T and 17 T (Nietz, 2003). It can finally be noticed that punctual apparatus are developed through the world allowing 1 s pulsed magnetic fields up to 25 T (Nojiri et al., 1998) or 40 T (Grössinger, et al., 2004), in order to achieve fields up to 70 T in the future. Thermal neutrons are thus largely used to refine magnetic structures in addition to the nuclear structure. Hot polarized neutron diffraction is also a very interesting technique used to obtain crucial information about the magnetic form factors (reflecting the state of the magnetic ions) and magnetisation distributions of para-, ferro- and some antiferro-magnetic single-crystal materials if measurements are performed under high magnetic field configuration. On the other hand, one can performed zero-field neutron polarimetry measurements using Cryopad (CRYOgenic Polarization Analysis Device) (Tasset et al., 1999; Lelièvre-Berna et al., 2005; Takeda et al., 2005) to refine complex antiferromagnetic structures or to perform accurate determination of antiferromagnetic distributions.

2.3.4. High pressure

High pressure condition is certainly the more used sample environment after low/high temperature. This field of research is also one of the most well-informed facing the number of published papers and reviews concerning experimental results or technical information (see for examples Paszkowicz, 2002; Boldyreva, 2008; Katrusiak, 2008; and references therein). Material studies under pressure are nowadays numerous due to the increase capabilities of modern X-ray diffractometer and the making of dedicated instruments on synchrotrons and neutron facilities. At the present time, about 30 synchrotron beamlines in the world (at 11 synchrotrons) allow high-pressure measurements to be performed, with possibility of high temperature too. It could be pointed out that 2012 marks the 35th anniversary of high pressure diffraction at synchrotron (Buras et al., 1977). The application of pressure on a material may induce major structural variations, particularly a modification in the weak intermolecular contacts and reorganization of the electronic structure, but also phase transitions or chemical reactions. Anyway, the application of pressure on a system induces more effects than the application of cryogenic temperatures, which is of interest to physicists, chemists, biologists and geologists. Pressure can give substantial information about structure-properties relationships and new thermodynamic phenomena (phase diagrams, polymorphism, cohesion forces, transformations...). Moreover, temperature variations can now be applied in combination to high pressure due to the high brilliance of modern sources at facilities which can bypass the unavoidable absorption of the beam by the cryostat or furnace and the pressure cell containing the tiny studied sample. Now, it is possible to perform experiments at 100 GPa and further thousands of Celsius degrees (Fiquet et al., 1999; Akahama et al., 2002; Guo et al., 2002; Hu et al., 2002; Orosel et al., 2012) using synchrotron radiations, this is achieved with the laser heating technique (Bassett, 2001). Due to the relatively large sample volume required for neutron diffraction, high pressure measurement are always a challenge and reached maximum pressures are not so high: it was recently reported a high-pressure single-crystal neutron diffraction to 10 GPa at ambient temperature taking advantage of the neutron diffractometer D9 at the Institut Laue-Langevin (France) (Bull et al., 2011). To succeed those measurements, teams generally use DAC (Diamond Anvil Cell) (Jamieson et al., 1959; Weir et al., 1959), MAC (Moissanite Anvil Cell) (Xu et al., 2002) or LAC (Large Anvil Cell, as Paris-Edinburgh cell) (figure 7c) (Ohtani et al., 1977; Besson et al., 1992; Mezouar et al., 1999; Le Godec et al., 2003) pressure cells. For low and moderate pressures up to about 3 GPa at ambient temperature, there are also piston-in cylinder clamped cells (McWhan et al., 1974) and continuously loaded cells (Paureau, 1975). From the firsts diffraction measurements under pressure (Barnett et al., 1963; Bassett et al., 1966) to the last ones (Basu et al., 2012; Orosel et al., 2012; Tulk et al., 2012) efforts still continue to increase the available pressure range in both static and dynamic conditions, and also to improve instruments (Utsumi et al., 2002) or allow new techniques (Bromiley et al., 2009).

3. New instrumental developments at dawn of the third millennium

Using extreme conditions, researchers can consider not only to extend the accessible parameters scale to measure some physical constants or to verify existing theories, but also to point out new behaviours and states which increase our knowledge of materials properties. With progresses in the domains of available sources as 3rd generation synchrotrons or powerful neutron sources and of new sample environment devices as explained in part 2., extreme conditions measurements have known since 10 years a meteoric rise in all scientific fields for fundamental studies, for examples, of complex fluids, high correlated electrons systems, geological and biological materials at the origin of earth or life. However, will this keen interest continue for a long time? Are technological innovations always possible in that scientific area to build more efficient instruments and devices? Could it be possible to go further the actual limits of high pressures, high and low temperatures, applied magnetic fields? Is it really necessary to cross these limits as numerous material properties are still unexplored? All these questions have one simple answer: yes. Going further in the field of crystallography under extreme conditions is also expanding our knowledge of the world and life by probing the matter at the border of the unknown. Scientist needs everyday new challenges and the one to push back those limits is a necessity for him, but mostly essential for mankind.

In this way, *extreme conditions beamline* projects have emerged throughout the world at Large scale facilities. It will be pointed out thereafter some examples of instrumental projects which were, or will be, realized to explore materials properties giving access to new non-ambient conditions. The proposed examples are a non-exhaustive list of the actual developments in crystallography under extreme conditions. These examples are chosen on the most powerful Large scale facilities around the world in terms of the most important parameters (energy, current...): SPring-8 (Japan), APS (USA) and ESRF (France) are the chosen synchrotron sources (table 1); ISIS (neutron spallation source, UK) and the ILL (neutron reactor, France) are the chosen neutron facilities.

Name	Location	Energy (GeV)	Perimeter (m)	Current (mA)	Emittance (mrad)	Number of straights
SPRING-8	Japan	8	1436	100	3	48
APS	USA	7	1060	100	3	40
ESRF	France	6	844	200	3.8	32
SSRF	China	3.5	396	300	4.8	20
ALBA	Spain	3	268.8	250	3.7	24
Australian Synchrotron	Australia	3	216	200	8.6	14
DIAMOND	UK	3	560	300	2.7	24
SPEARS3	USA	3	240	500	18	18
CLS	Canada	2.9	171	500	18	15
SOLEIL	France	2.85	354	500	3.1	24

Table 1. The most important parameters of the 10 world-wide most powerful 3rd generation light sources available in 2012.

During the last 10 years, numerous Large scale facilities begun an upgrade of their instruments and installations. This is partly due to the increase demand for high-brilliance X-ray and high-flux neutron beams by user communities requiring always an increase of the instrument performances and of the allocated beamtime. Users want to perform measurements on smaller samples mostly in *exotic* environments. One of the engineering goals for the upgrades is making available advanced sample environments with extreme conditions on the beamlines whilst fitting in the tight space around the sample also detectors and monitoring equipment. The challenge for the facilities will be to commonly supply extreme environments like pressures up to 100 GPa, low temperature below 1 K, high temperature above 3000 K, high magnetic field up to 50 T in pulsed mode and 30 T in continuous mode. To achieve these goals, facilities must also invest in the engineering of new dedicated instruments and particularly in the development of new detectors which have to be more sensitive, more efficient (especially for high-energy X-rays) and faster (notably in the sub-millisecond range to perform time resolved measurements). Upgraded a Large scale facility is difficult, long and very expensive (table 2).

Name	Type of facilities		Period	Total cost
SPRING-8	synchrotron	upgrade	2012-2019	375 M€
APS	synchrotron	upgrade	2010-2018	275 M€
ESRF	synchrotron	upgrade	2008-2017	290 M€
DIAMOND	synchrotron	construction	2002-2007	460 M€
SOLEIL	synchrotron	construction	2000-2006	403 M€
ILL	neutron reactor	upgrade	2001-2014	85 M€
ISIS- TS2	neutron spallation source	construction	2003-2009	200 M€

Table 2. Estimated time and total cost of the upgrade or construction of some synchrotron and neutron facilities around the world.

SPring-8 light source will be upgraded in order to advance promising science and to support industrial innovations that will improve our life and contribute to a more sustainable society (SPring-8, 2012). This upgrade programme is principally focused on the construction of a new ring in place of the actual one, retaining the existing insertion device beamlines. In 2019, the new ring (SPring-8 II) would produce 1000 times higher brilliance than the present SPing-8 (new stored current of 300 mA instead of 100 mA presently) with a short X-ray pulse option of around 1 ps. The frontiers of science will be definitely opened. Instruments like BL02B1 (designed for analyses of single crystal structures and for investigations of phase transitions under external fields), BL04B1 (dedicated to high pressures and high temperatures researches in geoscience), BL10XU (designed to perform X-ray diffraction structure analyses under high pressure - 300 GPa - and low temperature - 10 K - or high temperature - 3000 K -) or BL40XU (high flux beamline for time-resolved x-ray diffraction) will benefit of new opportunities.

On April 2010, the US Department of Energy approved the Advanced Photon Source (APS) upgrade project (APS, 2011). The upgrade will provide high energy, high brilliance, short pulse, new and upgraded instruments. The objective is to push the stored current to 150 mA, to upgrade 6 beamlines and to build at least 6 new ones between 2010 and 2018 (Mills, 2011). The new beamlines will especially deal with the overarching theme *Real materials under real conditions in real time*. It means an easier access for users to high pressures, low/high temperatures and in situ chemical reactions. One frontier for X-ray science in the 21st century is to combine atomic-level spatial and temporal information, a frontier which would be crossed after the upgrade. For extreme conditions experiments, the APS had plane to upgrade 4 instruments and to build 2 new ones. For examples, the outboard branch of 16-ID will be dedicated to high pressure spectroscopy using sub-µm probes (beam size of 100-500 nm with a flux $\geq 10^{12}$ photons.s^{-1} at sample position).

In this context, the ESRF (ESRF, 2007) has planned an upgrade programme for its instruments, like ID20 (beamline for magnetic and resonant X-ray scattering investigations under extreme conditions; Paolasini et al., 2007), ID22 (micro-fluorescence, imaging and diffraction beamline, Martinez-Criado et al., 2007), ID24 (Energy dispersive X-ray absorption spectroscopy beamline) or ID27 (high pressure beamline; Mezouar et al., 2005). The ID27 source was upgraded to the most recently developed cryogenic or superconducting undulator to obtain the maximum flux from the ESRF machine after its upgrade. This very high flux beamline will provide new approaches to very challenging problems such as the search for a new metallic superfluid state of matter predicted at pressures above 400 GPa, or to perform ultra-high pressure and temperature X-ray emission spectroscopy in combination with X-ray diffraction in the laser heated diamond anvil cells. In addition to this upgrade, the ESRF will build two new dedicated high pressure beamlines for diamond anvil cells and Paris-Edinburgh research. Time-resolved diffraction is also a priority for the ESRF which propose to build a dedicated beamline covering the timescale from 10 picoseconds to seconds in studies of physical, chemical and biological processes. The 4-circle diffractometer of this beamline will include Laue diffraction and classical diffraction, a very intense focused beam, a new 3 kHz chopper which isolates single pulses of X-rays from the timing modes and a fast readout (0.1 s) Frelon CCD based camera. An other upgrade example concerns the ID20 magnetic scattering beamline with an extend of its experimental conditions in order to provide new research opportunities. It is also planned to build a dedicated station for very low temperatures below 1 K, high magnetic fields up to 17 T and high pressure up to 5 GPa at 1.5 K.

The world's leading spallation neutron source, ISIS, has contributed significantly to many of the major breakthroughs in materials science, physics and chemistry since it was commissioned in 1985. To keep the UK at the forefront of neutron research, it was decided in 2003 to build a second spallation neutron source, ISIS Target Station 2, where three key areas of science are planned to be investigated in priority: soft matter, bio-sciences and advanced materials. ISIS TS-2 project was completed in 2009 on time and to budget. There are currently 7 available instruments (and 4 instruments under building) at ISIS TS-2 which has a capacity for a total of 18 instruments in the future, adding to the 20 instruments already available at ISIS Target Station 1. Concerning the new instruments dedicated to extreme condition measurements, the available WISH instrument (Chapon et al., 2011) is a long-wavelength diffractometer primarily designed for powder diffraction at long d-spacing in magnetic and large unit cell systems, with the option of enabling single-crystal and polarised beam experiments. The WISH sample environments include a dedicated low background cryostat (with ultra-low temperature inserts), a dedicated 13.6 T vertical magnet, but also the standard ISIS furnaces, pressure cells and gas rigs. Futhermore, there is the IMAT instrument (Kockelmann et al., 2007), in design phase and operating in 2015, which will be a neutron imaging and diffraction instrument for materials science, materials processing and engineering. IMAT will offer a unique combination of imaging and spatially resolved diffraction modes for *tomography-driven diffraction*: residual stresses inside engineering-sized samples can be more effectively analysed if the diffraction scans are guided by radiographic data. Finally, the future EXEED instrument (McMillan and Tucker, 2007) will be a neutron time-of-flight diffractometer optimised for extreme environment studies of materials which will complement the capabilities of WISH on Target Station 2 and PEARL on Target Station 1. The experiments will be performed at high pressure, above 50 GPa using diamond anvil cells, under combined very low temperatures (mK), very high temperatures (2000 K, including laser heating in cells) or intense magnetic field, up to 10 T.

Finally, the neutron reactor installations and instruments at the ILL are undergoing a modernisation phase called the *Millennium programme 2001-2014*, 30 years after the first experiments in 1972. During the first phase (2001-2008) of the ILL upgrade, significant advances had been provided. The efficiency of the instruments has been boosted by a factor of 19. Moreover, 6 new instruments were built and 8 others were upgraded. In the second phase (2008-2014), 7 new instruments are planned to be built and 4 actual instruments upgraded. For extreme conditions experiments, it was planned a new neutron diffractometer XtremeD (Rodriguez-Velamazan et al., 2011). This instrument will be optimized for high pressure, up to 30 GPa, and high magnetic field, up to 15 T in continuous mode, studies for both single crystals and powders. The diffractometer, mainly dedicated to study molecular chemistry and magnetism under extreme conditions, will include a large 2D position-sensitive detector and a radial oscillating collimator to suppress background.

It is clear, with all the above mentioned new instrumental developments at almost worldwide Large scale facilities, that the challenge for crystallography in the next years is to perform non-equilibrium and extreme condition measurements. All the domains of sciences are concerned in order to increase the knowledge and the understanding of materials behaviours and properties under external perturbations. It can be asserted without ambiguity that the present century is the one of *crystallography under extreme conditions*.

4. Conclusion

Since the first crystallographic experiment at the beginning of the 20th century, scientists have always whished to push back frontiers of measurements. Obtaining the structure-properties relationships in static and dynamic modes is one fundamental goal to increase our knowledge of materials and verify existing theories. This was largely made possible by the arrival of modern Large scale facilities allowing material investigations under extreme conditions. State-of-the-art crystallographic measurements discussed in this review and recent upgrades achieved at most facilities augur that limits will be increasingly crossed in the next years, giving rise to unsuspected highlight researches in all domains of sciences. The considered perspectives will offer new opportunities in crystallography.

Author details

Legrand Vincent
LUNAM Université - Université de Nantes - Ecole Centrale Nantes,
Institut de Recherche en Génie Civil et Mécanique (GeM), France
vincent.legrand@univ-nantes.fr

5. References

Akahama, Y., Kawamura, H. & Le Bihan, T. (2002). A new distorted body-centred cubic phase of titanium (δ-Ti) at pressure up to 220 GPa. *J. Phys.: Condens. Matter*, Vol.14, pp.10583-10588

APS (2011). Conceptual design report. Advanced Photon Source upgrade project. *http://aps.anl.gov/Upgrade/Documents/*

Bailey, I. F. (2003). A review of sample environments in neutron scattering. *Z. Kristallogr.*, Vol.218, pp.84-95

Ban, N., Nissen, P., Hansen, J., Moore, P. B. & Steitz, T. A. (2000). The complete atomic structure of the large ribosome subunit at 2.4 Å resolution. *Science*, Vol.289, pp.905-920

Banhart, J., Stanzick, H., Helfen, L. & Baumbach, T. (2001).Metal foam evolution studied by synchrotron radioscopy. *Appl. Phys. Letters*, Vol.78, pp.1152-1154

Barcza, A., Gercsi, Z., Knight, K. S. & Sandeman, K. G. (2010). Giant Magnetoelastic Coupling in a metallic helical metamagnet. *Phys. Rev. Letters*, Vol.104, pp.247202

Barnet, J. D. & Hall, H. T. (1964). High pressure – high temperature, X-ray diffraction apparatus. *Rev. Sci. Instrum.*, Vol.35, pp.175-182

Bassett, W. A., Takahashi, T. & Stook, P. W. (1967). X-ray diffraction and optical observations on crystalline solids up to 300 kbar. *Rev. Sci. Instrum.*, Vol.38, pp.37-42

Bassett, W. A. (2001). The birth and development of laser heating in diamond anvil cells. *Rev. Sci. Instrum.*, Vol.72, pp.1270-1272

Basu, A., Chandra, A. Tyagi, A. K. & Mukherjee, G. D. (2012). Reppearance of ferroelectric soft modes in the paraelectric phase of $Pb_{1-x}Ca_xTiO_3$ at high pressures : Raman and X-ray diffraction studies. *J. Phys.: Condens. Matter*, Vol.24, pp.115404

Beson, J. M., Nelmes, R. J., Hamel, G., Loveday, J. S., Weill, G. & Hull, S. (1992). Neutron powder diffraction above 10 GPa. *Physica B*, Vol.180-181, p.907-910

Bilderback, D. H., Elleaume, P. & Weckert, E. (2005). Review of third and next generation synchrotron light sources. *J. Phys. B: At. Mol. Opt. Phys.*, Vol.38, pp.S773-S797

Boldyreva, E. V. (2008). High-pressure diffraction studies of molecular organic solids. A personal view. *Acta Cryst. A*, Vol.64, pp.218-231

Born, M. (1933). Modified field equations with a finite radius of the electron. *Nature*, Vol.132, pp. 282

Born, M. (1935). Quantised field theory and the mass of the proton. *Nature*, Vol.136, pp. 952-953

Born, M. & Schrödinger, E. (1935). The absolute field constant in the new field theory. *Nature*, Vol.135, pp. 342

Bousseksou, A., Molnar, G., Salmon, L. & Nicolazzi, W. (2011). Molecular spin crossover phenomenon:recent achievements and pospects.*Chem. Soc. Rev.*, Vol.40, pp.3313-3335

Bragg, W. H. (1912). X-rays and crystals. *Nature*, Vol.90, pp. 219

Bragg, W. H. (1913a). X-rays and crystals. *Nature*, Vol.90, pp. 572

Bragg, W. H. (1913b). The reflection of X-rays by crystals. *Nature*, Vol.91, pp. 477

Bragg, W. H. (1915a). The distribution of the electrons in atoms. *Nature*, Vol.95, pp. 344

Bragg, W. H. (1915b). The structure of magnetite and the spinels. *Nature*, Vol.95, pp. 561

Bragg, W. H. & Bragg, W. L. (1913). The structure of diamond. *Nature*, Vol.91, pp. 557

Bragg, W. L. (1920). Crystals structure. *Nature*, Vol.105, pp. 646-648

Brochier, D. (1977). Cryostat à température variable pour mesures neutroniques ou optiques. *ILL Tech. Report 77/74*

Bromiley, G. D., Redfern, S. A. T., Le Godec, Y., Hamel, G. & Klotz, S. (2009). A portable high-pressure stress cell based on the V7 Paris-Edinburgh apparatus. *High pressure Research : an International Journal*, Vol.29, pp.306-316

Brown, P. J., Forsyth, J. B., Lelièvre-Berna, E. & Tasset, F. (2002). Determination of the magnetization distribution in Cr_2O_3 using spherical neutron polarimetry. *J. Phys.: Condens. Matter*, Vol.14, pp.1957-1966

Brown, P., Bushmelev, A., Butts, C. P., Cheng, J., Eastoe, J., Grillo, I., Heenan, R. K. & Schmidt, A. M. (2012). Magnetic control over liquid surface properties with responsive surfactants. *Angewandte Chemie Int. Ed.*, Vol. 51, pp.2414-2416

Bull, C. L., Guthrie, M., Archer, J., Fernandez-Diaz, M-T., Loveday, J. S., Komatsu, K., Hamidov, H. & Nelmes, R. J. (2011). High-pressure single-crystal neutron diffraction to 10 GPa by angle-dispersive techniques. *J. Appl. Cryst.*, Vol.44, pp.831-838

Buras, B., Olsen, J. S., Gerward, L., Will, G. & Hinze, E. (1997). X-ray energy-dispersive diffractometry using synchrotron radiation. *J. Appl. Cryst.*, Vol.10, pp.431-438

Chapon, L. C., Manuel, P., Radaelli, P. G., Benson, C., Perrott, L., Ansell, S., Rhodes, N. J., Raspino, D., Duxbury, D., Spill, E. & Norris, J. (2011). Wish: the new powder and single crystal magnetic diffractometer on the second target station. *Neutron News*, Vol.22, pp.22-25

Chasseau, D., Guionneau, P., Gaultier, J., Barrans, Y., Ducasse, L., Kepert, C. J., Day, P. & Kurmoo, M. (1997). Crystal structures of $(BEDT-TTF)_3CuBr_4$ at 10 K and 10 kbar. *Synthetic Metals*, Vol.86, pp.2045-2046

Chiba, A., Funamori, N., Nakayama, K., Ohishi, Y., Bennington, S. M., Rastogi, S., Shukla, A., Tsuji, K & Takenaka, M. (2012). Pressure-induced structural change of intermediate-range order in poly(4-methyl-1-pentene) melt. *Phys. Rev. E*, Vol.85, pp.021807

Clementi, E. & Roetti, C. (1974). Roothaan-Hartree-Fock atomic wavefunctions. *Atomic data and nuclear data tables*, Vol.14, pp.177-478

Cloetens, P., Ludwig, W., Baruchel, J., Van Dyck, D., Van Landuyt, J., Guigay, J. P. & Schlenker, M. (1999). Holotomography: quantitative phase tomography with mocrometer resolution using hard synchrotron radiation X-rays. *Appl. Phys. Letters*, Vol.75, pp2912-2914

Collet, E., Lorenc, M. Cammarata, M., Guérin, L., Servol, M., Tissot, A., Boillot, M-L., Cailleau, H. & Buron-Le Cointe, M. (2012a). 100 picosecond diffraction catches structural transients of laser-pulse triggered switching in a spin crossover crystal. *Chem. Eur. J.*, Vol.18, pp.2051-2055

Collet, E., Moisan, N., Baldé, C., Bertoni, R., Trzop, E., Laulhé, C., Lorenc, M., Servol, M., Cailleau, H., Tissot, A., Boilot, M-L., Graber, T., Henning, R., Coppens, P & Buron-Le Cointe, M. (2012b). Ultrafast spin-state photoswitching in a crystal and slower consecutive process investigated by femtosecond optical spectroscopy and picosecond X-ray diffraction. *Phys. Chem. Chem. Phys.*, Vol.14, pp.6192-6199

Coppens, P., Formitchev, D. V., Carducci, M. C. & Culp, K. (1998). Crystallographiy of molecular excited states. Transition metal nitrosyl complexes and the study of transient species. *J. Chem. Soc., Dalton Trans.*, pp.865-872

Coppens, P., Vorontsov, I. I., Graber, T., Gembicky, M. & Kovalensky, A. Y. (2005). The structure of short-lived excited states of molecular complexes by time-resolved X-ray diffraction. *Acta Cryst. A*, Vol.61, pp.162-172

Cristiglio, V., Hennet, L., Cuello, G. J., Pozdnyakova, I., Bytchkov, A., Palleau, P., Fisher, H. E., Zanghi, D., Saboungi, P-L. & Price, D. L. (2007). Structural study of levitated liquid $Y2O3$ using neutron scattering. *J. Non-Crystalline Solids*, Vol.353, pp.993-995

Guo, Q., Mao, H. K., Hu, J., Shu, J. & Hemley, R. J. (2002). The phase transitions of CoO under static pressure to 104 Gpa. *J. Phys.: Condens. Matter*, Vol.14, pp.11369-11374

Destro, R., Loconte, L., Lo Presti, L., Roversi, P. & Soave, R. (2004). On the role of data quality in experimental charge density studies. *Acta Cryst. A*, Vol.60, pp.365-370

Elder, F. R.; Gurewitsch, A. M.; Langmuir, R. V. & Pollock, H. C. (1947). A 70 MeV Synchrotron. *J. Appl. Phys.*, Vol.18, pp. 810-818

ESRF (2007). Science and technology programme 2008-2017. *http://www.esrf.eu/AboutUs/Upgrade/documentation/purple-book/*, Vol.1-2

Falconi, S., Lundegaard, L. F., Hejny, C. & McMahon, M. I. (2005). X-ray diffraction study of liquid Cs up to 9.8 GPa. *Phys. Rev. Letters*, Vol.94, pp.125507

Fermi, E. (1930). Magnetic moment of atomic Nuclei. *Nature*, Vol.125, pp. 16

Fermi, E. (1934). Radioactivity induced by neutron bombardment. *Nature*, Vol.133, pp. 757

Fermi, E. (1940). Reactions produced by neutrons in heavy elements. *Nature*, Vol.146, pp. 640-642

Fermi, E. (1947). Elementary theory of the chain-reacting pile. *Science*, Vol.105, pp. 27-32

Fertey, P., Argoud, R., Bordet, P., Reymann, J., Palin, C., Bouchard, C., Bruyère, R., Wenger, E. & Lecomte, C. (2007). A mini-goniometer for X-ray diffraction studies down to 4 K on four-circle diffrctometers equipped with two-dimensional detectors. *J. Appl. Cryst.*, Vol.40, pp.526-531

Fiquet, G. & Andrault, D. (1999). Powder X-ray diffraction under extreme conditions of pressure and temperature. *J. Synchrotron Rad.*, Vol.6, pp.81-86

Formitchev, D. V., Novozhilova, I. & Coppens P. (2000). Photo-induced linkage isomerism of transition metal nitrosyl and dinitrogen complexes studied by photocrystallographic techniques. *Tetrahedron*, Vol.56, pp.6813-6820

Garcia, P., Dahaoui, S., Fertey, P., Wenger, E. & Lecomte, C. (2005). Crystallographic investigation of temperature-induced phase transition of the tetrathiafulvalene-*p*-bromanil, TTF-BA charge transfer complex. *Phys. Rev B*, Vol.72, pp.104115

Garcia, P., Dahaoui, S., Katan, C., Souhassou, M. & Lecomte, C. (2007). On the accuracy of intermolecular interactions and charge transfer: the case of TTF-CA. *Faraday Discussions*, Vol.135, pp.217-235

Gembicky, M. & Coppens, P. (2007). On the desing of ultrafast shutters for time-resolved synchrotron experiments. *J. Synchrotron Rad.*, Vol.14, pp.133-137

Givord, F., Boucherle, J-X., Lelièvre-Berna, E. & Lejay, P. (2004). The cerium magnetic form facto rand diffuse polarization in $CeRh_3B_2$ as functions of temperature. *J. Phys.: Condens. Matter*, Vol.16, pp.1211-1230

Godfrin, H., Meschke, M., Lauter, H-J., Sultan, A., Böhm, H. M., Krotscheck, E. & Panholzer, M. (2012). Observation of a roton collective mode in a two-dimensional fermi liquid. *Nature*, Vol.483, pp.576-579

Goedhart, J., Stetten, D., Noirclerc-Savoye, M., Lelimousin, M., Joosen, L., Hink, M. A., Weeren, L., Gadella Jr, T. W. J. & Royant, A. (2012). Structure-guided evolution of cyan fluorescent proteins towards a quantum yield of 93%. *Nature Communications*, Vol.3, pp.751

Goujon, A., Gillon, B., Gukasov, A., Jeftic, J., Nau, Q., Codjovi, E. & Varret, F. (2003). Photoinduced molecular switching studied by polarized neutron diffraction. *Physical Review B*, Vol.67, pp.220401(R). DOI: 10.1103/PhysRevB.67.220401 http://prb.aps.org/abstract/PRB/v67/i22/e220401

Goujon, A., Gillon, B., Debede, A., Cousson, A., Gukasov, A., Jeftic, J., McIntyre, G. & Varret, F. (2006). Neutron Laue diffraction on the spin crossover crystal [Fe(1-n-propyltetrazole)$_6$](BF$_4$)$_2$ showing continuous photoinduced transformation. *Physical Review B*, Vol.73, pp.104413

Goward, F. K. & Barnes D. E. (1946). Experimental 8 MeV synchrotron for electron acceleration. *Nature*, Vol.158, pp. 413

Graafsma, H., Svensson, S. O. & Kvick, A. (1997). An X-ray charge density feasibility study at 56 keV of magnesium formate dihydrate using a CCD area detector. *J. Appl. Cryst.*, Vol.30, pp.957-962

Grimes, J. M., Burroughs, J. N., Gouet, P., Diprose, J. M., Malby, R., Zientara, S., Merten; P. P. C. & Stuart, D. I. (1998). The atomic structure of the bluetongue virus core. *Nature*, Vol. 395, pp.470-478

Grössinger, R., Sassik, H., Mayerhofer, O., Wagner, E. & Schrenk, M. (2004). Austromag: pulsed magnetic fields beyond 40 T. *Physica B*, Vol.346-347, pp.609-613

Guillot, R., Allé, P., Fertey, P., Hansen, N. K. & Elkaïm, E. (2002). Diffraction measurements from crystals under electric fields: instrumentation. *J. Appl. Cryst.*, Vol.35, pp.360-363

Guionneau, P., Brigouleix, C., Barrans, Y., Goeta, A. E., Létard, J-F. Howard, J. A. K., Gaultier, J. & Chasseau, D. (2001). High pressure and very low temperature effects on the crystal structures of some iron(II) complexes. *C. R. Acad. Sci. Paris, Chimie*, Vol.4, pp.161-171

Guionneau, P., Le Pévelen, D., Marchivie, M., Pechev, S., Gaultier, J., Barrans, Y. & Chasseau, D. (2004). Laboratory high pressure single-crystal X-ray diffraction – recent improvements and examples of studies. *J. Phys.: Condens. Matter*, Vol.16, pp.S1151-S1159. DOI: 10.1088/0953-8984/16/14/025 http://iopscience.iop.org/0953-8984/16/14/025

Gütlich, P. & Goodwin, H. A. (2004). Spin crossover in transition metal compounds. Eds. *Topics in Current Chemistry*, Springer-Verlag: Berlin, Vol.233-234-235

Hansen, N. K. & Coppens, P. (1978). Testing aspherical atom refinements on small-molecule data sets. *Acta Cryst. A*, Vol.34, pp.909-921

Hansen, N. K., Fertey, P. & Guillot, R. (2004).Studies of electric field induced structural and electron-density modifications by X-ray diffraction. *Acta Cryst. A*, Vol.60, pp.465-471

Hansen, T. C., Henry, P. F., Fischer, H. E., Torregrossa, J. & Convert, P. (2008). The D20 instrument at the ILL: a versatile high-intensity two-axis neutron diffractometer. *Meas. Sci. Technol.*, Vol.19, pp.034001

Hennet, L., Thiaudière, D., Gailhanou, M., Landron, C., Coutures, J-P. & Prince, D. L. (2002). Fast X-ray scattering measurements on molten alumina using a 120° curved position sensitive detector. *Rev. Scientific Instruments*, Vol.73, pp.124-129

Hennet, L., Thiaudière, D., Landron, C., Melin, P., Price, D. L., Coutures, J-P., Bérar, J-F. & Saboungi, M-L. (2003). Melting behaviour of levitated Y$_2$O$_3$. *Appl. Phys. Letters*, Vol.83, pp.3305-3307

Hennet, L., Pozdnyakova, I., Bytchkov, A., Cristiglio, V., Paleau, P., Fischer, H. E., Cuello, G. J., Johnson, M., Melin, P., Zanghi, D., Brassamin, S & Saboungi, M-L. (2006). Levitation apparatus for neutron diffraction investigations on high temperature liquids. *Rev. Scientific Instruments*, Vol.77, pp.053903

Hennet, L., Pozdnyakova, I., Cristiglio, V., Krisnan, S., Bytchkov, A., Albergamo, F., Cuello, G. J., Brun, J-F., Fischer, H. E., Zanghi, D., Brassamin, S., Saboungi, M-L. & Price, D. L. (2007). Structure and dynamics of levitated liquid aluminates. *J. Non-Crystalline Solids*, Vol.353, pp.1705-1712

Hennet, L., Pozdnyakova, I., Bytchkov, A., Cristiglio, V., Zanghi, D., Brassamin, S., Brun, J-F., Leydier, M. & Price, D. L. (2008). Fast X-ray scattering measurements on high temperature levitated liquids. *J. Non-Crystalline Solids*, Vol.354, pp.5104-5017

Hu, J., Xu, J., Somayazulu, M., Guo, Q., Hemley, R. & Mao, H. K. (2002). X-ray diffraction and laser heating: application of a moissanite anvil cell. *J. Phys.: Condens. Matter*, Vol.14, pp.10479-10481

Jamieson, J. C., Lawson, A. W. & Nachtrieb, N. D. (1959). New device for obtaining X-ray diffraction patterns from substances exposed to high pressure. *Rev. Sci. Instrum.*, Vol.30, pp.1016-1019

Katrusiak, A. (2008). High-pressure crystallography. *Acta Cryst. A*, Vol.64, pp.135-148

Katsumata, K. (2005). Synchrotron X-ray diffraction studies of magnetic materials under extreme conditions. *Physica Scripta*, Vol.71, pp.CC7-13

Kockelmann, W., Oliver, E. C. & Radaelli, P. G. (2007). IMAT - an imaging and materials science & engineering facility for TS-II. Draft proposal for discussion. *http://www.isis.stfc.ac.uk/Instruments/Imat/*

Kuhs, W. F., Archer, J. & Doran, D. (1993). A closed-shell furnace for neutron single-crystal diffraction. *J. Appl. Cryst.*, Vol.26, pp.730-733

Le Godec, Y., Dove, M. T., Redfern, S. A. T., Tucker, M. G., Marshall, W. G., Syfosse, G. & Klotz, S. (2003). Recent developments using the Paris-Edinburgh cell for neutron diffraction at high pressure and high temperature and some applications. *High Pressure Research*, Vol.23, pp.281-287

Legrand, V. (2005). Cristallographie et photo-cristallographie haute résolution de composés moléculaires à transition de spin : propriétés structurales, électroniques et mécanismes de conversion. *thesis, Univ. Henri Poincaré, Nancy 1 (France)*, n°1132

Legrand, V., Carbonera, C., Pillet, S., Souhassou, M., Létard, J-F., Guionneau, P. & Lecomte C. (2005). Photo-crystallography: from the structure towards the electron density of metastable states. *Journal of Physics: Conf. Series*, Vol.21, pp.73-80

Legrand, V., Pillet, S., Souhassou, M., Lugan, N. & Lecomte C. (2006). Extension of the experimental electron density analysis to metastable states : a case example of the spin crossover complex Fe(btr)2(NCS)2.H2O. *J. Am. Chem. Soc.*, Vol.126, pp.13921 13931

Legrand, V., Pillet, S., Weber, H-P., Souhassou, M., Létard, J-F., Guionneau, P. & Lecomte C. (2007a). On the precision and accuracy of structural analysis of light-induced metastable states. *J. Appl. Cryst.*, Vol.40, pp.1076-1088

Legrand, V., Pillet, S., Carbonera, C., Souhassou, M., Létard, J-F., Guionneau, P. & Lecomte C. (2007b). Optical, magnetic and structural properties of the spin crossover complex [Fe(btr)2(NCS)2].H2O in the light-induced and thermally quenched metastable states. *European Journal of Inorganic Chemistry*, pp.5693-5706

Legrand, V., Le Gac, F., Guionneau, P. & Létard., J-F. (2008). Neutron powder diffraction studies of two spin transition Fe(II)-complexes under pressure. *Journal of Applied Crystallography*, Vol.41, pp.637-640

Lelièvre-Berna, E., Bourgeat-Lami, E., Fouilloux, P., Geffray, B., Gibert, Y., Kakurai, K., Kernavanois, N., Longuet, B., Mantegezza, F., Nakamura, M., Pujol, S., Renault, L-P., Tasset, F., Takeda, M., Thomas, M. & Tonon, X. (2005). Advances in spherical neutron polarimetry with Cryopad. *Physica B*, Vol.356, pp.131-135

Lorenz, G., Neder, R. B., Marxreiter, J., Frey, F. & Schneider, J. (1993). A mirror furnace for neutron diffraction up to 2300 K. *J. Appl. Cryst.*, Vol.26, pp.632-635

Mak, A. N-S., Bradley, P., Cernadas, R. A., Bogdanove A. J. & Stoddard B. L. (2012). The crystal structure of TAL effector PthXo1 bound to its DNA target. *Science*, Vol.335, pp.716-719

Martinez-Criado, G., Steinmann, R., Alén, B., Labrador, A., Fuster, D., Ripalda, J. M., Homs, A., Labouré, S. & Susini, J. (2007). New cryogenic environment for beamline ID22 at the European Synchrotron Radiation Facility. *Rev. Sci. Instrum.*, Vol.78, pp.025106

McMillan, P. & Tucker, M. (2007). EXEED - an extreme sample environment diffractometer. Draft proposal for discussion. *http://www.isis.stfc.ac.uk/instruments/Exeed/*

McWhan, D. B., Bloch, D. & Parisot, G. (1974). Apparatus for neutron diffrction at high pressure. *Rev. Sci. Instrum.*, Vol.45, pp.643-646

Mezouar, M., Le Bihan, T., Libotte, H., Le Godec, Y. & Häusermann, D. (1999). Paris-Edinburgh large volume cell coupled with a fast imaging-plate system for structural investigation at high pressure and high temperature. *J. Synchrotron Rad.*, Vol.6, pp.1115-1119

Mezouar, M., Crichton, W. A., Bauchau, S., Thurel, F., Witsch, H., Torrecillas, F., Blattmann, G., Marion, P., Dabin, Y., Chavanne, J., Hignette, O., Morawe, C. & Borel, C. (2005). development of a new state-of-the-art beamline optimized for monochromatic single-crystal and powder X-ray diffraction under extreme conditions at the ESRF. *J. Synchrotron Rad.*, Vol.12, pp.659-664

Mignot, J-M. (2008). Diffusion neutronique sous conditions extrêmes. *Collection SFN 9. EDP Sciences, Les Ulis*. DOI:10.1051/sfn:2008012

Mills, D. M. (2011). The Advanced Photon Source - where we are and where we are going. *Nuclear Instruments and Methods in Physics Research A*, Vol.649, pp.22-24

Muchmore, S. W. (1999). Experiences with CCD detectors on a home X-ray source. *Acta Cryst. D*, Vol.55, pp.1669-1671

Neumaier, K., Heidemann, A. & Mageri, A. (1984). A dilution refrigerator insert for standard ILL cryostat. *Rev. Phys. Appl. (Paris)*, Vol.19, pp.773-774

Nietz, V. (2003). Prospects for the use of the pulsed fields in neutron research of condensed matter. *J. Magnetism and Magnetic Materials*, Vol.260, pp.84-104

Nojiri, H., Takahashi, K., Fukuda, T., Jujita, M., Arai, M. & Motokawa, M. (1998). 25 T repeating pulsed magnetic fields system for neutron diffraction experiments. *Physica B*, Vol.241-243, pp.210-212

Nozawa, S., Adachi, S., Takahashi, J., Tazaki, R., Guérin, L., Daimon, M., Tomita, A., Sato, T., Chollet, M., Collet, E., Cailleau, H., Yamamoto, S., Tsuchiya, K., Shioya, T., Sasaki, H.,

Mori, T., Ichiyanagi, K., Sawa, H. & Koshihara, S. (2007). Developing 100 ps-resolved X-ray structural analysis capabilities on beamline NW14A at the Photon Factory Advanced Ring. *J. Synchrotron Rad.*, Vol.14, pp.313-319

Ohtani, A., Mizukami, M., Katayama, M., Onodera, A. & Kawai, N. (1977). Multi-anvil apparatus for high pressure X-ray diffraction. *Jpn. J. Appl. Phys.*, Vol.16, pp.1843-1848

Okada, J. T., Sit, P. H-L., Watanabe, Y., Wang, Y. J., Barbiellini, B., Ishikawa, T., Itou, M., Sakurai, Y., Bansil, A., Ishikawa, R., Hamaishi, M., Masaki, T., Paradis, P-F., Kimura, K., Ishikawa, T., & Nanao S. (2012). Persistence of covalent bonding in liquid silicon probed by inelastic x-ray scattering. *Phys. Rev. Lett.*, Vol.108, pp.067402

Orosel, D., Dinnebier, R. E., Blatov, V. A. & Jansen, M. (2012). Structure of a new high-pressure – high temperature modification of antimony(III)oxyde, γ-Sb$_2$O$_3$, from high-resolution synchrotron powder diffraction data. *Acta Cryst. B*, Vol.68, pp.1-7

Paolasini, L., Detlefs, C., Mazzoli, C., Wilkin, S., Deen, P. P., Bombardi, A., Kernavanois, N., de Bergevin, F., Yakhou, F., Valade, J. P., Brslavetz, I., Fondacaro, A., Pepellin, G. & Bernard, P. (2007). ID20: a beamline for magnetic and resonant X-ray scattering investigations under extreme conditions. *J. Synchrotron Rad.*, Vol.14, pp.301-312

Paszkowicz, W. (2002). High pressure powder X-ray diffraction at the turn of the century. *Nucl. Instr. And Meth. In Phys. Res. B*, Vol.198, pp.142-182

Paureau, J. (1975). Nouveau dispositif d'étanchéité pour hautes pressions. *Revue de Physique Appliquée*, Vol.10, pp.475-478

Pillet, S., Legrand, V., Weber, H-P., Souhassou, M., Létard, J-F., Guionneau, P. & Lecomte C. (2008). Out-of-equilibrium charge density distribution of spin crossover complexes from steady-state photocrystallographic measurements: experimental methodology and results. *Z. Kristallogr.*, Vol.223, pp.235-249

Price, T. C., Grant, D. M., Legrand, V. & Walker G. S. (2010). Enhanced kinetics for the LiBH$_4$:MgH$_2$ multi-component hydrogen storage system - the effects of stoichiometry and decomposition environment on cycling behavior. *International Journal of Hydrogen Energy*, Vol.35(9), pp.4154-4161

Riekel, C., Müller, M. & Vollrath, F. (1999). In situ X-ray diffraction during forced silking of spider silk. *Macromolecules*, Vol.32, pp.4464-4466

Riekel, C. & Vollrath, F. (2001). Spider silk fibre extrusion: combined wide- and small-angle X-ray microdiffraction experiments. *Inter. J. Biological Macromol.*, Vol.29, pp.203-210

Rodriguez-Velamazan, J. A., Campo, J., Rodriguez-Carvajal, J. & Noguera, P. (2011). XtremeD - a new neutron diffractometer for high pressures and magnetic fields at ILL developed by Spain. *J. Phys.: Conference Series*, Vol.325, pp.012010

Sakai, I., Murai, K., Jiang, L., Umesaki, N., Honma, T. & Kitano, A. (2005). Aerodynamic levitation apparatus for structure study of high temperature materials coupled with Debye-Scherrer Camera at BL19B2 of Spring-8. *J. Electron Spectroscopy and Related Phenomena*, Vol.114-147, pp.1011-1013

Scanlon, D. O., Kehoe, A. B., Watson, G. W., Jones, M. O., David, W. I. F., Payne, D. J., Egdell, R. G. Edwards, P. P. & Walsh, A. (2011). Nature of the band gap and origin of the conductivity of PbO$_2$ revealed by theory and experiment. *Phys. Rev. Letters*, Vol.107, pp.246402

Seta, J., Ma, J., Davis, S. A., Meldrum, F., Gourrier, A., Kim, Y-Y., Schilde, U., Sztucki, M., Burghammer, M., Maltsev, S., Jäger, C. & Cölfen H. (2012). Structure-preoperty relationships of a biological mesocrystal in the adult sea urchin spine. *PNAS*, Vol.109, pp.3699-3704

SPring-8 (2012). SPring-8 upgrade plan preliminary report. *http://www.spring8.or.jp/en/about_us/whats_sp8/spring-8_II/.*

Sun, L., Chen, W-J., Guo, J., Gao, P., Huang, Q-Z., Wang, H., Fang, M., Chen, X., Chen, G., Zhang, C., Gu, D., Dong, X., Wang, L., Yang, K., Li, A., Dai, X., Mao, H-K. & Zhao, Z. (2012). Re-emerging superconductivity at 48 kelvin in iron chalcogenides. *Nature*, Vol.483, pp.67-69

Suzuki, H., Naher, S., Shimogushi, T., Mizuno, M., Ryu, A. & Fujishita, H. (2002). X-ray diffraction measurement below 1 K. *J. of Low Temperature Physics*, Vol.128, pp.1-7

Szilárd, L. (1935). Absorption of residual neutrons. *Nature*, Vol.136, pp. 950-951

Takeda, M., Nakamura, M., Kakurai, K., Lelièvre-Berna, E., Tasset, F. & Renault, L-P. (2005). Cryopad on the triple-axis spectrometer TAS-1 at JAERI. *Physica B*, Vol.356, pp.136-140

Tasset, F. (1989). Zero-field neutron polarimetry. *Physica B*, Vol.156-157, pp. 627-630

Tasset, F., Brown, P. J., Lelièvre-Berna, E., Roberts, T., Pujol, S., Allibon, J. & Bourgeat-Lami, E. (1999). Spherical neutron polarimetry with Cryopad-II. *Physica B*, Vol.267-268, pp.69-74

Tulk, C. A., Klug, D. D., dos Santos, A. M., Karotis, G., Guthrie, M., Molaison, J. J. & Pradhan, N. (2012). Cage occupancies in the high pressure structure H methane hydrate: a neutron diffraction study. *J. Chem. Phys.*, Vol.136, pp.054502

Utsumi, W., Funakoshi, K. I., Katayama, Y., Yamakata, M., Okada, T. & Shimomura, O. (2002). High-pressure science with a multi-anvil apparatus at Spring-8. *J. Phys. Condens. Matter*, Vol.14, pp.10497-10504

Van Kan Parker, M., Sanloup, C., Sator, N., Guillot, B., Tranche, A. J., Perrillat, J-P., Mezouar, M., Rai, N. & Westrenen, W. (2012). Neutral buoyancy of titanium-rich melts in the deep lunar interior. *Nature Geoscience*, Vol.5, pp.186-189. DOI: 10.1038/ngeo1402 http://www.nature.com/ngeo/journal/vaop/ncurrent/full/ngeo1402.html

Vettier, C. & Wright, A. (1984). Workshop "sample environments in neutron and X-ray experiments". *Revue de Physique Appliquée*, Vol.19, pp.643-836

Walker, G. S., Grant, D. M., Price, T. C., Yu, X. & Legrand, V. (2009). High capacity multicomponent hydrogen storage materials: investigation of the effect of stoichiometry and decomposition conditions on the cycling behaviour of $LiBH_4$-MgH_2. *Journal of Power Sources*, Vol.194(2), pp.1128-1134

Weir, C. E., Lippincott, E. R., Van Valkenburg, A. & Bunting, E. N. (1959). Infrared studies in the 1- to 15-micron region to 30,000 atmospheres. *J. Res. Nat. Bur. Stand. A*, Vol.63, pp.55-62

White, M. A., Pressprich, M. P. & Coppens, P. (1994). Apparatus for the measurement of the electronic excited-state structure of single crystals using X-ray diffraction. *J. Appl. Cryst.*, Vol.27, pp.727-732

Wimberly, B. T., Brodersen, D. E., Clemons, W. M., Morgan-Warren, R. J., Carter, A. P., Vonrhein, C., Hartsch, T. & Ramakrishnan, V. (2000). Structure of the 30S ribosomal subunit. *Nature*, Vol.407, pp.327-339

Xu, J., Mao, H. K., Hemley, R. J. & Hines, E. (2002). The moissanite anvil cell: a new tool for high-pressure research. *J. Phys.: Condens. Matter*, Vol.14, pp.11543-11548

Zhurov, V. V., Zhurova, E. A. & Pinkerton, A. A. (2008). Optimization and evaluation of the data quality for charge density studies. *J. Appl. Cryst.*, Vol.41, pp.340-349

Mineralization of Lipid A-Phosphates in Three- and Two-Dimensional Colloidal Dispersions

Henrich H. Paradies, Peter Quitschau, Hendrik Reichelt,
Chester A. Faunce and Kurt Zimmermann

Additional information is available at the end of the chapter

1. Introduction

Crystal growth and crystal nucleation has attracted interest for centuries and goes back to Johannes Kepler [1] in 1611. Though progress in the understanding of crystal nucleation and crystal growth were theoretically developed, exact quantitative prediction of the nucleation rates and their kinetics still remains unresolved [2]. As early in 1959, the protocol of preparation of virus crystals and their physical analysis revealed an interparticle spacing of 250 nm [3, 4]. The crystal growth from microcrystalline material to single crystals of viruses marked another event for crystal nucleation. It was concluded for this *Tipula* iridescent virus that the hydrated virus particles in the crystal are not in contact but are separated by large distances of water (~ 50 nm) showing soft modes in the direction of a lattice site and the lattice displacements at a given time is caused by a longitudinal phonon with a wave vector at the zone boundary [5, 6]. These crystals are probably held together by long-range forces operating at a distance comparable with the size of the particles themselves. This is very similar to the recently discovered *autovaccines* obtained from non-pathogenic *E. coli* as liquid colloidal and solid nanocrystals [7-11]. The former virus crystals as well as the discussed liquid *autovaccine* crystals seem to present the well-established instance of a unique ordering of iso-dimensional colloidal particles in three dimensions in solutions and in the solid state. This holds also for the prediction of polymorphic crystal forms like liquid crystals, especially for complex crystalline solid forms originating from colloidal dispersions of e.g. chiral Lipid A-phosphates [12], cationic lipids [13-17], anionic surfactants [18], diblock copolymers [19], or surfactant-water complexes [20]. Crystallization, phase transitions and crystal growth remain a central topic of condensed matter physics. A detailed understanding of the controlled formation of crystalline materials is of great importance for numerous applications, especially for colloidal particles [21-23], biomimetic approaches to mineralization [24], curved crystalline shapes that emerged from mixtures of barium or strontium carbonates and silica in alkaline media [25], and usually from devices derived from self-assembly of colloidal materials [26, 27].

2. Overview

In view of the above mentioned and widely accepted explanations and morphological descriptions of crystal, nuclei formation and phase transformations, the focus of this contribution relies on recent advances in the physical understanding of the disorder-order, fluid-crystal-fluid transitions of Lipid A-phosphates, the formation of a re-entrant phase, particularly for Lipid A-diphosphate e.g freezing and melting, and the molecular mimicry of hierarchical self-assembly of Lipid A-phosphates. This was achieved using a variety of experimental techniques e.g. scanning electron microscopy (SEM), high resolution transmission electron microscopy (HRTEM) and small-area electron diffraction (SAED), small-angle X-ray diffraction & solution scattering (SAXD & SAXS), static & quasi-elastic light scattering (LS & QELS). The contribution is outlined as follows: In remainder of the brief introduction, a short description is given which is devoted to the chemistry and importance of this class of molecules ("nano-medicine") in the day-to-day life. In Section 3, the phase transition of Lipid A-diphosphate clusters, the formation of various crystal forms as a function of particle number density (n) or volume fraction (ϕ), ionic strength (I) and T will be introduced since they affect the crystallization in 2d and 3d. Observations on freezing and melting of Lipid A-diphosphate clusters on a physical molecular level are also presented. The occurrence of the re-entrant fluid phase of Lipid A-diphosphate clusters upon addition of μM NaOH after a crystalline BCC phase will be compared with the various crystalline phases upon decreasing I and increase of n for NaCl. The self-assembly of Lipid A-diphosphates clusters composed of different "subunits" e.g. six-hexaacylated chains and the corresponding Lipid A-diphosphate with four acylated fatty acid chains with the same phosphorylated disaccharide will be discussed in more detail (crystals, symmetry packing). These crystalline Lipid A-diphosphate cluster complexes may be called *"autovaccines"* (*Note:* An immunizing agent is composed of a selected or modified chemical entity of an original microorganism or virus, which does not cause clinical signs associated with the parent microorganism or virus, but still infects & multiplies in the host so as to induce immunity). Section 4 reports on the formation of Lipid A-diphosphate crystals by surface-induced gradient-induced crystallization in monolayers, their thermodynamics with relation to surface tension and their morphologies. In Section 5 the conclusions are presented.

2.1. Chemistry and the biological role of Lipid A-diphosphate

The lipopolysaccharides (*LPS*) are a group of diverse lipid-containing carbohydrates that exhibit a wide variety of biological activities. They occur naturally on the outer cell membranes of Gram-negative bacteria such as *Escherichia coli*. Although the lipopolysaccharides are large molecules, most of their biological activities result from the activity of a small portion of the molecule known as Lipid A-diphosphate. The structure of Lipid A consists of two β (1,6)-linked D-glucosamine units with polar phosphate groups at 1 and 4' positions (Fig. 1) [28]. This problem caused approximately 21 000 mortalities in 1996 in the U.S. alone [29].

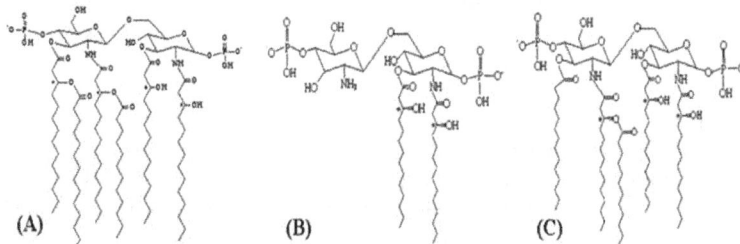

Figure 1. Chemical structures of Lipid A-diphosphate (A) and two Lipid A-diphosphate derivatives (B & C) with strong biological implication on the anti-inflammatory mediator level [28, 30]. Lipid A-diphosphate from *E. coli* is a 1,4-di-phosphorylated β-1,6-linked D-glucosamine disaccharide with four residues of amide-and O-esterified R-(-)-3-hydroxy fatty acids (* denotes the chiral centers in the hydroxy fatty-acid esters), apart from the chiral and epimeric carbons in the glucosamine moieties which are not marked. The antagonistic Lipid A-diphosphate molecules shown in B & C contain the same disaccharide as in (A); however, they differ in the number anchored carbohydrate positions and the number of chiral fatty-acid chains but the chain lengths is the same. The monophosphate of Lipid A is only phosphorylated at the reducing end of the disaccharide (C-1).

The Lipid A-diphosphate is associated with lethal endotoxicity, pyrogenicity and specific immune response. It is also responsible for triggering a cascade of cellular mediators, e.g. tumor necrosis factor α, interleukins, leukotrienes, thromboxane A2 from monocytes and macrophages. The Lipid A-diphosphate and their analogues are distinct from normal lipids with respect to structure, chirality and chemical building units (Fig. 1) [30]. The 2 and 2' amino positions and the 3 and 3' hydroxyl groups are esterified with hydroxy fatty acids. It is known that natural Lipid A-phosphates and approximants are potent immunostimulants which induce a number of desirable effects but also some undesirable ones [30]. Various analogues of Lipid A-diphosphate have been developed to avoid such unwanted effects as toxicity and pyrogenicity, and therefore, they are very distinct from other lipids and surfactants.

Lipid A-diphosphate and approximants possess beneficial effects in clinical therapy against chronic inflammatory diseases and are capable of decreasing resistance to antibiotics and cationic antimicrobial peptides (CAMP) [31]. In this context the CAMP play also a significant role in the immune reaction to gut commensals (inflammatory bowel disease (Crohn), ulcerative colitis) and possibly in antibiotic resistance [30, 31]. This infers to an increased bacterial invasion of the surface of the respective tissues accompanied by the loss of the protective barrier. This accounts for bacterial contamination of the intestinal surface where host and invader are physically in close contact. Accordingly, this view strongly supports the production of *"intestinal autovaccines"* and its therapeutic potential e.g for protection of CAMP synthesis and sustaining remissions. The chemical structure of the pro-inflammatory component of *LPS*, Lipid A (Fig. 1), varies between bacteria of different species where the Gram-negative bacteria modulate the structure of their *LPS*.

3. Fluid-solid phase transition of self-assembled Lipid A-diphosphate cluster

By applying freeze-fracture electron microscopy and X-ray diffraction techniques a qualitative phase relationship of Lipid A-diphosphate has been reported [32, 33]. These results were taken from samples of Lipid A extracted from Salmonella Minnesota, E. coli rough mutant LPS and from Salmonella enterica serovar Minnesota. In their qualitative study the aqueous specimens have been examined in the presence of mM phosphate buffers by synchrotron radiation and analyzed by their diffraction profiles e.g. cubic, hexagonal or simple cubic structures. However, there ionic strength was a magnitude higher than the one used for preparing electrostatically stabilized Lipid A-diphosphate dispersions with $I = 10^{-4}$ to 10^{-6} M [12]. Especially, SAXD and SAXS are very suitable to characterize the 3-d structure or ordering in solution of the self-assembled Lipid A-phosphate clusters and probe their corresponding long-range order parameters. For comparison with light scattering, the most important advantage of the applied X-rays is their low refractive index contrast (the difference in index of refraction is of the order of 10^{-6}), so the occurrence of multiple scattering and incoherent background scattering is significantly reduced and a much wider range of scattering vectors are available. It was observed that upon reaching a nanocluster size of 500 – 600 nm the suspensions became iridescent to visible light: thus, the iridescence acts as a visual marker of nanocluster size. Recently, the iridescent solutions have been physically analyzed and interpreted in terms of mass, surface charge, size and shape. The influence of polydispersity in charge, size and mass has been elucidated and included in all further experiments and simulations [7-13]. The most compelling observations of the colloidal crystallization and also of Lipid A-diphosphates in aqueous solutions at very low ionic strength conditions ($< 10^{-6}$ M) are the order-disorder transition and the structural transition from a body-centered cubic (BCC) to a face-centered (FCC) structure [34, 35]. In these dispersions it is essential to consider the co-occurrence of phase separation and crystalline ordering, where it has been suggested that the crystalline phase is a supercooled liquid phase with some liquid retained in a metastabile state for a certain period of time, even at $\phi > 0.58$. Experimental phase diagrams of Lipid A-diphosphate dispersions in NaCl or NaOH as shown in Fig. 2 are very helpful in searching for crystal formation, order–disorder as a function of ionic strength I, n and T. Moreover, they are useful for comparison with the theoretical predictions [36, 37]. Normally the crystalline arrays form spontaneously through self-assembly of charged colloidal Lipid A-phosphate spheres in low ionic strength but at low polydispersity in size, mass and charge. The spontaneous formation of self-assembled Lipid A-diphosphate crystallization is mainly driven by the excluded-volume entropy. The decrease in entropy in the colloidal crystals is associated with a nonuniform mean density, however, a greater local volume that each particle can independently explore compensates for these phenomena. If the amount of base (c_s) were increased, a new fluid re-entrant disordered phase of self-assembled Lipid A-diphosphate clusters was encountered followed by a fluid ordered-crystalline BCC phase for low n when using NaOH as an electrolyte in the crystallization process.

3.1. Freezing and melting

It was possible to prepare stable aqueous colloidal dispersions of Lipid A-diphosphate and their approximants with low polydispersity in shape, size, and charge over a discrete range of volume fraction, ϕ [9-11]. The phase transitions were a correlated liquid phase, a cubic FCC and a body-centered cubic crystalline phase [35]. These phases were detected in the presence of mM NaCl for different volume fractions, ϕ; and various crystal forms (BCC and FCC) could be obtained. It was found that these assemblies were consistent with an assembly for a BCC lattice (*Im3m*) with a = 35.5 nm. However, a mixture of equimolar concentrations of the two antagonistic molecules revealed a SAXS-powder diffraction pattern a light scattering profile for crystals of sizes of 1 µm that could be indexed for a much larger face-centered (*Fd3m*) unit cell, with a = 58.0 nm. However, when employing the similar conditions for colloidal Lipid A-diphosphate dispersions but through addition of µM NaOH rather than by removing NaCl [35] a very different phase behavior was observed (Fig. 2). By varying the amount of added NaOH (or NaCl) it was possible to determine the effective charge, Z_{eff} for various n values and the screening parameter, k, for the excess electrolyte (c_s). For sufficiently large values of n, Lipid A-diphosphate crystallized through an increase in Z_{eff} at a constant c_s when adding µM NaOH. When the c_s were increased, the crystals melted with little change in $\overline{Z_{eff}}$. An increase in the c_s enhanced interparticle interaction and attraction due to many-body effects (NaOH), but influences long-range interactions where particles repel one another at large separation distances in the presence of NaCl; thus repulsion increases with decreasing particle-particle separation and decreasing ionic strength. The effective charge and $k = 4\pi k_B T \lambda_B$ (n Z_{eff} + 2 $\cdot 10^3/N_A \cdot c_{se})^{1/2}$, with λ_B = 0.735 nm, the Debye screening length, account for counterion condensation and many-body effects. - If the effective charge determined from scattering measurements was used in the simulations, the equilibrium phase boundaries were consistent with predicted universal melting-line simulations [36, 37]. It seems that the presence of NaOH in the aqueous dispersions of Lipid A-diphosphate stabilized a two-state structure. The stabilizing effect was promoted by long-range (counter-ion mediated) attractive interactions between the crystalline clusters and the many-body effects. This Lipid A-diphosphate cluster is able to undergo both a fluid-order and an order-fluid transition.

When the NaOH concentration and the particle-number density of Lipid A-diphosphate is increased, the length scale of the repulsion decreased, because of many-body effects and the disorder-order-transition occurred at a particle-number density close to the freezing transition. When the NaOH concentration and the particle-number density of Lipid A-diphosphate is increased, the length scale of the repulsion decreased, because of many-body effects and the disorder-order-transition occurred at a particle-number density close to the freezing transition. At lower particle-number densities, as the length scale of the repulsive forces increased, the fluid-crystalline transition gave rise to BCC-type crystals.

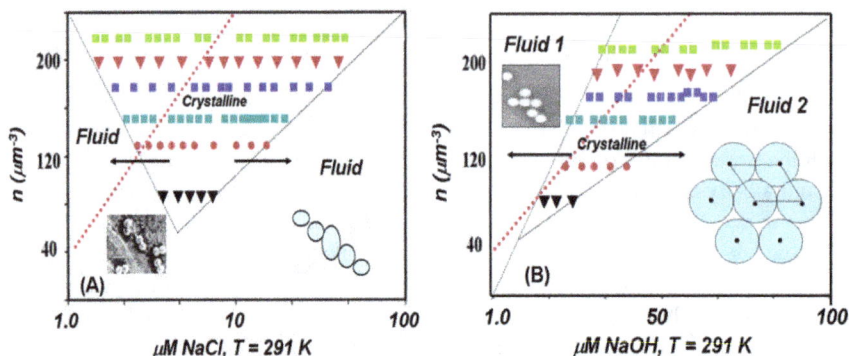

Figure 2. Experimental phase diagrams of self-assembled charged Lipid A-diphosphate and their approximants at constant temperature (T = 291 K) as a function of particle number density, n, and ionic strength (*I*). **(A)** Ordered crystal phases appeared after considerable reduction of (*I*) with a large Debye length (NaCl) due to a stabilized electrostatic repulsion between the various self-assembled Lipid A-diphosphate clusters: ▼ Lipid A-diphosphate (A in Fig. 1) crystals; • self-assembled Lipid A-diphosphate clusters comprising of Lipid A-diphosphate (A) and (C) shown in Fig. 1; ■ self-assembly of Lipid A-diphosphate clusters with (A) and (B) components; ■ self-assembly of components B and C; ▼ self-assembly of component (A) with six chains & with two double chained Lipid A-diphosphate at the non-educing end of the disaccharide; ■ self-assembly of the Lipid A-diphosphate approximant C (Fig. 1). *Inset:* Single spherical Lipid A-diphosphate clusters with d = 70.0 nm (SEM image). They form because of low shape and charge polydispersity (≤ 10%). **(B)** Phase diagram of charged spherical Lipid A-diphosphate clusters with the same composition as in **(A)**, however, in the presence of μM NaOH. The re-entrant melting lines are shown at the left boundary (charging) and at the right boundary (screening), respectively. The two-phase regions are indicated as horizontal arrows. The red dotted lines indicate the equivalent titration points obtained from conductometric titrations. The •••lines are theoretical fits to the experimental data applying an effective charge of Z_{eff} = 345 ± 76 for the left boundary, for the right boundary at maximum interaction the effective charge was Z_{eff} = 320 ± 50. The corresponding values for NaCl **(A)** were Z_{eff} = 470 ± 61 and Z_{eff} = 500 ± 50, respectively. **Insets:** An ordered hexagonal columnar phase of non-crystalline Lipid A-diphosphate in the aqueous dispersion (Fluid 2) in (B) at high μM NaOH, contrary to (A). Lipid A-diphosphate clusters of diameter d = 6.0 nm are only present in the Fluid 1 Phase (SEM image) in (B), whereas the Lipid A-clusters in the hexagonal Fluid Phase 2 exhibit a nearest neighbor distance of 35.1 nm and a packing fraction of 0.68.

This implies that self-screening was much smaller than in previous studies and very different for Lipid A-monophosphate phases [38]. The experimental observations and the simulation support the existence of a transition from a fluid to a BCC structure rather than to an expected FCC structure for Lipid A-diphosphate clusters in e.g 5.0 mM NaCl. However, there was no FCC structure present for a certain NaOH concentration, but there was a crystalline BCC phase present between two clearly defined fluid phases with no crystals. It was rather unusual to encounter BCC structures for ionic strengths of different magnitude (5.0 μM NaCl vs. 50.0 μM NaOH) and for the same n, therefore, the OH- ions must contribute to the scenario. It is also feasible that the structural transition path was influenced by topological dissimilarities but differently forms the corresponding Lipid A-monophosphate crystalline phase, because of altered elastic deformations between the crystalline BCC phase which formed and the fluid phases from which it originated. The nearest-neighbor interparticle distance, $2d_{exp} = 2\pi/Q_{110} = 32.2$ nm was estimated from the experimental peak positions and compared with the average theoretical distance, $2\ d_{th} = \sqrt{3}/\sqrt[3]{4n}$, for various n, 20 μm^{-3} ≤ n ≤ 400 μm^{-3}. For this limited particle-number density range the double-logarithmic plot of $2\ d_{exp}$ and $2\ d_{th}$ vs. n revealed a straight line with a slope of -0.32. The nearest-neighbor distance in the re-entrant fluid phase was determined as $d_{N-N} = 33.0$ nm, surprisingly for all Lipid A-diphosphate clusters including those with different "subunits". The nearest-neighbor distance for the fluid phase before freezing was found to be $d_{N-N} = 32.2$ nm. The nearest-neighbor distance for the crystalline phase revealed a value of $d_{N-N} = 30.8$ nm. This value was expected to be smaller than the average interparticle distance $2d_{th} = 32.2$ nm when a homogeneous particle distribution was assumed with cubic bcc symmetry.

The observed intersphere spacing was normally close to the calculated mean sphere distance except at the interfacial region of the dispersions, where there was contact with the air or a wall. Melting is initiated when the amplitude of vibration becomes sufficiently large for the occurrence of partly shared occupancy between adjacent particles. It occurs when the root mean square of the vibration amplitude of a crystal exceeds a threshold value (~ 0.15 d_{N-N}) according to Lindemann [39] and it would result in a movement of ~ 6.0 nm, which is a distance significantly smaller than the spacing between the surfaces of any of the spherical colloidal Lipid A-diphosphate clusters and a distance much less than the screening length. The Lindemann rule does not hold for crystals in 2d which have quasi-long-range instead of long-range translational order [40]. Measurements of the short-time and the long-time diffusion constants by quasi-elastic light scattering yielded a ratio of ~ 11 in the freezing region for the Lipid A-diphosphate cluster [41]. Furthermore, the possibility exists that the charge deduced from the melting line was also essential to center the particle within the cubic BCC unit cell. Note: According to the rule of Verlet and Hansen [42], crystallization occurs when the structure factor of ordinary liquids exceeds a value of 2.85 for 3d and ≅ 4 in 2d. Furthermore, when plotted S(Q) and n on a double-log graph a value of d_f = 1.75 was obtained, where S(Q) is the effective structure factor, Q is the scattering vector, d_f is the fractal dimension and close to the one found for a diffusion-limited cluster aggregation of 1.80.

This result may explain why the Lipid A-diphosphate basic arrays adopted the form of colloidal clusters at low particle-number densities and low NaOH concentrations. The number of seeding Lipid A-diphosphate clusters of $\bar{d}$ = 7 nm required to form a fractal nucleus of a given size was considerably lower than the number required for a compact nucleus. These parameters determined whether or not small clusters or sub-critical nuclei developed. Once formed, further cluster growth took place if a sufficient supply of Lipid A-diphosphate clusters of $\bar{d}$ = 7 nm were available. Therefore, cluster growth depended upon n, the size and the charge polydispersity and the availability of residual Lipid A-diphosphate clusters. A further increase in the NaOH concentration resulted in crystal melting and a re-entry into the fluid phase. The morphology of the clusters detected in such dispersions was similar to those of the Lipid A-diphosphate clusters observed before freezing took place. At a sufficiently high concentration of μM NaOH, where strong interactions occurred, freezing can take place and a two-phase region was formed. If the μM NaOH was increased there was an increase in charge on the self-assembled Lipid A-diphosphate clusters according to R–H$_2$PO$_4$H + OH$^-$ → R–HPO$_4^-$ + H$_2$O. At low μM NaOH, the Lipid A-diphosphate clusters retained their BCC structure. With additional increases in the μM NaOH the interactions became screened with neither an increase nor a decrease in charge and the system began to melt. Although the Lipid A-diphosphate samples investigated covered a limited pH-range (4.5 ≤ pH ≤ 7.5), there was a sufficient excess of NaOH to attribute the re-entrant occurrence to added screening produced by the excess. This is completed different form the situation using NaCl. Because of the Lipid A-diphosphate cluster counterions, the screening parameter, κ increased steadily if $n^{-1/3}$ remained constant. As a result, a decrease in the equilibrium-state lines will be observed and the melting line was crossed. This accounted for the maximum interaction equilibrium state line when the surface charges approached their highest value. Consequently, any further addition of μM NaOH screened the Lipid A-diphosphate particle surface charge and initiated a reduction in the cluster-cluster interaction. The screening parameter, κ, then depended only on the screening electrolyte (NaOH or NaCl). The result was an increase in the equilibrium state lines which again crossed the melting line.

3.2. BCC and FCC structures of Lipid A-phosphates

Some new crystalline Lipid A-diphosphate clusters and their approximants have been developed since the protocol for obtaining electrostatically stabilized solutions of Lipid A-diphosphates or for the corresponding monophosphates at various n was available [11-13]. The well-ordered Lipid A-diphosphate clusters and the presence of higher order diffraction peaks corroborated the existence of crystalline Lipid A-diphosphate material documented for the BCC and FCC structures assigned to the space groups $Im\bar{3}m$ & $Fd\bar{3}$ m [43] depicted in Figs 3 and 4. It was observed that the (211) peak at c$_S$ = 3.15 μM NaOH and the height of this reflection increased with n holding c$_B$ constant (Fig. 3). In addition the sensitivity of Lipid A-diphosphate clusters on pH is illustrated in Fig. 5B. The absence of the specific reflections (Figs. 3B, 4 & 5) of the X-ray diffraction patterns and small-area electron diffraction pattern reinforced the argument that the lattice type was face-centered cubic, therefore, two space groups were possible, namely $Fd\bar{3}$ and $Fd\bar{3}m$.

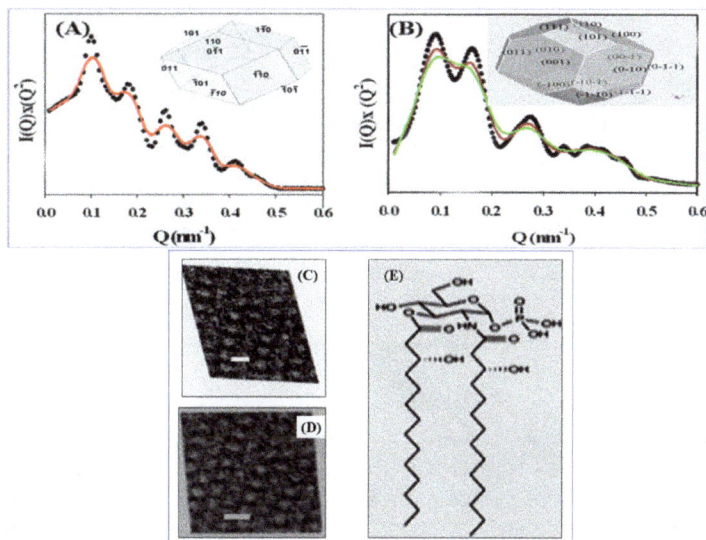

Figure 3. (A) SAXS profiles I(Q) vs. Q, with Q = (4π/λ)·sinθ/2), of BCC type colloidal crystals (λ = 1.54 nm) with a = 37.6 nm composed of Lipid A-diphosphate and "subunit" C (Fig. 1) . The assigned space group was *Im3m*, origin at center m3m, and equivalent positions 0, ½, ½; ½, 0, ½; ½, ½, 0 (Q^{229}) based on the molecular composition and the assumed spherical diameter of *d* = 7.0 nm which was consistent with form-factor scattering at higher Q. The black dotted scattering profile is for the Lipid A-diphosphate phase at φ = 3.5 x 10⁻⁴, *I* = 0.5 mM NaCl, the solid-red line is the profile for an equimolar mixture of the antagonistic molecule depicted in B of Figure 1 for Lipid A-diphosphate with *a* = 35.5 nm. (**B**) SAXS profiles of colloidal crystals of the FCC type (*Fd 3 m*), the black-dotted line corresponds to Lipid A-diphosphate with *a* = 57.5 nm comprising of the components A and B (Fig. 1). The red-solid line is for the colloidal mixture of Lipid A-diphosphate and antagonistic Lipid A-diphosphate [47] (Fig. 1B) both are at φ = 5.4 x 10⁻⁴, I = 0.5 mM NaCl. The green solid SAXS profile represents the results from a mixture of Lipid A-diphosphate with the corresponding monosaccharide of Lipid A-diphosphate at φ = 3.4 x 10⁻⁴, I = 0.5 mM NaCl. **Insets:** Crystal morphologies as they appear in SEM images and simulated with the *Accelrys Software Materials Studio 4.4 Module Morphology Version 6.0,* San Diego (USA). The corresponding TEM images are shown in (C) and (D); the scale bar is 10 nm; (D) illustrates the chemical structure of the antagonistic Lipid A-diphosphate [47] composed only of a diphosphorylated glucosamine residue and two fatty-acid chains.

The two space groups were also centrosymmetric and belong to the Laue classes $m\overline{3}$ and $m\overline{3}m$. *Note: Fd $\overline{3}$ m* corresponds to special positions of *Fd $\overline{3}$*. The observed and calculated *hkl* sets of Bragg reflections were consistent with a combination of different sites in the *Pm $\overline{3}$ n* (*a* and *d*), or of sites *a* and *d* in the *Fd $\overline{3}$ m* (Fig. 5). High-resolution transmission electron microscopy, SAED on crystalline Lipid A-diphosphate rods of the order of 2-3 μm in length and diameters of several nm obtained at pH 8.0 [45] revealed that the rods were held to the truncated polyhedra with a five-fold symmetry (Fig. 6).

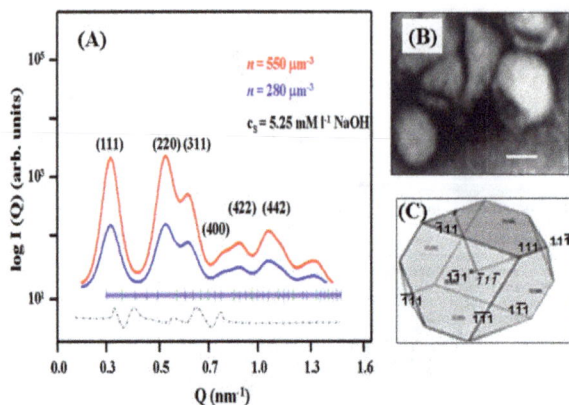

Figure 4. (A) S(Q) vs. Q profile for colloidal FCC-type crystals of Lipid A-diphosphate clusters with a = 57.0 nm for n = 140 μm^{-3} and cs = 2.05 μM NaOH. The lower curve (-----) reveals the differences between observed (•••) and calculated (—) intensities of the refined parameters form Rietveld analysis and X-ray powder diffraction pattern ($Fd\,3\,m$, site: 8a: 1/8, 1/8, 1/8 and 16-hedra: site 16d: ½, ½, ½, 12-hedra). **(B)** An SEM image is shown of μm sized single crystals (bar scale ≈1.0 μm) and the overall morphology of this crystal type and the Miller indexed faces are shown in (C).

By using computer simulations with multi-slice calculations, the Lipid A-diphosphate structure was obtained. The image multi-slice image simulations were carried out for the 100kV TEM Joel Microscope (T 3010) with imaging facilities for hollow-cone illuminations and using the Cowley algorithm [46, 47]. In addition, after successful indexing of the X-ray diffraction profiles and the selected area diffraction pattern (SAED), after the successful indexing of the powder-diffraction patterns and the selected-area diffraction patterns, a Pawley refinement was performed, taking the following parameters into consideration: cell parameters, peak-profile parameters, background and zero shifts. Following this refinement, a structure solution was imitated using a direct-space Monte Carlo-simulated annealing approach and a full-profile comparison was implemented. By employing a global optimization algorithm, trial structures were continuously generated by modifications in the specified degrees of freedom, i.e. three translations, three rotations and the dihedral angles.

It was possible to show that most of the Lipid A-diphosphate particles were orientated in the [001] direction with respect to the substrate for one of the five deformed tetrahedral subunits, i.e., the fivefold axis was parallel to the surface of the substrate [48]. Due to the presence of Lipid A-diphosphate and the surface tension, the only growth in the direction of the fivefold axis of decahedra was possible, resulting in long rods.

Since $n = n^*$ (n^* = particle/length³) the orientational entropy and the electrostatic repulsion of the free energy expression for these charged rods favor antiparallel alignment of the rods [45], this will give rise to a cubic lattice-like interparticle structures as noticed in Figure 5**(C)**. For $n \gg n^*$ and 10 μM NaOH a hexagonal packing of parallel rods can be anticipated as an appropriate description. If one considers the correlation of nearest-neighbor rods of Lipid A –diphosphate, then a parallel alignment of nearest neighbor rods is observed with a local ordering parameter $S = 0.07$ at $n = 2.8\ n^*$. Individual rod-shaped particles are noticed in Figure 5B whereas in Figure 5A N-regions of Lipid A-diphosphate particles are observed, where the particles exhibit parallel orientations, which may correspond to Sm precursors. When n is significantly increased, the Lipid A-diphosphate clusters grow laterally, their contours become clearer, and more layering of the clusters appear. The particle packing fraction ϕ_P for the Sm phase was estimated to 0.28. The Sm layer period and in-layer separation were calculated to be 2.8 and 2.4 nm, respectively, for $n = 5.5\ n^*$.

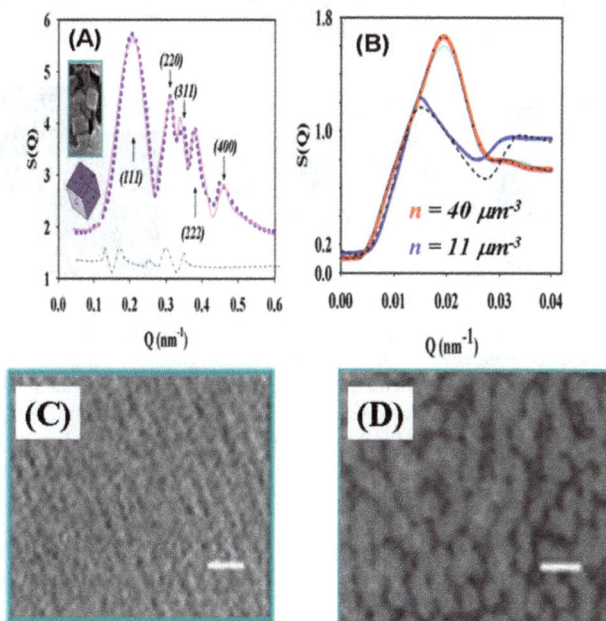

Figure 5. (A) S(Q) vs. Q profile for colloidal FCC-type crystals with $a = 57.0$ nm for $n = 140\ \mu m^{-3}$ and $c_s = 2.05\ \mu M$ NaOH. The lower curve (-----) reveals the differences between observed (•••) and calculated (—) intensities of the refined parameters. The inset depicts a SEM image of single crystals (size ≈1.0 μm) and the morphology of this crystal type. (B) S(Q) vs. Q profiles as a function of n obtained by static light scattering for rods (—) and SAXS (—) at $c_s = 7.6\ \mu M$ NaOH, pH 7.85. These materials do not exhibit any iridescence. The dotted lines (---) represent the calculated S(Q) values for an isotropic solution of rods with $L = 800$ nm and $d = 5.6$ nm as a function of n; (·····) for a calculated dodecahedral Lipid A-diphosphate rod-model for $n = 40\ \mu m^{-3}$ (T = 295 K). (C) shows a high resolution electron micrograph (HTEM) image of highly ordered and crystalline Lipid A-diphosphate nanorods in the [100] direction observed at pH 7.85 approaching the melting line for $n = 55\ \mu m^{-3}$ (T = 295 K), the scale bar is 100 nm. (D) shows a HRTEM image of Lipid A-diphosphate observed at pH 7.85, crossing the melting line for $n = 55\ \mu m^{-3}$ (T = 295 K), and at 5 μM NaOH. The scale bar is also 100 nm.

3.3. Morphologies of Lipid A phosphates

The various morphologies of Lipid A-diphosphate and Lipid A-monophosphate nanocrystals as observed by SEM are shown in Figure 7.

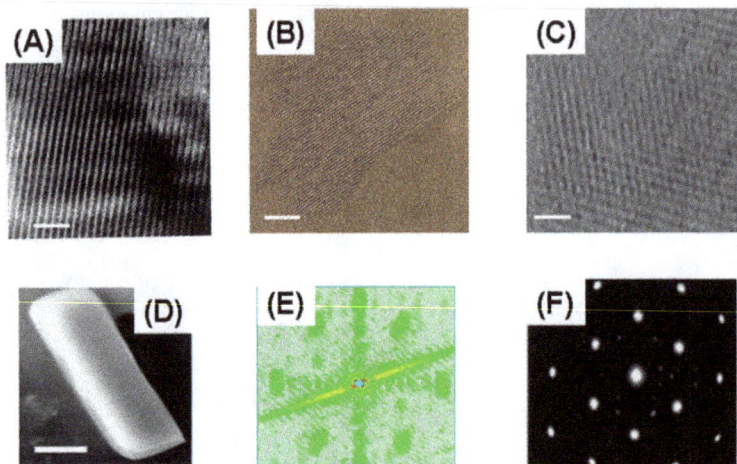

Figure 6. (A) – (C) show overviews of HRTEM images of crystalline Lipid A-diphosphate rods at pH 7.85 (295 K) along the [111] plane. The bar in (A) is 50 nm, 100 nm in (B) and 200 nm in (C). (D) Shows a long rod of Lipid A-diphosphate as a SEM image, the bar is 1 μm. This crystalline rod reveals FCC stacking faults along the [111] plane. A calculated Fourier transform image for models of truncated dodecahedral Lipid A-diphosphate particles is shown in (E); they are obtained form SAED pattern (F) from the lattices shown in (A) to (C). Model calculations support a five-fold axis in the [1-10] direction, and parallel to the (001) plane, i.e. 0° and 18°, but perpendicular to the fivefold axis (E) accounting for the experimental SAED pattern shown in (F).

These crystals are obtained at different particle number densities n as indicated and at constant T but at very low ionic strength, ($I \sim 10^{-6}$ M in NaCl or $I \sim 10^{-5}$ M in NaOH. Although the obtained nanocrystals at 10 μM NaOH show a better 3d order than those grown in very low NaCl concentrations according to the corresponding SAED and SAXS pattern quality, they generally reveal SEM images of icosahedral or dodecahedral morphology. Typical sizes of these crystals are of the order of 0.3-1 μm. Moreover, there was no congruent metastable other phases found for the icosahedral phase for the crystals grown in the presence of 10 μM NaOH, but there was in the presence of 1.0 μM NaCl, or 10 nM Ca^{2+} [11].]. Therefore, some small portion of FCC Lipid A-diphosphate was always present. Rhombic triacontahedral has been observed for Lipid A-monophosphate nanocrystals and is not the only icosahedral quasicrystal.

The antagonistic mediated incorporation into cubic crystalline Lipid A-diphosphate assemblies is more effective if the colloidal particles are oblate ellipsoids rather than curved. The assumed deformable soft spheres incorporate flexibility on a simple level [47]. At mechanical equilibrium the soft Lipid A-diphosphate sphere may be approximated into a prolate or oblate ellipsoid of revolution while preserving its volume at $\pi d^3/6$ where d is the hydrodynamic diameter of the soft sphere (~ 7 nm). The distortion is given through the aspect ratio $\rho = a/b$ of the self-assembly, a is the symmetry semiaxis length of the ellipsoid and b is its orthogonal semiaxis length.

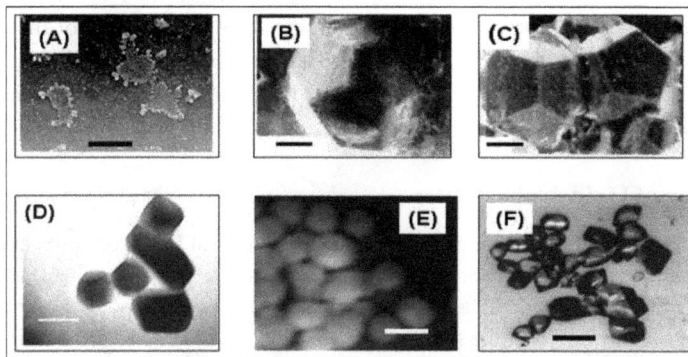

Figure 7. SEM images of the morphologies of Lipid A-diphosphate (**A – C**) and Lipid A-monophosphate crystals (**D - E**). These crystals are usually grown from aqueous dispersions containing either 10^{-5} M NaCl (A, B, E) or 10 μM NaOH (D, F). The icosahedral phase exhibits nanocrystals with pentagonal dodecahedral faceting. The scale bar is 100 nm in (A) - (C) and (E). The scale bar in (D) is 0.5 μm. In (**F**) faceted Lipid A-monophosphate nanocrystals are shown, the scale bar is 1.0 μm. The particle number density, n, was 350 μm⁻³ in (A) –(C), 150 μm⁻³ in (D) – (E), and 450 μm⁻³ in (F).

Furthermore, some Lipid A-diphosphate and approximants were formed by a peritectic reaction from the solid Lipid A-diphosphate phase and the liquid Lipid A-diphosphate (Form C of Fig. 1) at T = 295 K. These nanocrystals possess pentagonal dodecahedral solidification morphologies, but with exclusively pentagonal faces. Since this observation is associated with molecular motion and rearrangements of the positions in the Lipid A-phosphate nanocrystals lattices above a certain temperature (T > 295 K) or slowly cooling form T = 295 K to T = 288 K, which was deduced from broadening of the X-ray diffraction lines and SAED peaks, these modes belong to the phason long-wave-length phason. Relaxation of the phason strain is a diffusive process and therefore intimately related to the crystal growth process and is much slower than the phonon strain.

Quasicrystals exhibited non-crystallographic packing of non-identical Lipid A-phosphate spheres and a spatial packing of these spheres in either a cuboctahedron or an icosahedron were representative of sound physical models. Following an increase in temperature, a BCC phase was revealed and this structure gave rise to dodecagonal quasicrystals, which formed from spherical particles. Electron diffraction patterns of the dodecahedrons were recognized from the magnitudes of non-identical intensities of Lipid A-diphosphate and antagonistic Lipid A-diphosphate, which contained only 4 acyl-chains. It should be noted that the observed (3.3.4.3.4) was a crystalline analogue of the above-mentioned icosahedral quasicrystal with a different length scale. The tiling pattern of triangles (N_3) and squares (N_4) where the vertices were surrounded by a triangle-square-triangle tiling pattern possessed a *p4gm* plane group. Another coded Lipid A-diphosphate approximant showed an 8/3 ratio with 6-fold symmetry and plane group *p6mm*. Both (3.3.4.3.4) and dodecagonal phases revealed a N_3/N_4 ratio of approx. 2.34; the ratio of the p6mm plane group was 8/3. Because bond orientational order existed, the direction of domains was classified into three orientations for the (3.3.4.3.4) tiling, but only two for the 8/3 approximants. The average magnitude of the prominent scattering vectors, $|Q| = 0.121$ nm^{-1}, and the length of sides of the triangles and of the squares was 50 nm.

3.4. Packing of Lipid A-phosphates

Under the assumption that the macroscopic shape of a crystal is related to its microscopic symmetry and taking the various X-ray diffraction patterns, the SAED's and the morphology into consideration, the Lipid A-diphosphate structures and their approximants can be reconciled by lowering the symmetry from cubic, $Im\overline{3}m$ or $Ia\overline{3}d$ (a = 4.55 nm) for the diphosphate or monophosphate, respectively, to rhombohedral $R\overline{3}m$, and finally to monoclinic (P2$_1$ or C2) which is acceptable if the Lipid A-phosphate anions were completely orientationally ordered [12]. This could be attributed to two different space-filling packings: (i) two dodecahedra on site **a** and six tetrakaidecahedra on site **d**, forming a $Pm\overline{3}n$ lattice; (ii) or sixteen dodecahedra on site **d** and eight hexadecahedra on site **a,** forming an $Fd\overline{3}m$ lattice (Fig. 8). The space-filling network of Lipid A-diphosphate consisting of slightly distorted polyhedra is reminiscent of the known basic frameworks of gas-hydrates and sodium silicon sodalites. The final geometry of the "spheres", i.e. whether they were rounded or faceted in shapes, was established. The equilibrium separation distance between the interfaces of the "spheres" suggested that they repel each other as a result of electrostatic, steric, van der Waals forces as well as water layer surrounding the spheres. The small cubic $Pm\overline{3}n$ (a = 6.35 nm) structure observed for the Lipid A-monophosphate clusters [38, 48], but different from the large cubic unit cell with a = 49.2 nm materialized as a result of a space-filling combination of two polyhedra, a dodecahedra and a tetrakaidodecahedra. This is in contrast to the tetrakaidecahedra ($Im\overline{3}m$) or rhombodo-decadecahedra ($Fd3m$) packings observed for the Lipid A-diphosphate assemblies (Fig. 8). The small cubic $Pm\overline{3}n$ (a = 6.35 nm) structure observed for the Lipid A-monophosphate clusters [38, 48], but different form the large cubic unit cell with a = 49.2 nm materialized as a result of a space-filling combination of two polyhedra, a dodecahedra and a tetrakaidodecahedra. This is in contrast to the tetrakaidecahedra ($Im\overline{3}m$) or rhombodo-decadecahedra ($Fd3m$) packings observed for the Lipid A-diphosphate assemblies. **Note:** The limited number of reflections observed for this large cubic unit cell in case of the Lipid A-monophosphate was not sufficient to discriminate between primitive and body-centered-cubic lattices, or more precisely between Pn-n and I- - - extinction groups. The two differ by the reflection condition *hkl*: *h=k+l=2n*, which satisfied the BCC lattice, but the first set of reflections on which this condition could be tested would be the {421} reflection. This reflection was not observed because of the low resolution of the diffraction pattern.

Space groups such as $Im\bar{3}m$, $I4\bar{3}m$, $I4\bar{3}2$, etc. are all possible and may not be separated by extinction alone. Thus, it may be argued that favorable space filling packings will be achieved when satisfactory geometrical conditions were fulfilled (Plateaus's law). For the Lipid A-phosphates this resulted from minimizing the surface area between the aqueous films and so achieving a high homogeneous curvature. However, in the case of the Lipid A-monophosphate, rhombodo-decadecahedra ($Fd\bar{3}m$) packing seems to be suppressed a result of instability in the mean curvature between the tetrahedral and the octahedral nodes. Tetrakaidodecahedra packing shows only tetrahedral nodes. However, the tetrahedral angle (109.47°) can only be restored between all of the edges if the hexagonal faces of the truncated octahedron change and generate change and generate planer surfaces with no mean curvature and form Kelvin's minimal polyhedra [49,50].

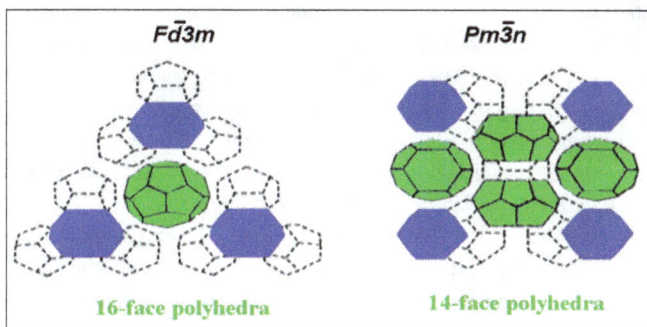

Figure 8. Packing of $Fd\bar{3}m$ and $Pm\bar{3}n$ tetrahedrally close-packed Lipid A-diphosphate and Lipid A-monophosphate structures for two cubic crystalline phases using Wigner-Seitz cells. For both structures, aqueous bilayer compartmentalizes the hydrophobic portion of the Lipid A-phosphates into tetrahedrally networks. This network is a combination of a pentagonal dodecahedron (blue) with 14 – face polyhedra (green) in $Pm\bar{3}n$ and with 16-face polyhedra (green) in $Fd\bar{3}m$.

4. Faceted crystal growth

4.1. Two-phase coexistence regions in monolayers

Normally equilibrium thermodynamics prevent faceting in two dimensions (2d) because the one-dimensional perimeter of a two-dimensional crystal exhibits no long-range order at any non-zero temperature. However, the formation of stable facets during crystallization need not prevent faceted crystal growth in two dimensions, which is supported both experimentally [18, 51-53] and by computer simulations [54-56]. This possibility is extremely useful in studies of cell surface recognition in the presence of Lipid A-diphosphate e.g. surface patterning, mechanical properties and cell mechanics with optical tweezers. Moreover, surface tension effects become important as the interface is no longer planar and it introduces a length scale of the order of a few nm, influencing crystal shape, morphology and stability as well as biomineralization [57].

4.2. Surface-tension-gradient-induced crystal formation in monolayers

We observed a surface-tension gradient induced crystal growth phenomena in Lipid A-diphosphate and Lipid A-diphosphate approximants layers comprising of the same chemical composition as studied for the re-entrant phase BCC and FCC networks. Briefly, the surface tension gradient was created by heating the trough (1 cm^3, 5.0 mins/°C), measuring simultaneously the surface tension, γ, for various n, either in the presence of a fluorescent dye (Alexa Fluor 488, Molecular Probes USA) which was spread on the aqueous surface, or by direct observation with a Scanning Electron Microscope (SEM) when the samples were withdrawn from the container under a light microscope (Olympus BX 60) and transferred to a substrate, coated with Pt (1min) and studied in a SEM (JEOL 6400). A video recording system was hooked up to the light microscope for monitoring the morphology changes with time. Double-chained lipids reveal saturation coverage of ~1.0 molecule/0.5nm^2 in aqueous media, which are magnitudes different from Lipid A-diphosphate where ordering of Lipid A-diphosphate occurs at concentrations that are less than 1.33×10^{-11} mbar·s or saturation monolayer coverage of $\sim 10^{-5}$ L [58, 59]. The surface density of the dye is $\sim 10^{-2}$ molecule/nm^2. Since n is much lower than the CMC of Lipid A-diphosphate (14.0 µg/mL (3.5×10^{-5} mM) at 10°C and 7.5 µg/mL at 20°C (2.0×10^{-5} mM; Kraft point 5.8°C), the dye does not dissolve in the bulk phase, and because the measurements are far from the stability region of the Lipid A-diphosphate 3d crystals, the surface film is 2-dimensional. By compressing the monolayer film a liquid phase-liquid condensed phase coexistence space is reached. The coexistence phase is characteristic of the formation of dark and fractal like liquid condensed phase domain when viewed through the fluorescent microscope. The morphology of the various 2d Lipid A-diphosphate images and their evolution are depicted in Figure 9.

Faceted domains built up on the tip of the fractal branches, but these tips are not stable and after continuous crystal growth a dendritic pattern evolved. The domains are squeezed in-between the dendritic stems promote to grow. However, the external tips of the domains have a higher probability to expand into a dendritic pattern. It appears that the hexagonal shaped Lipid A-diphosphate crystals grew in the liquid-crystalline boundary (Fig. 9). It was also observed that the hexagonal domains enlarged to the dendritic pattern, where the corners of the hexagons are very sharp and the dendrites revealed stable tips and strong stems are clearly observed. Quantitatively, the Lipid A-diphosphate is higher than in the middle of the straight edge at the corners of the faceted pattern. This implies that the main transfer rate to the corners is higher than at other places in space causing an instability region. As a result the crystal edge corners grew more rapidly than the center region and (curled) dendrites appeared on the corners of the hexagon (Fig. 9g).

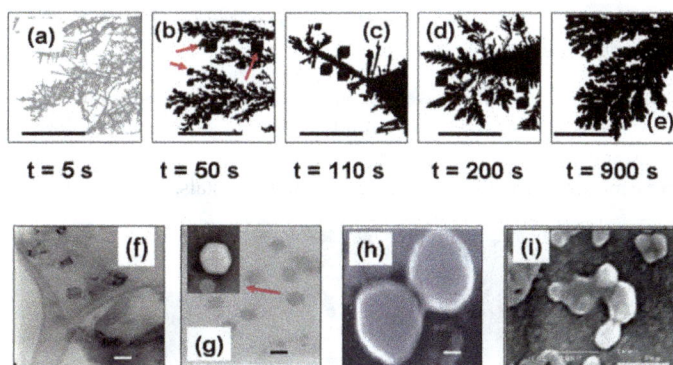

Figure 9. Development of fractal crystalline pattern of Lipid A-diphosphate domains as a function of time. The video-recorded light-optical microscopy images were directly converted to black and white images by using the Adobe Photoshop (version 8.0.1.) **(a)** Observed fractal patter when the monolayer of Lipid A-diphosphate is compressed to the liquid-crystalline to liquid-expanded coexistence region. **(b) – (e)** show the time evolution of the same Lipid A-diphosphate pattern from fractal to dendritic behavior under the microscope. At the tips (red arrows) in **(b & c)** become thicker and reveal facets; normally these tips grow into thick dendrites with clearly main stem, strong and stable tips **(d & e)**. The bar size is 50 μm. The morphologies of the liquid-crystalline domains of Lipid A-diphosphate are shown **(f-i)**. In **(f)** a HRTEM image of faceted cubic-like and unconnected Lipid A-diphosphate crystals are shown (the bar is 100 nm). **(g)** depicts the observed hexagon domain which grows usually near the cubic-like crystal domain; but this is dependent on T, however, at constant I; the bar size is 10 nm. **(h)** Shows a SEM image of fusing Lipid A-diphosphate crystals observed when the monolayer expands to a fluid state, or compresses to a certain density d_c where the crystals contact one another and fuse. The bar is 10 μm, T = 295 K and c = 6.0 μg/mL. **(i)** Shows single Lipid A-diphosphate nanocrystals grown at T = 295 k but in c = 60.0 μg/mL and in the presence of 20 μM NaOH. These nanocrystals exhibit icosahedral faces and also rhombic triacontahedral faceting; the bar size is 1 μm.

For this instability, a field gradient existed where crystals grew faster as they reached deeper into the gradient. As a result of this instability was an invasion of the more viscous phase by the less viscous phase without ordering or a characteristic length scale. It was also a process which operated during diffusion-limited aggreagtion, in which case the diffusive instability led to a fractal structure. Moreover, in the growth process far from equilibrium the aggregation of the fractal-like domains can be faster than the relaxation process of the Lipid A-diphosphate clusters. Hence the Lipid A-diphosphate clusters form rectangular lattices. The 2d-hexagonal domains nucleated and grew under lower force, where in the meantime the Lipid A-diphosphate clusters gain enough time to relax to minimum lattice energy positions. Therefore, the difference in macroscopic morphology together with the caused instability implies that, the structure of the liquid crystal domains depend on the driving force. Also what should be mentioned is the influence of different chiral conformers in the bulk solution whose presence has been supported by circular dichroism experiments and molecular simulations (unpublished results, 2011). The observed hexatic phase (Fig. 10), which separates the isotropic liquid Lipid A-diphosphate cluster phase and the liquid crystalline phase has short-range translational and quasi long-range orientational order.

In accord with theory [22, 60-64] we are able to detect three phases: a liquid, hexatic and a crystal Lipid A-diphosphate phase, but no fluid-crystal or hexatic-crystal phase coexistence phase. This is supported from surface tensiometry measurements as a function of T and μg/mL bulk concentration revealing hexagonally shaped crystals.

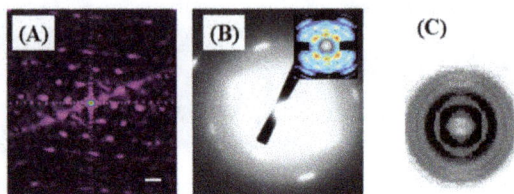

Figure 10. SAED's (Joel, T 3010, 300 kV, 15 cm) and SAXS images for the various crystalline Lipid A-diphosphate clusters at different μM NaOH and Lipid A-diphosphate concentrations (c). **(A)** SAED for a 2d-pattern, a 10 nm thick layer obtained from very small Lipid A-diphosphate crystallites (sizes ∼ 0.1-0.2 μm, c = 3.5 μg/ml, T = 295 K and γ_c = 25.5 mN/m in 5 μM NaOH. This 2d pattern indicates a hexagonal or centered trigonal unit cell with a = 3.70 nm. The bar size is 0.5 nm^{-1}, and the image was obtained with a CCD camera. **(B)** Enlarged SAED pattern of a 2d crystallite for c = 8.5 μg/ml, T_c = 295 K and γ_c = 25.5 mN/m but for 5.75 μM NaOH. This crystalline Lipid A-diphosphate material is maintained for approximately 1-1.5 h., and is indicative for the hexatic phase (inset). **(C)** SAXS pattern for the fluid phase where only diffuse rings but no dots are visible and close to the melting line; a change from solid to liquid occurred ($d\gamma/dT$ < 0, S_s >S_b) where S_s is larger than the bulk entropy (S_b) and where the density is very similar to that of the bulk (for T > T_c).

Normally, lipids or surfactant concentrations are well in the range of mM (mg/mL) when forming a liquid-like monolayer, but in the case of Lipid A-diphosphate we are in the μg/ml range or lower and strongly dependent on polydispersity in charge and size distribution! This is significantly different from insoluble surfactant monolayers. Preliminary theoretical fitting results of the thermodynamics, $\gamma(T)$ vs. T (T< T_c), with T_c is the critical temperature where the slope changed ($d\gamma/dT)_c$, yielded values for the chain interactions of $b \cong 41.8$ k$_B$T and the effective sugar-phosphate headgroup attraction of $a \cong -6.6$ k$_B$T [58].

Different phase can be distinguished from 2d structure factors (Fig. 10). The functional form of the angular intensity profile of S(Q) is the square root Lorenzian of the hexatic phase and Lorenzian of the crystal phase [65, 66]. In the hexatic phase both the ring and the six spots (Fig. 10A & B) are clearly noticeable, where the six spots indicate a small ordered patchy like located in the dense liquid Lipid A-diphosphate dispersion (Fig. 10C). Using the disorder parameter ⟨D⟩ [67, 68], which is a function of T, the solid hexatic phase and the liquid phase reveal different slopes [69]. In the hexatic phase both ⟨D⟩ and the variance sharply increases with configurational temperature. Approaching the melting region close to the liquid phase the average and the variance of ⟨D⟩ approaches ∼5.0 and 4.0, respectively, where ⟨D⟩$_j$ = 0 corresponds to a perfect triangular lattice and becomes larger for more disordered structures. The thermal fluctuation and entropic interactions lower the free energy (Eq. 1) of the interface when two Lipid A-diphosphate particles approach each other in a sort of Casimir type of effective attraction [70-72]. The polydispersity influences the coexistence region of the fluid and crystal to higher volume fractions. At fixed supersaturation, the height of the nucleation barrier is not affected by a polydispersity up to ~ 6% in our experimental studies; while for larger polydispersities the barrier increases sharply; thus an increase in surface tension γ with T and n is noticed.

But, when melting of Lipid A-diphosphate crystals commenced above a critical temperature (T_c), γ_c as a function of n, the evaporation rate becomes slower when the crystallites shrink. These crystallites may sediment rapidly when transforming into the bulk liquid with a higher density than from the onset of the crystallization, where a lower density of the bulk solution is met than the actual crystal and evaporates swiftly. This melting scenario is reminiscent to the coexistence of a dense and expanded crystal phase. This Lipid A-diphosphate crystal phase also depends strongly on polydispersity in size, mass and charge for T and n is constant. The total number of Lipid A-diphosphate crystals depends on the kinetics during the monolayer compression and is a function of n, T and γ. Once the early seed have developed the number of domains is fixed and does not change with subsequent compression unless the monolayer expands to a fluid state, or compresses to a certain density d_c where the crystals contact one another and fuse (Fig. 8h). Evidently this depends on the rate of compression (or evaporation) or "impurities" seen as "various Lipid A-diphosphate conformers" present in the bulk Lipid A-diphosphate dispersion due to the conformational changes of the disaccharide as noticed from the crystal structures of Lipid A-monophosphate [73]. The cause of this resulting instability originates from an increase of lipid A-phosphate conformers over another conformer on a characteristic length scale rather than on impurities [74]. It is also a process which operated during diffusion-limited aggreagtion, in which case the diffusive instability led to a fractal structure. Furthermore, the shape of the Lipid A-diphosphate crystals (Fig. 8A) is also affected by the interfacial free energy between the solid and liquid phase. Particularly, the interface influences the free energy penalty $A\gamma$, which is proportional to the area of the interface and the surface energy γ. Consequently, the free energy difference of the crystal and the fluid is:

$$\Delta G = A \cdot \gamma - V \cdot n_{crystal} \Delta \mu \qquad (1)$$

Where V is the volume of the crystal nucleus, $n_{crystal}$ is the particle number density in the crystal and $\Delta \mu = \mu_{Fluid} - \mu_{crystal}$ is the difference of the chemical potential of fluid and crystal, respectively.

Furthermore, the 2d order quality of the Lipid A-diphosphate crystals is influenced by grains due to mixed sizes and shapes (Fig. 9C). Complete uniformity cannot be expected. Assuming the 2d lattice and the force between adjacent spheres (or ellipsoidal particles with a low axial ratio) are identical may yield under compression (sedimentation, gravity, capillary forces) another packing than hexagonal resulting in change to a cubic lattice. This cubic lattice has actually been found for both lipids. Consequently the ordered 2d hexagonal structure which is present in the fluid state in the presence of μM NaOH (Fig. 2B) has a minimum density, whereas the intermediate density rests with the detected cubic structure and finally the maximum density will be closer to the hexagonal close packing than to the simple cubic structure. This is contrary for the Lipid A-diphosphate invariant to the particle number density, n, and T=constant, but at very low I of μM to mM NaCl or nM Ca^{2+}, respectively [9, 11].

The influence of the Lipid A-diphosphate crystal-crystal-coexistence within the crystallization process has also to be considered. Since there is a noticeable variation in large and small μm-sized crystals as a function of polydispersity in size and charge, the ratio may be important so that the expanded Lipid A-diphosphate crystal phase is metastabile and a function of ϕ, T and I. As a result the density of the Lipid A-diphosphate nuclei is increasing with ϕ (n), which is contrary of the classical nucleation theory (CNT) [60]. According to this theory the density of the crystal is the same as the bulk density, or the density decreases with increasing of the Young-Laplace pressure, $\Pi_c = \gamma/L_c = \gamma \cdot g$, where L_c is the capillary length and depends on the curvatures inside and outside of the nuclei, due to capillary forces [74, 75], which takes also the surface tension and the chemical potential (μ) into account. Thus the assumption of CNT is independent on γ is no longer valid. This result in a decrease of the nucleation rate with an increase in supersaturation is governed by the increase in γ rather by slowing down the kinetics. There is also strong evidence that the 2 d crystals of Lipid A-diphosphate do not melt in a first-order transition but may be in second-order transition. This behavior follows the theory developed by Kosterlitz, Thouless, Halperin, Nelson, and Young (KTHNY), which predicts that a third phase, namely the hexatic phase with short-range translational order and quasi-long-range order exists between crystal and liquid [60-64].

5. Conclusion

The successful production of Lipid A-phosphate crystals makes it extremely useful to study various Lipid A-diphosphate assemblies of e.g. four, three, and penta- or hexaacylated Lipid A-phosphate approximants including those of modified disaccharide or monosaccharide moieties. This still remains to be elucidated. It was possible to construct different Wigner-Seitz polyhedra that make up the overall volume of the Frank-Kasper type unit cells with complexes comprised of Lipid A-diphosphate, antagonistic and non-toxic Lipid A-phosphate analogues depending on volume fraction, ϕ ($\phi = \upsilon_2 \cdot c$), the nature of the counterions and temperature. They form by spontaneous self-assembly and appear to obey the principles of thermodynamically reversible self-assembly but once self-assembled strongly resist disassembly. Base on these principles, Lipid A-phosphate assemblies can be designed which form large unit cells by containing more than hundreds of Lipid A-phosphates. The range of Lipid A-phosphate structures may also be increased further by employing various different ("non-identical subunits") and identical subunits of Lipid A-phosphate in analogy with block copolymers. The rational design of such assemblies and the nucleation and creation of polymorphic Lipid A-phosphates production of mesoscopic suitable cellular networks, and structure-function relationships will be impacted by a theoretical and practical understanding of the spherical assemblies, rod-like assemblies and the mixtures thereof. Furthermore, the unit cell found for a four-single-chained Lipid A-phosphate approximant contained four honeycomb cells: two triangular and two quadrangular. However, the corresponding monophosphate contained 16 cells, of which either 10% or 66% were quadrangular. Given the theoretical and practical importance of this system, we expect that the attention given to it will substantially increase our knowledge on Lipid A-di-and monophosphates and the driving forces for the ordered assemblies. Furthermore, the structure of the Lipid A-diphosphate rod can be explained as truncated large dodecahedra.

The crystallization and phase behaviour of the Lipid A-diphosphate in two-dimensional (2d) and three-dimensional (3d) systems has been elucidated in more detail than before and analyzed as a function of ϕ, T, γ, morphology, and structure stability with the application of the CNT and KTHNY theories. But the experimental situation appear to be more complicated, because no real long-range translational order exists in 2d crystals and the phase behaviour close to freezing has been found to be richer than in 3d systems. We discovered for the Lipid A-diphosphate system a hexatic phase with short-range-translational order and quasi-long-range orientational order between crystal and liquid.

Author details

Henrich H. Paradies, Hendrik Reichelt and Chester A. Faunce
The University of Salford, Joule Physics Laboratory, Manchester, United Kingdom

Peter Quitschau
Fachhochschule Südwestfalen, University of Applied Sciences, Biotechnology & Physical Chemistry, Iserlohn, Germany

Kurt Zimmermann
The Symbio Herborn Group Inc., Institute for Microecology, Herborn, Germany

Acknowledgement

The authors thank Professors S. E. Donnelly, Drs. N. M. Boag, S. Simpson (Manchester, U.K.) for very critical discussions on electron microscopy, X-ray diffraction and chirality, and for using the High Resolution Electron Microscope (Joel T 3010) and the X-ray diffraction equipments. Professor K. Stadtlander (Iserlohn, Germany) for lending us the Olympus light microscope and the video camera. Financial support from the Biomaterials Project (Bruxelles, Grant BMH4-CT-96-0013), Chemical-Biotechnological Laboratories Inc (Iserlohn) and from Orthomol (Langenfeld, Germany) are gratefully acknowledged.

6. References

[1] Kepler, J. (1611). *De Nive Sexangula Godfrey Tampach*. Frankfurt am Main.
[2] Tiller, W. A. (1991). *The Science of Crystallization*. Cambridge University Press, Cambridge. (1991.

[3] Klug, A, Franklin, R. E., Humphreys-Owen, S. P. F. (1959). The Crystal Structure of Tipula Iridescent Virus As Determined By Bragg Reflection of Visible Light Biochim. *Biophys. Acta*, 32, 203-219.

[4] Holmes, K. C., Finch, J. T. (1967). Structural Studies of Viruses. *In Methods of Virology*, Vol. III, 351-474.

[5] Jones, R. A. L. (2002). *Soft Condensed Matter*. Oxford Master Series In Condensed Matter Physics, Oxfords University Press.

[6] Kittel, Ch. (1996). *Introduction to Solid State Physics*, John Wiley & Sons, Inc. New York.

[7] Thies, M., Quitschau, Zimmermann, K., Rusch, V., Faunce, A. C., Paradies, H. H (2002). Liquid-like ordered colloidal suspensions of Lipid A: The influence of Lipid A particle concentration, *J. Chem. Phys.* 116, 3471-3483.

[8] Faunce, C. A., Quitschau, P., Thies, M. Scheidt, T, Paradies, H. H. (2002). Bacteria and Bacterial fragments as Immunomodulatory Agents, in Probiotics, *Old Herborn University Seminar* Vol. 15, 95-120.

[9] Faunce, C. A., Quitschau, P., Paradies, H. H. (2003). Solution and structural properties of colloidal charged Lipid A (diphosphate) dispersions. *J. Phys. Chem. B.* 107, 2214-2227.

[10] Faunce, C. A., Paradies, H. H. (2007). Density fluctuation in Coulombic colloid dispersion: self-assembly of Lipid A-phosphates. *Mater. Res. Soc. Symp. Proc.* Vol. 947, 0947-A03-11.

[11] Faunce, C. A., Reichelt, H., Quitschau, P. Zimmermann, K., Paradies, H. H. (2005). The liquidlike ordering of Lipid A-diphosphate colloidal crystals: The influence of Ca^{2+}, Mg^{2+}, Na^+, and K^+ on the ordering of colloidal suspensions of Lipid A-diphosphate in aqueous solutions. *J. Chem. Phys.* 122, 214727-214750.

[12] Faunce, C. A., Reichelt, H., Paradies, H. H., Quitschau, P. Rusch, V., Zimmermann, K. (2003). The formation of colloidal crystals of Lipid A-diphosphate: Evidence for the formation of nanocrystals at low ionic strength. *J. Phys. Chem. B.* 107, 9943-47.

[13] Paradies, H. H., Habben, F. (1993). Structure of N-hexadecylpyridinium chloride monohydrate. *Acta Crystallograph.* C 49, 744-747.

[14] Alonso, B., Massiot, D., Florian, P., Paradies, H. H., Gaveau, P., Mineva, T. (2009). ^{14}N and ^{81}Br. Quadrupolar nuclei as sensitive NMR probes of n-alkyltrimethylammonium bromide crystal structures. An experimental and theoretical study. *J. Phys. Chem. B.* 113, 11906-11920.

[15] Paradies, H. H., Clancy, S. F. (2000). Crystalline Polymorphism of Cetyltrimethyl-ammonium bromide and Distearyldimethyl ammonium (DSDMA) Compounds, *Rigaku J.* 17, 20.

[16] De Gennes, P., and Prost, J. (1993). *The Physics of Liquid Crystals*. International Series of Monographs on Physics. Oxford Science Publications (Clarendon Press, Oxford).

[17] Faunce, C. A., Paradies, H. H. (2009). Two new colloidal crystal phases of Lipid A-monophosphate: Order-to-order transition in colloidal crystals. J. Chem. Phys., 131, 244708-244728.

[18] Berge, B., Faucheux, L., Schwab, K., Libchaber, A. (1991). Faceted crystal growth in two dimensions, *Nature* 350, 322-324.

[19] Hamley (2000). *Introduction to Soft Matter* (2nd Edit.) J. Wiley, Chichester, U.K.

[20] Israelachvili, J. (1992). *Intermolecular and Surface Forces* (2nd Edit.), Academic Press.

[21] Hynninen, A.P., Thijssen, J. H. J., Vermolen, E. C. M. , Dijkstra, M., van Blaaderen, A. (2007). Self-assembly route for photonic crystals with a bandgap in the visible region. *Nat. Materials* 6, 202-205.

[22] Gasser, U. (2009). Crystallization in three-and two-dimensional colloidal suspensions. *J. Phys.: Condes. Matter* 21, 203101-2031025.

[23] Pieranski, P. (1980). Two-dimensional Interfacial Colloidal Crystals. *Phys. Rev. Lett* 45, 569-572.

[24] Sommerdijk, W. A., Cölfen, H. (2010). Lessons from Nature-Biomimetic Approaches to Mineral with Complex Structures. *MRS Bulletin*, 35 (2) 116-121.

[25] Garcia-Ruiz, J. M., Hyde, S. T., Carnerup, A. M., Christy, A.G., van Kranendonk, M. J., Welham, N. J. (2003). Self-assembled-carbonate Structures and detection of Ancient Microfossil. *Science* 302, 1194-1197.

[26] Crocker, J.C., Grier, D.G. (1998) From Dynamics to Devices: Directed Self-Assembly of Colloidal Material. *MRS Bulletin* Vol.23 (10) 24-50.

[27] Weitz, D. A., Russel, W.B. (2004). New Developments in Colloid Science. *MRS Bulletin* Vol. 9 (2) 82-105.

[28] Raetz, C. R., Whitefield, C. (2002). *Lipopolysaccharide endotoxins*. Ann. Rev. Biochem., 71, 635-700.

[29] Morbidity and Mortality Weekly Report; Technical Report No: 46 (1997); Center for Disease Control: Atlanta, GA, 941-944.

[30] Christ, W. J., Hawkins, L. D., Lewis, M. D., Kishi, Y. (2003). Synthetic Lipid A antagonist for sepsis treatment. In *"Carbohydrate-based Drug Discovery"*, Vol. 1, pp 341-355; Ed. C.-H. Wong, Wiley-VCH Verlag GmbH & Co, KGA.

[31] Wehkamp, J., Bals, R., Kreft, B., Schroeder, J.-M., Stange, E. F. (2007). Innate Immunity. *Dtsch. Ärztebl, 104 (5)*, A 257 – 262.

[32] Brandenburg, K., Koch M. H., Seydel, U. (1990). Phase diagram of Lipid A from Salmonella minnesota and Escherichia coli rough mutant lipopolysaccharide. *J. Struct. Boil.* 105, 11-21.

[33] Brandenburg, K., Richter, W., Koch M. H., Seydel, U (1998). The nonlamellar cubic and HII structures of Lipid A from Salmonella enterica serovar Minnesota by X-ray diffraction and freeze-fracture electron microscopy. *Chem. Phys. Lipids* 91, 53-69.

[34] Gast, E.P., Russel, W.B., Hall, C.K. (1986). An experimental and theoretical study of phase transitions in the polystyrene latex and hydroxyethylcellulose system. *J. Colloid Interface Sci* 109, 161-171.

[35] Reichelt, H., Faunce, C. A., Paradies, H. H. (2008). The phase transition of charged colloidal Lipid A-diphosphate. *J. Phys. Chem.* 112, 3290-3293.

[36] Robbins, M...O., Kremer, K., Grest, G.S. (1988). Phase diagram and dynamics of Yukawa systems. *J. Chem. Phys.* 88, 3286-3312.

[37] Meijer, E. J., Frenkel (1991). Melting line of Yukawa system by computer simulation. *J. Chem. Phys.* 94, 2269-2271.

[38] Faunce, C. A., Reichelt, H., Quitschau, P., Paradies, H. H. (2007). Ordering of Lipid A-monophosphate clusters in aqueous solutions. *J. Chem. Phys.* 127, 115103-11525.

[39] Lindemann, F. A. (1910). Über die Berechnung molekularer Eigenfrequenzen. *Phys. Z.* 14, 609-612.

[40] Strandburg, K.J. (1988). Two-dimensional melting. *Rev. Mood. Phys.* 60, 161-207.

[41] Löwen H., Palberg, T., Simon, R. (1993). Dynamic criterion for freezing of colloidal liquids. *Phys. Rev. Lett.* 70, 1557-1470.

[42] Verlet, L, Hansen, J.-P. (1969). Phase transitions of Lennard-Jones Systems. *Phys. Rev.* 184, 151-161.

[43] International Tables for X-ray Crystallography, Vol. 1, published for The International union of Crystallography, Birmingham, England 1965.

[44] Aschauer, H., Grob, A., Hildebrandt, J., Schuetze, E., Stuetz, P. (1990). Highly purified lipid X is an LPS-antagonist only. I. Isolation and characterization of immunostimulating contaminants in a batch of synthetic lipid X. *J. Biol. Chem.* 265, 9159-9164.

[45] Faunce, C. A., Paradies, H. H. (2008). Observations of liquid like order of charged rodlike Lipid A-diphosphate assemblies at pH 8.5. *J. Chem. Phys.* 128, 065105-065113.

[46] Cowley, J. M., Moodie, A. F., The scattering of electrons by atoms and crystals (1957). *Acta Crystallograph.* 10, 609-619.

[47] Goodman, P., Moodie, A. F., Numerical evaluation of N-beam wave functions in electron scattering by multi-slice method (1974). *Acta Crystallograph.* A30, 280-290.

[48] Paradies, H. H., Faunce, C. A., Thompson, R. (2005). Probing Complex Fluid Membranes and Films with Neutron Spin-Echo, Indiana University, August 14-17, (2005):
http://www.iucf.indiana.edu/events/neutronmembrane/presentations/Paradies_lipid _diphosphates.pdf.

[49] Batista, V. M. O., Miller, M. A. (2010) Crystallisation of deformable spherical colloids. *Phys. Rev. Lett.* 105, 088305-088309.

[50] Faunce, C. A., Paradies, H. H. (2009). Two new colloidal crystal phases of Lipid A-monophosphate: Order-to-order transition in colloidal crystals. *J. Chem. Phys.* 244708-244728.

[51] Ziherl, P. and Kamien, R. D. (2001). Maximizing Entropy by Minimizing Area: Towards a New Principle of Self-Organization. *J. Phys. Chem. B* 105, 10147-10158.

[52] Anderson, D. M., Gruner, S. M., Leibler S. (1988). Geometrical aspects of the frustration in the cubic phases of lyotropic liquid crystals. *Proc. Natl. Acad. Sci.* USA, 85, 5364-5369.

[53] Miller, A. Knoll, Möhwald, H (1986). Fractal growth of Crystalline Phospholip Domains in Monomolecular Layers. *Phys. Rev. Lett* 56, 2633-36.

[54] Wang, M, Wildburg, G., van Esch, J.H., Bennema, P., Nolte, R.J.M., Ringsdorf (1993). Surface-tension-gradient-induced pattern Formation in Monolayers. *Phys. Rev. Lett.* 71, 4003-4006.

[55] McConnell, H.M., Keller, D, Gaub, H. (1986). Thermodynamic Models for the Shapes of Monolayer crystals. *J. Phys. Chem.* 90, 1717-1721.

[56] Meakin, P. (1988). Noise-reduced and anisotropy-enhanced Eden and screened-growth models. *Phys. Rev. A* 38, 418-426.

[57] Ben-Jacob, E., Garik, P. (1990). The formation of patterns in non-equilibrium growth. *Nature* 343, 523-530.

[58] Xiao, R.-F, Alexander, J.I.D., Rosenberger, F. (1988). Morphological evolution of growing crystals: A Monte Carlo simulation. *Phys. Rev. A* 38, 2447-2456.

[59] Faunce, C.A., Paradies, H.H. (2008). Nucleation of Calcium Carbonate as Polymorphic Crystals in the Presence of Lipid A-diphosphate. *J. Phys. Chem. B.*, 112, 8859-8862.

[60] Faunce, C.A., Paradies, H.H. (2012). Observations of solution growth of nanocrystals of charged lipid A-diphosphate. *Phys. Chem. Lett.* Submitted.

[61] Lin, I.-H, Miller, D.S., Murphy, C.J., de Pablo, J. Abbott, N.L. (2011). Endotoxin-induced-Structural Transformations in Liquid Crystalline Liquids. *Science* 332, 1297-1300.

[62] Volmer, M., Weber, A., (1926). Germ formation in oversaturated figures. *Z. Phys. Chem.* 119, 277-301.

[63] Becker, R., Döring, W. (1935). Kinetic treatment of germ formation in supersaturated vapour. *Ann. Phys.* 24, 719-752.

[64] Young, A.P. (1979). Melting and the vector Coulomb gas in two dimensions. *Phys. Rev. B* 19, 1855-1866.

[65] Nelson, d. R., Halperin, B.I. (1979). Dislocation-mediated melting in two dimensions. *Phys. Rev. B.* 19, 2457-2484.

[66] von Grünberg, H.H., Keim, P., Maret, G. (2007). Phase transitions in Two-dimensional colloidal systems. *Soft Matter, Colloidal Order: Entropic and Surface Forces*, Vol. 3, eds. G. Gomper & M. Schick (Weinheim: Wiley-VCH) 40-83.

[67] Marcus, A.H., Rice, S.A. (1996). Observations of First-Order Liquid-to-Hexatic and Hexatic- to-Solid Phase Transitions in Confined Colloidal Suspensions. *Phys. Rev. Lett.* 77, 2577-2580.

[68] Davey, S.C., Budai, J., Goodby, J.W., Pindak, R., Moncton, D. E. (1984). X-ray study of the Hexatic-B-Smectic A Phase Transition in liquid Cyrstal Films. *Phys. Rev. Lett.* 53, 2129-2132.

[69] Qi, W.-K., Wang, Z.-R., Han, Y., Chen, Y. (2010). Melting in two-dimensional Yukawa systems: A Brownian dynamics simulation. *J. Chem. Phys.* 133, 234508-234520.

[70] Larson, a. E., Grier, D.G., (1996). Melting of Metastabile Crystallites in Charge-Stabilized Colloidal Suspensions. *Phys. Rev. Lett.* 76, 3862-3865.

[71] Shiba, H., Onuki, A., Araki, T. (2009). Structural and dynamical heterogeneities in two-dimensional melting. *Europhys. Lett.* 86, 66004-66010.

[72] Nikolaides, M.G., Bausch, A.R., Hsu, M.F., Dinsmore, A.D., Brenner, M.P., Gay, C., Weitz, D.A. (2002) Electric-field-induced capillary attraction between like-charged particles at liquid interfaces. *Nature* 420, 299-301.

[73] Goulian, M., Bruinsma, R., Pincus, P (1993). Long-range forces in heterogeneous fluid membranes. *Europhys. Lett.* 22, 145-150.

[74] Golestanian, R., Goulian, M., Kardar, M (1996). Fluctuation-induced interactions between rods on a membrane. *Phys. Rev. E.* 54, 6725-6734.

[75] Faunce, C.A., Paradies, H.H. (2011). Studies on Structures of Lipid A-monophosphate clusters. *J. Chem. Phys.* 134, 104902-149016.

[76] Mullins, W.W. (1984). Thermodynamic equilibrium of a crystalline sphere in a fluid. *J. Chem. Phys.* 81, 1436-1442.

[77] Cacciuto, A. Frenkel. D. (2005*).* Stresses inside critical nuclei. *J. Phys. Chem. B* 109, 6587-6594.

Crystal Engineering

Chapter 6

Crystals in Materials Science

Dinesh G. (Dan) Patel and Jason B. Benedict

Additional information is available at the end of the chapter

1. Introduction

Intense and sustained research into single crystals and polycrystalline assemblies continues to produce advanced materials crucial to the function of modern and future electronic and photonic devices. Device function is not only based upon the optical and electronic properties of the active material (for example, the light absorbing layer in molecular solar cells), but also the molecular and supramolecular ordering. Herein we will focus on devices and technologies where single crystalline materials are employed, and where appropriate, discuss competing materials and devices where polycrystalline or amorphous materials are used. We acknowledge that there is a sustained push to use solution processable conjugated polymer films for several similar technological applications; however, these films are typically disordered and suffer from impurities and chemical defects inherent to synthetic macromolecules making definitive characterization using X-ray diffraction difficult or impossible. Small molecule crystalline materials, on the other hand, can be obtained in high purity with a high level of organization in the solid state.

Additionally, our discussion will be largely limited to organic materials as there is a drive to replace conventional inorganic active materials (for example the semiconductor channel layers in transistors) with organic compounds. The reasoning for this lies in the fact that organic compounds can be synthesized and purified relatively easily using bench-top techniques without the requirement of high temperatures and energy intensive processes. As an example, to obtain electronics and solar cell grade silicon several high temperature steps are required to reach the necessary purities in excess of 99.9999%.[1] Organic compounds can be highly purified using conventional solution crystallization or vapor growth[2] methods yielding materials which exhibit impressive performance, in some cases rivaling their inorganic counterparts.[3]

For this review we selected to focus on two areas: (i) photochromic molecules that can be used for molecular scale data storage and as actuators capable of turning molecular level changes into macroscopic motion, (ii) organic field effect transistors and organic photovoltaics that make use of single crystals for charge transport. Other important areas of research such as catalysis, nonlinear optics, magnetic materials, and biology also make effective use of crystalline materials but are beyond the scope of this review.

2. Photochromic materials

Photochromism is the light induced reversible change in the absorption spectrum of a molecule and is necessarily accompanied by changes in both electronic and molecular structure and properties.[4] A generic photochromic system in which component A can be transformed into B via irradiation with light of energy $h\nu_1$ is illustrated in Figure 1. Species B reverts back to A upon exposure to light with energy $h\nu_2$. As photochromic molecules have the ability to optically cycle between two spectroscopically distinct states, they have been studied for possible applications in molecular switching, variable transmission lenses, and molecular scale optical data storage.[5] In some cases a system may also have a thermally activated pathway for the interconversion of the chemical species. Given the close packing and perceived rigidity of crystalline lattices, photochromism in the crystalline state was thought to be rare compared to observation of this phenomenon in solution[6] or in polymer matrices[7] where the medium is less constrained and thus permits the associated changes in molecular shape and volume.

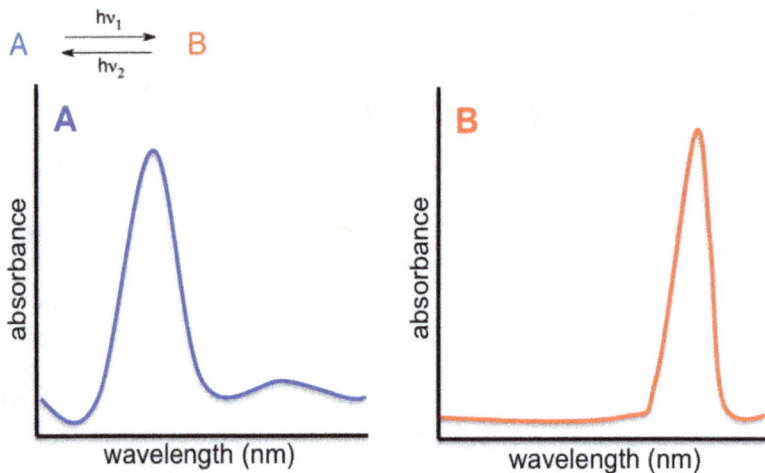

Figure 1. Photochromic compounds can interconvert by irradiation with light of different wavelengths. The two components, A and B, of a system have different absorption spectra. Though not specifically shown, a thermal pathway to interconversion may be present.

The spirooxazine (SO) and spiropyran (SP) class of photochromic materials are typified by a heterocyclic ring-closed isomer, the spiro-form, which is usually colorless and dominates the equilibrium. The isomerization of representative SO and SP compounds is shown in Scheme 1. Irradiation with high-energy UV light leads to cleavage of the C-O bond at the spirocyclic carbon giving a ring-opened form, the photomerocyanine form, that is strongly colored owing to the extended conjugation afforded by planarity and the donor-acceptor character of the molecule.[5, 8] Removal of the UV irradiation source or irradiation with visible light affects the reverse reaction. While photochromic spiropyrans[9] and spirooxazines[10] have been known for some time, photoisomerization has almost exclusively been observed in solution and in polymer matrices, though recent research suggests that crystalline state photoisomerization may be more general and thus make SOs and SPs amenable to use in solid state devices. Understanding and visualizing the structural and electronic changes that occur upon photoisomerization would be indispensable in aiding the design and fabrication of SO and SP containing devices and guide the synthesis of new materials.

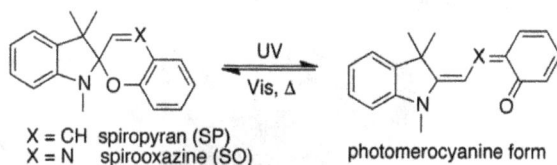

Scheme 1. General structure of spirooxazines and spiropyrans. Photoisomerization is initiated with UV light; visible light or heat affects the ring closing reaction.

More than ten years ago, Benard and Yu used UV-Vis absorption measurements to show that photoisomerization occurred in microcrystalline powders of cationic SO **1**[11] and SPs **2** and **3**[12] shown in Figure 2. This was attributed to the presence of a large iodide counter ion giving the necessary free space in the crystal lattice along with stabilization of the iminium cation that was afforded upon ring opening. When examined by X-ray diffraction, however, the powder patterns showed only small changes suggesting that the observed isomerization was likely a surface effect and did not occur in the bulk.

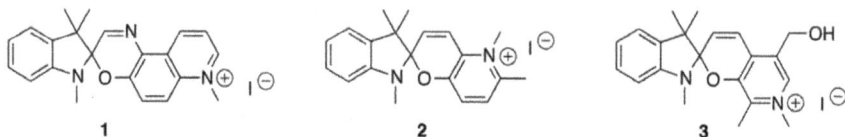

Figure 2. Molecular structures of cationic spirooxazines and spiropyrans that show photochromism in microcrystalline powders.

Arguably, the single crystalline phase is more constrained than microcrystalline powders giving less room for the large structural changes that are necessary upon ring opening and planarization in SOs and SPs. When the indoline group typically used for spirooxazines is replaced with an azahomoadamantyl group as in SO **4** (Figure 3), isomerization was found to occur at room temperature in prism shaped single crystals grown from a hexanes solution.[13] The initially colorless, block shaped crystals turned dark purple upon irradiation with UV light. Unlike in solution, the thermal reaction leading to color bleaching was not observed in the single crystal, and the dark purple color persisted for several months compared to seconds or minutes in solution. Analysis of the electron density difference map

(Figure 3c) showed only small changes indicating that there was likely much less than 10% photoconversion within the crystal.[14] The polarized absorption spectra before and after irradiation (Figure 3d) illustrate the dramatic changes in the absorption spectra that occur upon photoisomerization. These results demonstrate that crystalline SO materials could be used as filter materials and possibly as data storage media where the differences in absorption could correspond to the 1's and 0's of the binary computer language. It should be noted that changes in the polarizer orientation do not lead to significant changes in absorption because of the orthogonal arrangement of molecules in the crystal lattice.[13] The exact structure of species responsible for the purple color was never discussed, although it may be some ring opened form that is not completely planar through the conjugated portion and that may not be able to revert to the closed form owing to crystal packing forces.

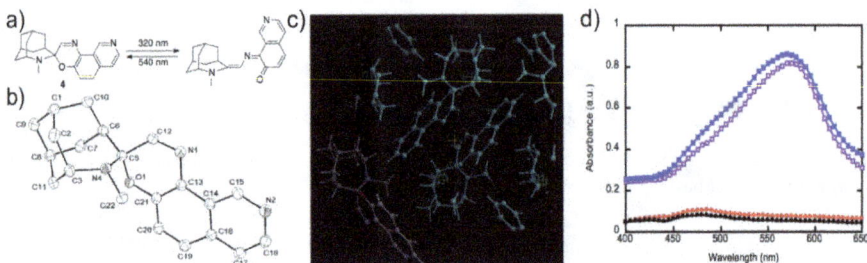

Figure 3. (a) Molecular structure and photoisomerization of SO **4**, (b) crystal structure of SO **4**, (c) the electron density difference map of a single crystal of **4** before and after irradiation with UV light, and (c) polarized absorption spectra of **4** before UV irradiation with the polarizer at 0° (red circles) and 90° (black circles), and after irradiation with the polarizer at 0° (purple squares) and 90° (blue squares) relative to the [011] plane. The electron density difference map is defined as the difference between the electron density map of an irradiated single crystal and that of the same crystal before irradiation. Thermal ellipsoids are shown at the 50% probability level with hydrogens omitted for clarity. (Figure adapted from reference 13. Reproduced by permission of The Royal Society of Chemistry).

Harada and coworkers[15] reported similar results for the irradiation of single crystals, obtained from methanol or acetone solution, of a series of structurally related SOs and SPs at various temperatures. UV irradiation at room temperature resulted in little or no color changes. In contrast, irradiation of the single crystals at 90 K led to strong color changes that persisted. The population of the photogenerated form was low, however, as it could not be detected by X-ray diffraction. The supposition was made that the thermal reversion to the colorless spirocyclic form is fast at room temperature and thus cryogenic temperatures essentially prevent this reaction. The authors note that it is unclear whether these single crystal photoreactions occur at defect sites or whether lattice control is directly responsible and investigations to answer this question are currently underway.

In contrast to the spirooxazine and spiropyran class of photochromic materials, diarylethenes require only modest structural changes upon photoisomerization. The observed isomerization is typically not thermally reversible if aryl groups with low aromatic stabilization energy such as thiophene, benzothiophene, furan, and thiazole are used.[18]

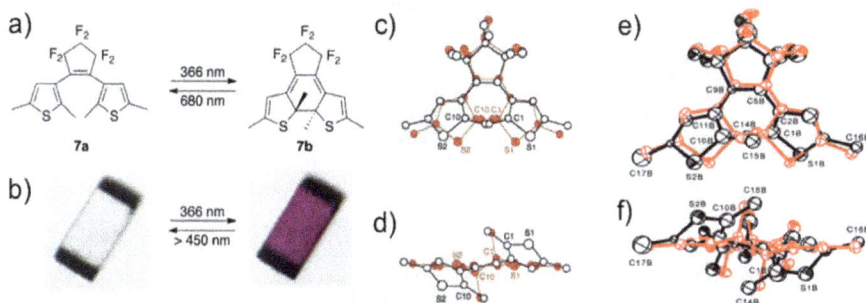

Figure 4. (a) Molecular structure of dithienylethene **7** that is photochromic in the single crystal. (b) Photograph of a single crystal of **7a** under polarized light before (left) and after (right) UV irradiation. (c) Top view and (b) side view of the movement of each atom of **7a** in the crystal during photoisomerization and relative atom positions of **7a** (O, open circles) and **7b** (b, red circles). (e) Top view and (f) side view of the molecular structure of a crystal of **1b** before and after irradiation with 680 nm light for 100 h: The closed-ring form **1b** and open-ring form **1a** resulting from irradiation were drawn by red and black lines, respectively. The ellipsoids are drawn at the 50% probability level. Hydrogen atoms were omitted for clarity. (Adapted with permission from references 16 and 17. Copyright 1999 and 2000 American Chemical Society).

Initial work by Irie and Uchida[18] showed that a single crystal of a thiophene substituted perfluorinated diarylethene exhibited the photoisomerization phenomenon in the single crystalline phase. A single crystal of this compound, which was initially clear and colorless, became reddish upon irradiation with UV light. As a proof of concept that these crystals could be used for data storage, the authors inscribed the Chinese character for "light" on the crystal with micron dimensions.[18]

Subsequent work has shown that photoisomerization in diarylethene single crystals is a rather general phenomenon.[16-17, 19] The compound 1,2-bis(2,5-dimethyl-3-thienyl) perfluorocyclopentene (compound **7**), shown in Figure 4 in its ring opened (**7a**) and ring closed (**7b**) forms, serves as a quintessential example.[16] Single crystals of **7a** grown from hexanes are rhomboid in shape and colorless. Irradiation with 366 nm light causes the crystal to turn reddish purple in color (Figure 4). Polarized absorption spectra showed that the changes in absorption occurred without loss of order in the crystal and X-ray diffraction experiments confirmed that ring closed isomer **7b** was formed. Figures 4c and 4d show an overlay of the molecular structure of **7a** and **7b** and it can be clearly seen that the majority of motion occurs in the region of the sulfur and reacting carbon atoms (atoms C1 and C10). To show that the ring opening reaction was also possible, crystals of **7b** were also grown and irradiated with visible light.[17] The conversion (**7b** to **7a**) was only about 9% but did prove that this reverse transformation was also possible. When an organometallic diarylethene **8** (Figure 5) was synthesized, it exhibited even higher photoconversion factors in the single crystalline phase. The isomerization was found to proceed through a triplet sensitized pathway leading the authors to conclude that crystals of such compounds could easily find use as molecular switches.[20]

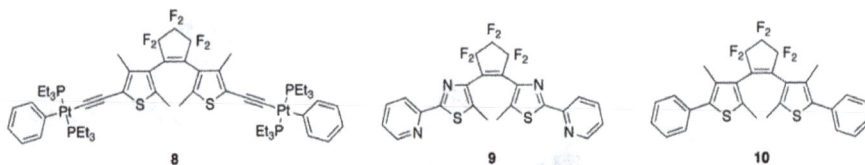

Figure 5. Molecular structures of ring open diarylethenes that photoisomerize to the ring closed form in the single crystalline state. Compounds **9** and **10** demonstrate additional photomechanical properties.

The ability to efficiently photoisomerize in the single crystalline phase endows diarylethenes with the unique ability to use light to carry out a change on the molecular level that then leads to action on the macroscopic scale. Colombier and coworkers observed irradiated diarylethene crystals of **9** (Figure 5) "jump" distances of 500 μm; when crystals are immobilized and unable to "jump" cracks appear on the surface.[21] While the jumping phenomeon was not observed in compound **10** (Figure 5), these crystals showed reversible surface morphology changes as determined by atomic force microscopy (AFM) when alternately irradiated with UV and visible wavelength light. This observation led Irie and coworkers to examine the possibility of using suitably sized diarylethene crystals as actuators to perform mechanical work using light activation.[22]

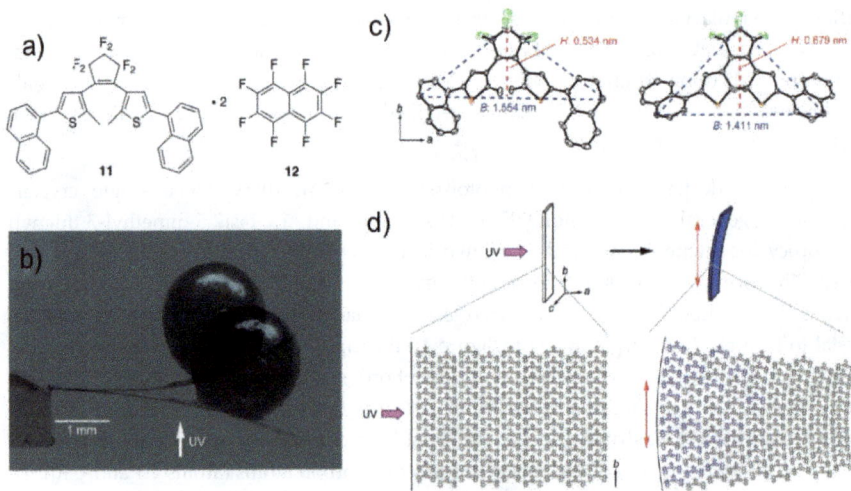

Figure 6. (a) Diarylethene **11** and perfluoronaphthalene (**12**) components of a co-crystal capable of photomechanical work. (b) Image of a rod-like co-crystal of **11** and **12** lifting a 2mm diameter, 46.77 mg lead ball after the application of UV light. (c) ORTEP diagram of **11** in the ring-opened and ring-closed forms with dimensions given for the long and short axis of the molecule. (d) Crystal packing in the co-crystal of **11** and **12** showing how photoisomerization at the molecular level leads to deformation in the crystal. The blue molecules are the ring-closed forms obtained after UV irradiation. (Adapted with permission from reference 23. Copyright 2010 American Chemical Society).

Initial attempts using single component rod-like crystals[22] or co-crystals consisting of a 1:2 molar ratio of diarylethene **11** and perfluonaphthalene (**12**)[23] demonstrated that photomechanical work was possible as the crystals were able to move a small load upon irradiation (Figure 6). When a rod-like crystal holding a small lead bearing was irradiated with UV light, the crystal bent away from the light source lifting the attached weight in the process. This effect is attributed to the change in the length of the short axis of the molecule (Figure 6c); when the molecules on the surface of the crystal are irradiated and photoisomerize to their ring closed state the net effect is an expansion on one side of the crystal (Figure 6d) which manifests as the macro-scale bending observed. The co-crystals proved quite durable by withstanding a few hundred cycles without degrading compared to less than one hundred cycles for the single component crystals.

Recent literature reports indicate that macroscopic motion due to molecular scale photoisomerization reactions is not unique to the diarylethene class of photochromes. A dimethylamino-subsistuted azobenzene, compound **13** in Figure 7, was shown to exhibit crystal bending despite the fact that a large change in molecular geometry is required upon photoisomerization, namely a *trans* → *cis* rearrangement.[24] Another compound worth mentioning is salicylideneaniline **14**, which has been known to undergo tautomerization in single crystals.[14] Small plate-like microcrystals (73 x 4.5 x 1.1 µm) were grown by sublimation, and when irradiated with 365 nm light bent in a direction away from the light source.[25] Irradiation with visible light returned the crystals to their former position. X-ray diffraction experiments confirmed that approximately 10% of the molecules had photoisomerized. Thus, it has been hypothesized that photochromic actuators could find use in artificial muscles and devices where light initiated photoisomerization is effectively translated into macroscopic motion.

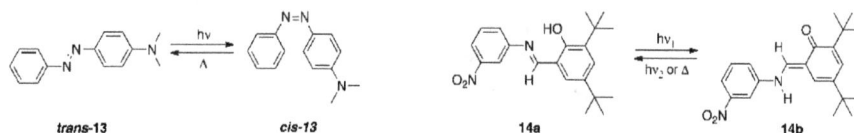

Figure 7. The molecular structures of photochromic azobenzene **13** and salicylideneaniline **14**. Both compounds show photomechanical effects in the single crystalline state.

In addition to mechanical and optical effects, the change in bonding pattern associated with the photoisomerization process has been explored as related to its influence on magnetic properties. In particular, the differences in conjugation between the two components of a photochromic system containing magnetic centers can lead to switchability in magnetic properties. Magnetism arises from the motion of charge, in particular unpaired electrons. Therefore, metal ions or organic radicals are typically employed and the resulting magnetic properties explored using single crystalline or microcrystalline materials.

Using the highly versatile diarylethene molecular framework, Irie showed that coupling between two terminal nitronyl nitroxide radical groups depended on whether the diarylethene was in its ring opened or ring closed form.[26] Microcrystalline powders of the ring opened form showed little coupling between the two nitronyl nitroxide radicals, while the ring closed form showed strong anti-ferromagnetic coupling. This was attributed to the ability of the radicals to delocalize over the π-system, which was in turn dependent on whether the molecule was in its ring opened or ring closed state.

A relatively new means of controlling the magnetic properties of a material relies on the interaction between a metal center with unpaired electrons and a photoisomerizable ligand.[27] The ligand field, which in turn affects the magnetic behavior of the metal electrons, is predicted to change upon isomerization from one photochromic state to another. This has been clearly demonstrated by Nishihara[28] and one example illustrated in Figure 8. Irradiation of a microcrystalline styrylpyridine iron(II) based complex led to an increase in magnetization (measured on the y-axis) for a given temperature with the associated $Z \rightarrow E$ isomerization. X-ray crystallographic analysis of the isomers confirmed the observation of a change in magnetization.

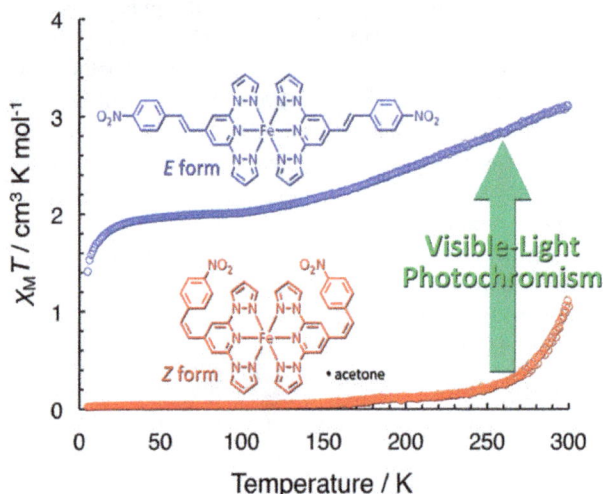

Figure 8. Photoisomerization of a styrylpyridine ligand leads to changes in magnetization of an iron complex at room temperature. (Adapted from reference 28. Copyright 2012 American Chemical Society).

3. Crystalline molecular semiconductors

No class of materials has been more important to the evolution of modern electronic devices than semiconductors. Inventions including the transistor, photovoltaics, and light emitting diodes all rely upon the semi-conductors unique ability to convert from electronically insulating to a fully conductive material. These devices typically require inorganic active materials that must undergo extensive and expensive processing to give materials of the required purity. Recently, there has been a push to replace traditional inorganic components with organic materials in the form of organic field effect transistors (OFETs) and organic photovoltaic (OPV) devices. While the subject of OPV and OFET fabrication and operation is beyond the scope of this review, it is worth mentioning that one of the most important properties of a semiconducting material is its ability to conduct charge (referred to as the mobility and given in units of $cm^2/V \cdot s$), either electrons or holes, and this is highly dependent on electronic structure, energy level (HOMO/LUMO) position, and on supramolecular organization.[29] Both π-conjugated polymers and small molecule based compounds have been examined, but only the molecular materials can be obtained in the required purity with the required supramolecular order that gives rise to the necessary charge transport properties.[30] The acene class of molecules shown in Figure 9 are among the most widely studied and best performing materials with regard to charge conductivity.[31]

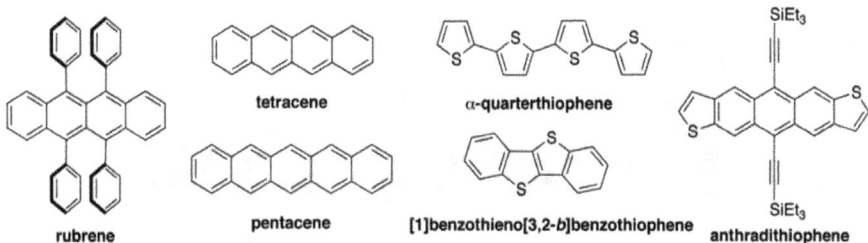

Figure 9. The structure of commonly used acene small molecules.

Facile processing is one of the greatest advantages organic materials hold over their inorganic counterparts. Organic molecules can be deposited on substrates in a number of ways including spin coating from solution, dip coating, vapor deposition methods, and ink-jet printing. The best performing materials are grown using the vapor deposition method,[2b] illustrated in Figure 10a, which can afford high quality single crystals while removing unwanted contaminants.[2a, 32] The single crystals are then manually harvested and manipulated into the appropriate device architecture. In contrast, dip-coating and other solution processing methods typically give lower quality materials relative to vapor-grown crystals (Figure 10b).

Although pentacene (structure show in Figure 9) was a popular molecule for OFET studies, rubrene (Figure 9) is now growing in popularity due to its superior stability in air as pentacene is easily oxidized.[34] Rubrene is known to exhibit several polymorphs depending on crystal growth conditions with packing also determined by competition between quadrupolar and π-stacking interactions.[35] It has the ability to form high quality crystals with an electronic band structure[36] as opposed to discrete energy states and was first examined as a single crystal OFET by Podzorov and coworkers who fabricated devices with mobilities as high as 1 $cm^2/V \cdot s$[37] and up to 20 $cm^2/V \cdot s$ with an optimized device.[38] Soon after, Brédas and coworkers compared the crystalline state structures of rubrene to unsubstituted acenes both computationally and experimentally and found that, contrary to expectation, the bulky phenyl side groups force molecular displacements that promote large intermolecular electronic coupling.[39] Furthermore, mobility was shown to be axially dependent.[40] However, when an "in channel" device was fabricated, where charge carriers moved through the bulk of the crystal as opposed to near the surface, mobility values of ~40 $cm^2/V \cdot s$ were obtained

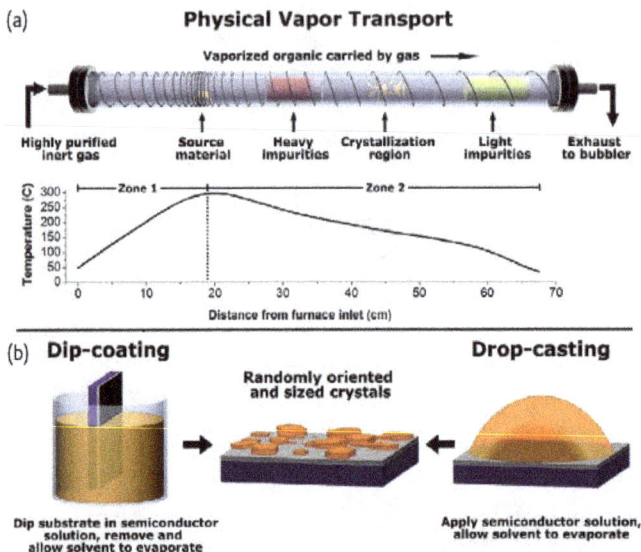

Figure 10. A schematic showing (a) the physical vapor transport method of single crystal growth contrasted with the (b) growth of crystals by solution methods. The physical vapor transport method offers crystals of high purity and greater uniformity. (Figure taken from reference 33 with permission).

thus showing that surface traps and adsorbed impurities greatly impact OFET performance. More importantly, the high mobilities measured in these materials indicate that organic single crystals are suitable for use in electronic devices.

Andrienko and coworkers[41] have suggested, that single crystal based OFETS may not be practical, likely owing to the delicate manipulations needed to make the devices coupled with the fragile nature of the crystals. Anthony and coworkers[31a, b, 42] have tried to synthesize acenes with solubilizing groups that can be deposited using solution methods while maintaining a high level of crystalline order. Examples of these soluble acenes are shown in Figure 11. Some of the derivatives show good mobilities (on the order of 1.0 cm^2/V•s) for solution-processed materials. Interestingly, small changes in the solubilizing groups lead to pronounced changes in the crystal packing as determined by X-ray diffraction. A triethylsilyl substituted acene (compound **16**) exhibited a brick-work[31c] packing motif that is ideal for maximized π-orbital overlap between neighboring molecules which translated into mobilities of 1.0 cm^2/V•s. A simple switch from the triethyl to the trimethylsilyl substituted acene (compound **17**) resulted in crystal structure that exhibited herringbone packing, negligible intermolecular orbital overlap, and mobilities that were too low to be measured.

Given the interest in making high quality single crystals or crystalline films of organic semiconductors for OFETs, new methods have emerged for the deposition of such structures. Inkjet printing techniques[43] have been shown to give microcrystals capable of

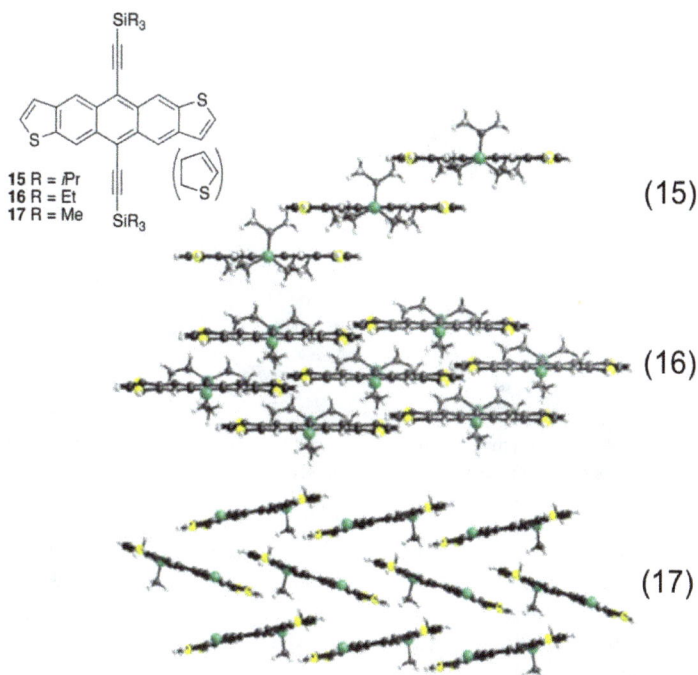

Figure 11. Crystal structure of anthradithiophenes **15 – 16** and their corresponding crystal structures. The alkyl groups on the front-most silicon atoms have been omitted for clarity. (Adapted with permission from reference 42c. Copyright 2005 American Chemical Society).

diffracting X-rays and giving diffraction patterns similar to bulk materials. This was achieved by inkjet printing a solution of the semiconductor in close proximity to a droplet of an antisolvent leading to slow crystallization as the solvents mix and evaporate. The crystals of a benzothiophene-based molecule obtained using this technique exhibit average mobilities of 16.4 cm^2/V•s which are similar to that of rubrene single crystal devices.

The Bao group has actively been pursuing OFET technologies and has introduced patterned growth techniques that have the potential to grow crystals directly on devices obviating the need for manual manipulation of delicate crystals as is required in the vapor growth method.[44] Patterned growth is achieved by using a pre-patterned, self assembled monolayer as a growth template. Takeya and coworkers have introduced an oriented growth method[3, 45] where an inclined substrate and a "liquid-sustaining piece" allow high quality crystals to form on a SiO$_2$ coated substrate. The resulting materials were single crystalline as confirmed by X-ray diffraction and gave mobilities on the order of 2 – 6 cm^2/V•s. Thus, it can be seen that high quality single crystals with good mobility characteristics can be obtained by solution methods.

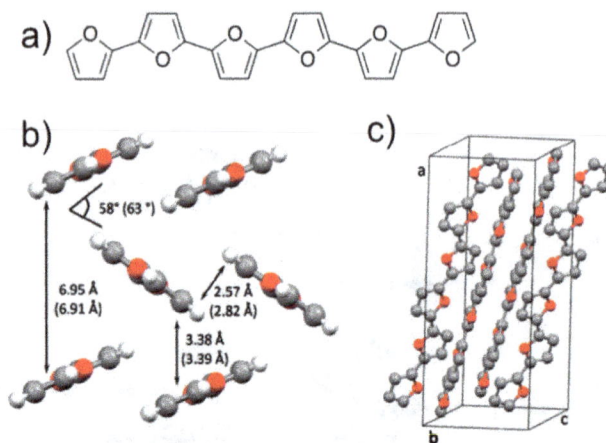

Figure 12. (a) The molecular structure of α-furan hexamer **10** and its crystal packing (b) and (c). The values given in parenthesis are for the thiophene analogue. (Adapted with permission from reference 46. Copyright 2010 American Chemical Society).

While significant effort has gone into new methods of growing high quality crystals and improving the solubility of acenes known to exhibit good performance in OFETs, effort is also directed toward the design of new materials with improved properties, such as smaller intramolecular distances in the solid state, that would lead to improved mobilities. While attention has been focused largely on thiophene based compounds[47] a recent report by Bendikov and coworkers[46] used X-ray crystallography, among other techniques, to reexamine the solid state structure of furan based materials. In particular, they synthesized and studied a series of furan oligomers including α-furan hexamer **10**, the structure and crystal packing of which is shown in Figure 12. When compared to the thiophene analogue,[48] **10** was found to exhibit closer packing (2.57 Å for **10** versus 2.89 Å for the thiophene analogue) and shorter inter-ring C-C bond distances indicating less aromatic character and more quinoidal character, both of which could facilitate better charge transport in furan-based materials. The Bendikov group has also explored OFETs from oligofuran materials,[49] and a recent study showed that solution grown crystals of a naphthodifuran gave mobilities of 3.6 cm²/V•s despite exhibiting herringbone packing, which leads to less than optimal intermolecular π-orbital overlap.[50] Thus, it appears that furan based crystalline semiconductors, with further optimization and crystal engineering, could eventually surpass their thiophene analogues.

Given the high mobilities intrinsic to single crystalline acene type semiconductors, it would seem that single crystalline materials would be well suited for OPVs, which similar to OFETs, require high mobility materials, but also require materials able to efficiently absorb solar energy. There are, however, few examples of single crystal organic photovoltaics[51] as most research is devoted to polymeric materials,[52] solution processable small molecule semiconductors,[53] and dye-sensitized solar cells.[54] Recently, Yang and coworkers assembled a solar cell based on a single crystal of tetracene (shown in Figure 9) as the light absorbing layer and a C_{60} electron transport layer.[55] The device showed efficiencies that were less than 1%, and efficiency depended critically on the thickness of the crystal used in addition to the crystal face that was irradiated. The authors hypothesized that the device suffered from low performance because of the poor absorption of visible light by tetracene. Campione and coworkers similarly, examined single crystal photovoltaic devices based on rubrene, perylene, α-quarterthiophene, and tetracene without employing a C_{60} layer. While the voltage generated by the cell was adequate, the current generated was very low giving efficiencies that were on the order of 10^{-5}%. The authors attribute the low performance to a high rate of exciton recombination, which they say does not necessarily preclude the use of single crystals in solar cells, but may necessitate the use of electron accepting materials to ensure a higher propensity for exciton separation in the active layer. Higher efficiencies have been obtained using organic materials; Kippelen and coworkers achieved an efficiency of ~2.7 % using a cell consisting of pentacene and C_{60}, however, the active layer was polycrystalline.[56]

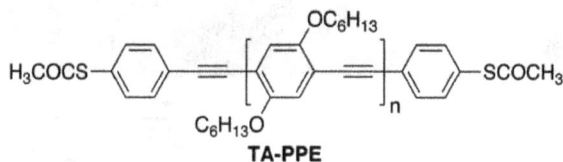

Figure 13. The molecular structure of **TA-PPE**.

Single crystal based devices show impressive mobilities, yet are not suitable for use when larger surface area coverage is desired such as in manufacturing displays and some solar cells, both of which are intensely pursued by academic and industrial groups seeking to commercialize these technologies. Polymeric and molecular materials amenable to spin coating and spray coating methods are ideally suited for coverage of large areas, but the mobilities resulting from amorphous and polycrystalline morphologies are orders of magnitude lower than those obtained for small single crystals. In an effort to close this gap, self-assembly processes have been examined for the generation of morphologies conducive to good charge transport and concomitant application in large area devices. In particular, nano-scale wires based on organic materials have been shown to be highly crystalline in nature, thus translating into effective charge transport materials[57] using simple and inexpensive solution methods.

Conjugated polymer nanowires with the potential for high mobility have been observed, although polymeric materials are typically spin coated or spray coated from dilute solutions giving amorphous films after solvent evaporation. In a prototypical example, however, careful and controlled deposition of the polymer **TA-PPE** (Figure 13) gave crystalline nanowires.[58] These nanowires were obtained by slow evaporation (over a period of days) of a tetrahydrofuran solution of polymer with concentrations ranging from 0.05 – 0.10 mg/mL. When incorporated into an OFET, the nanowires exhibited mobilities on the order of 0.01 –

0.1 cm^2/V•s, which were 3 – 4 orders of magnitude higher than the mobilities obtained from an amorphous film of **TA-PPE** obtained by spin coating. The slow evaporation ensured good ordering of the polymer wires as shown in Figure 14. Selected Area Electron Diffraction (SAED) gave sharp diffraction peaks indicating a high level of order as manifested in the high mobility values. Thus, conditions that allow polymer units to organize into higher order morphologies may be broadly applicable.

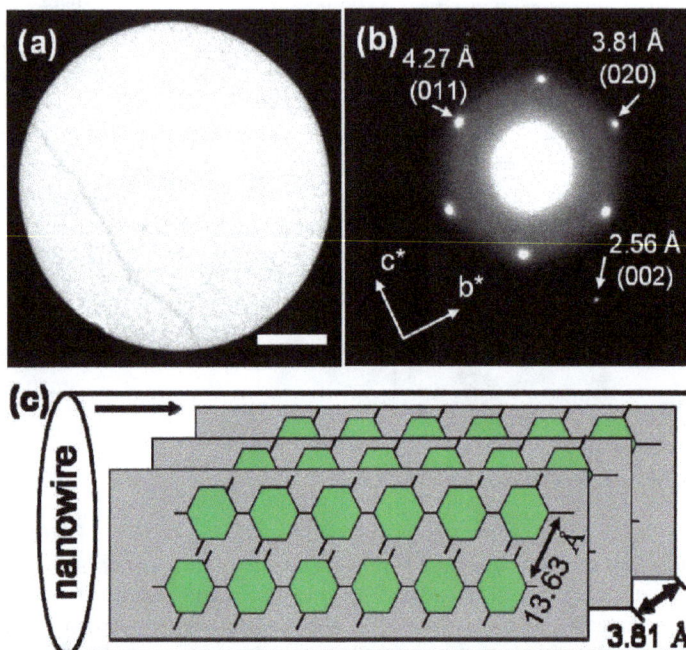

Figure 14. (a) TEM image of an individual TA-PPE nanowire (scale bar: 150 nm) and (b) its corresponding SAED pattern. (c) Schematic diagram of possible molecular packing in the one-dimensional nanowires; for clarity the end-capping groups have been omitted. (Adapted with permission from reference 58. Copyright 2009 American Chemical Society).

Molecular materials with a propensity to form ordered, π-stacks have also demonstrated the ability to form crystalline molecular wires. A dianthracene diester gave crystals that packed with good π-overlap between neighboring anthracene units.[59] Crystalline nanorods, when irradiated with UV light, underwent a [2+2] photopolymerzation reaction in a crystal-to-crystal type transformation. While the resulting polymer was not well suited to charge transport, the authors clearly demonstrated how single crystalline nanorods could be used to give other materials with application in organic electronics. In a similar study, a dicyanovinyl substituted anthracene molecule formed nanowires when films were cast from dichloromethane solution. These nanowires gave remarkably clear SAED patterns proving the single crystalline organization of the component molecules. Moreover, while the mobility properties were not measured, the nanowires exhibit photoswitchable conductivity potentially allowing their use in photodetectors.

For most electronic applications, crystalline organic compounds still lag behind their inorganic counterparts in terms of device performance; however, the gap has been steadily decreasing. Given the continuous rapid advances of organic crystalline semiconductors, these materials may soon replace traditional inorganics and usher in an era of carbon-based electronic devices.

Author details

Dinesh G. (Dan) Patel and Jason B. Benedict
Department of Chemistry, State University of New York at Buffalo, Buffalo, New York, USA

4. References

[1] (a) Braga, A. F. B.; Moreira, S. P.; Zampieri, P. R.; Bacchin, J. M. G.; Mei, P. R., *Sol. Energy Mater. Sol. Cells* 2008, *92* (4), 418-424; (b) Mei, P. R.; Moreira, S. P.; Cardoso, E.; Cortes, A. D. S.; Marques, F. C., *Sol. Energy Mater. Sol. Cells* 2012, *98*, 233-239; (c) Taft, E. A.; Horn, F. H., *J. Electrochem. Soc.* 1958, *105* (2), 81-83.

[2] (a) Kloc, C.; Simpkins, P. G.; Siegrist, T.; Laudise, R. A., *J. Cryst. Growth* 1997, *182* (3-4), 416-427; (b) Laudise, R. A.; Kloc, C.; Simpkins, P. G.; Siegrist, T., *J. Cryst. Growth* 1998, *187* (3-4), 449-454.

[3] Uemura, T.; Hirose, Y.; Uno, M.; Takimiya, K.; Takeya, J., *Applied Physics Express* 2009, *2* (11), 111501.

[4] Dürr, H.; Bous-Laurent, H., *Photochromism: Molecules and Systems*. Elsevier: Amsterdam, 1990.

[5] Crano, J. C.; Guglielmetti, R. J., *Organic Photochromic and Thermochromic Compounds*. Plenum Press: New York, 1999.

[6] Dürr, H.; Bouas-Laurent, H., *Photochromism : molecules and systems*. Rev. ed.; Elsevier: Amsterdam ; Boston, 2003; p liii, 1044 p.

[7] (a) Evans, R. A.; Hanley, T. L.; Skidmore, M. A.; Davis, T. P.; Such, G. K.; Yee, L. H.; Ball, G. E.; Lewis, D. A., *Nat. Mater.* 2005, *4* (3), 249-253; (b) Evans, R. A.; Such, G. K., *Aust. J. Chem.* 2005, *58* (12), 825-830; (c) Such, G. K.; Evans, R. A.; Davis, T. P., *Macromolecules* 2004, *37* (26), 9664-9666; (d) Such, G. K.; Evans, R. A.; Davis, T. P., *Macromolecules* 2006, *39* (4), 1391-1396.

[8] Patel, D. G.; Paquette, M. M.; Kopelman, R. A.; Kaminsky, W.; Ferguson, M. J.; Frank, N. L., *J. Am. Chem. Soc.* 2010, *132* (36), 12568-12586.

[9] Hirshberg, Y., *J. Am. Chem. Soc.* 1956, *78* (10), 2304-2312.

[10] Chu, N. Y. C., *Can. J. Chem.* 1983, *61* (2), 300-305.

[11] Bénard, S.; Yu, P., *Chem. Commun.* 2000, (1), 65-66.

[12] Bénard, S.; Yu, P., *Adv. Mater.* 2000, *12* (1), 48-50.

[13] Patel, D. G.; Benedict, J. B.; Kopelman, R. A.; Frank, N. L., *Chem. Commun.* 2005, (17), 2208-2210.

[14] Harada, J.; Uekusa, H.; Ohashi, Y., *J. Am. Chem. Soc.* 1999, *121* (24), 5809-5810.

[15] Harada, J.; Kawazoe, Y.; Ogawa, K., *Chem. Commun.* 2010, *46* (15), 2593-2595.

[16] Kobatake, S.; Yamada, T.; Uchida, K.; Kato, N.; Irie, M., *J. Am. Chem. Soc.* 1999, *121* (11), 2380-2386.

[17] Yamada, T.; Kobatake, S.; Muto, K.; Irie, M., *J. Am. Chem. Soc.* 2000, *122* (8), 1589-1592.

[18] Irie, M.; Uchida, K., *Bull. Chem. Soc. Jpn.* 1998, *71* (5), 985-996.

[19] (a) Kobatake, S.; Hasegawa, H.; Miyamura, K., *Cryst. Growth. Des.* 2011, *11* (4), 1223-1229; (b) Kobatake, S.; Kuma, S.; Irie, M., *J. Phys. Org. Chem.* 2007, *20* (11), 960-967; (c) Liu, G.; Liu, M.; Pu, S.; Fan, C.; Cui, S., *Tetrahedron* 2012, *68* (10), 2267-2275; (d) Liu, G.; Pu, S.; Wang, X., *Tetrahedron* 2010, *66* (46), 8862-8871; (e) Matsuda, K.; Takayama, K.; Irie, M., *Inorg. Chem.* 2003, *43* (2), 482-489; (f) Ohara, H.; Morimoto, M.; Irie, M., *Photochem. Photobiol. Sci.* 2010, *9* (8), 1079-1081; (g) Pu, S.; Liu, G.; Shen, L.; Xu, J., *Org. Lett.* 2007, *9* (11), 2139-2142.

[20] Brayshaw, S. K.; Schiffers, S.; Stevenson, A. J.; Teat, S. J.; Warren, M. R.; Bennett, R. D.; Sazanovich, I. V.; Buckley, A. R.; Weinstein, J. A.; Raithby, P. R., *Chem. Eur. J.* 2011, *17* (16), 4385-4395.

[21] Colombier, I.; Spagnoli, S.; Corval, A.; Baldeck, P. L.; Giraud, M.; Léaustic, A.; Yu, P., *Mol. Cryst. Liq. Cryst.* 2005, *431* (1), 495-499.

[22] Kobatake, S.; Takami, S.; Muto, H.; Ishikawa, T.; Irie, M., *Nature* 2007, *446* (7137), 778-781.

[23] Morimoto, M.; Irie, M., *J. Am. Chem. Soc.* 2010, *132* (40), 14172-14178.

[24] Koshima, H.; Ojima, N.; Uchimoto, H., *J. Am. Chem. Soc.* 2009, *131* (20), 6890-6891.

[25] Koshima, H.; Takechi, K.; Uchimoto, H.; Shiro, M.; Hashizume, D., *Chem. Commun.* 2011, *47* (41), 11423-11425.

[26] Matsuda, K.; Irie, M., *J. Am. Chem. Soc.* 2000, *122* (30), 7195-7201.

[27] (a) Boillot, M.-L.; Chantraine, S.; Zarembowitch, J.; Lallemand, J.-Y.; Prunet, J., *New J. Chem.* 1999, 179-183; (b) Boillot, M.-L.; Roux, C.; Audiere, J.-P.; Dausse, A.; Zarembowitch, J., *Inorg. Chem.* 1996, *35*, 3975-3980; (c) Roux, C.; Zarembowitch, J.; Gallois, B.; Granier, T.; Claude, R., *Inorg. Chem.* 1994, *33* (10), 2273.

[28] Takahashi, K.; Hasegawa, Y.; Sakamoto, R.; Nishikawa, M.; Kume, S.; Nishibori, E.; Nishihara, H., *Inorg. Chem.* 2012, *51* (9), 5188-5198.

[29] Mas-Torrent, M.; Rovira, C., *Chem. Rev.* 2011, *111* (8), 4833-4856.

[30] Hoeben, F. J. M.; Jonkheijm, P.; Meijer, E. W.; Schenning, A., *Chem. Rev.* 2005, *105* (4), 1491-1546.

[31] (a) Anthony, J. E., *Chem. Rev.* 2006, *106* (12), 5028-5048; (b) Anthony, J. E., *Angew. Chem. Int. Ed.* 2008, *47* (3), 452-483; (c) Würthner, F.; Schmidt, R., *ChemPhysChem* 2006, *7* (4), 793-797.

[32] Jurchescu, O. D.; Baas, J.; Palstra, T. T. M., *Appl. Phys. Lett.* 2004, *84* (16), 3061-3063.

[33] Reese, C.; Bao, Z., *Mater. Today* 2007, *10* (3), 20-27.

[34] Hasegawa, T.; Takeya, J., *Science and Technology of Advanced Materials* 2009, *10* (2).

[35] Jurchescu, O. D.; Meetsma, A.; Palstra, T. T. M., *Acta Cryst. B* 2006, *62* (2), 330-334.

[36] Ding, H.; Reese, C.; Makinen, A. J.; Bao, Z.; Gao, Y., *Appl. Phys. Lett.* 2010, *96* (22), 3.

[37] Podzorov, V.; Pudalov, V. M.; Gershenson, M. E., *Appl. Phys. Lett.* 2003, *82* (11), 1739-1741.

[38] Podzorov, V.; Menard, E.; Borissov, A.; Kiryukhin, V.; Rogers, J. A.; Gershenson, M. E., *Phys. Rev. Lett.* 2004, *93* (8), 086602.

[39] da Silva Filho, D. A.; Kim, E. G.; Brédas, J. L., *Adv. Mater.* 2005, *17* (8), 1072-1076.

[40] Sundar, V. C.; Zaumseil, J.; Podzorov, V.; Menard, E.; Willett, R. L.; Someya, T.; Gershenson, M. E.; Rogers, J. A., *Science* 2004, *303* (5664), 1644-1646.

[41] Vehoff, T.; Baumeier, B.; Troisi, A.; Andrienko, D., *J. Am. Chem. Soc.* 2010, *132* (33), 11702-11708.

[42] (a) Anthony, J. E.; Eaton, D. L.; Parkin, S. R., *Org. Lett.* 2002, *4* (1), 15-18; (b) Payne, M. M.; Odom, S. A.; Parkin, S. R.; Anthony, J. E., *Org. Lett.* 2004, *6* (19), 3325-3328; (c) Payne, M. M.; Parkin, S. R.; Anthony, J. E.; Kuo, C.-C.; Jackson, T. N., *J. Am. Chem. Soc.* 2005, *127* (14), 4986-4987; (d) Purushothaman, B.; Bruzek, M.; Parkin, S. R.; Miller, A.-F.; Anthony, J. E., *Angew. Chem. Int. Ed.* 2011, *50* (31), 7013-7017.

[43] Minemawari, H.; Yamada, T.; Matsui, H.; Tsutsumi, J.; Haas, S.; Chiba, R.; Kumai, R.; Hasegawa, T., *Nature* 2011, *475* (7356), 364-367.

[44] (a) Briseno, A. L.; Aizenberg, J.; Han, Y.-J.; Penkala, R. A.; Moon, H.; Lovinger, A. J.; Kloc, C.; Bao, Z., *J. Am. Chem. Soc.* 2005, *127* (35), 12164-12165; (b) Briseno, A. L.; Mannsfeld, S. C. B.; Ling, M. M.; Liu, S.; Tseng, R. J.; Reese, C.; Roberts, M. E.; Yang, Y.; Wudl, F.; Bao, Z., *Nature* 2006, *444* (7121), 913-917.

[45] Soeda, J.; Hirose, Y.; Yamagishi, M.; Nakao, A.; Uemura, T.; Nakayama, K.; Uno, M.; Nakazawa, Y.; Takimiya, K.; Takeya, J., *Adv. Mater.* 2011, *23* (29), 3309-3314.

[46] Gidron, O.; Diskin-Posner, Y.; Bendikov, M., *J. Am. Chem. Soc.* 2010, *132* (7), 2148-2150.

[47] Perepichka, I. F.; Perepichka, D. F., *Handbook of thiophene-based materials.* Wiley: Hoboken, 2009.

[48] Horowitz, G.; Bachet, B.; Yassar, A.; Lang, P.; Demanze, F.; Fave, J. L.; Garnier, F., *Chem. Mater.* 1995, *7* (7), 1337-1341.

[49] Gidron, O.; Dadvand, A.; Sheynin, Y.; Bendikov, M.; Perepichka, D. F., *Chem. Commun.* 2011, *47* (7), 1976-1978.

[50] Mitsui, C.; Soeda, J.; Miwa, K.; Tsuji, H.; Takeya, J.; Nakamura, E., *J. Am. Chem. Soc.* 2012, *134* (12), 5448-5451.

[51] (a) Bredas, J.-L.; Durrant, J. R., *Acc. Chem. Res.* 2009, *42* (11), 1689-1690; (b) Clarke, T. M.; Durrant, J. R., *Chem. Rev.* 2010, *110* (11), 6736-6767.

[52] Günes, S.; Neugebauer, H.; Sariciftci, N. S., *Chem. Rev.* 2007, *107* (4), 1324-1338.

[53] (a) Dang, X.-D.; Tamayo, A. B.; Seo, J.; Hoven, C. V.; Walker, B.; Nguyen, T.-Q., *Adv. Funct. Mater.* 2010, *20* (19), 3314-3321; (b) Roncali, J., *Acc. Chem. Res.* 2009, *42* (11), 1719-1730; (c) Tamayo, A. B.; Tantiwiwat, M.; Walker, B.; Nguyen, T.-Q., *J. Phys. Chem. C* 2008, *112* (39), 15543-15552; (d) Walker, B.; Tamayo, A.; Duong, D. T.; Dang, X. D.; Kim, C.; Granstrom, J.; Nguyen, T. Q., *Advanced Energy Materials* 2011, *1* (2), 221-229.

[54] (a) Hagfeldt, A.; Boschloo, G.; Sun, L.; Kloo, L.; Pettersson, H., *Chem. Rev.* 2010, *110* (11), 6595-6663; (b) Nazeeruddin, M. K.; Kay, A.; Rodicio, I.; Humphry-Baker, R.; Müller, E.;

Liska, P.; Vlachopoulos, N.; Grätzel, M., *J. Am. Chem. Soc.* 1993, *115* (14), 6382-6390; (c) O'Regan, B.; Grätzel, M., *Nature* 1991, *353* (6346), 737-740.

[55] Tseng, R. J.; Chan, R.; Tung, V. C.; Yang, Y., *Adv. Mater.* 2008, *20* (3), 435-438.

[56] Yoo, S.; Domercq, B.; Kippelen, B., *Appl. Phys. Lett.* 2004, *85* (22), 5427-5429.

[57] Briseno, A. L.; Mannsfeld, S. C. B.; Jenekhe, S. A.; Bao, Z.; Xia, Y., *Mater. Today* 2008, *11* (4), 38-47.

[58] Dong, H.; Jiang, S.; Jiang, L.; Liu, Y.; Li, H.; Hu, W.; Wang, E.; Yan, S.; Wei, Z.; Xu, W.; Gong, X., *J. Am. Chem. Soc.* 2009, *131* (47), 17315-17320.

[59] Al-Kaysi, R. O.; Dillon, R. J.; Kaiser, J. M.; Mueller, L. J.; Guirado, G.; Bardeen, C. J., *Macromolecules* 2007, *40* (25), 9040-9044.

Chapter 7

Epitaxial Electrodeposition of Chiral Films Using Chiral Precursors

Rakesh Gudavarthy and Elizabeth A. Kulp

Additional information is available at the end of the chapter

1. Introduction

Chiral drugs have played an important role in driving the market for past few decades. Currently, more than half of the drugs marketed are chiral. [1-4] It is well established that chiral drugs often differ in their pharmacological, toxicological and pharmacokinetic properties. Historically, the pharmaceutical industry has relied on using enzymatic reactions to produce enantiospecific molecules. The chiral drugs are often synthesized in the racemic form, and are then resolved into pure enantiomer. [5] From an Industry perspective, the process is neither cost effective nor completely safe. The enantiomer should be characterized in detail in order to develop a safe and reliable formulation. One way to get around that would be to develop enantiospecific catalyst that can be reused. Chiral surfaces offer this possibility. They have been produced by decorating the surfaces of achiral substrates by chiral molecules, or by introducing defects in the single crystals which exposes chiral kink sites. [6-17] It was shown that chiral surfaces can also be produced by electrodeposition technique. [18, 19] Unlike other vacuum techniques, electrodeposition is cheaper and can be carried out at ambient conditions. [20, 21] In this method, chiral molecules present in the electrodeposition solution determine the final Chirality of the electrodeposited thin film. In this regard, electrodeposition resembles biomineralization, where organic molecules adsorbed on surface, reduce the symmetry of surfaces, resulting in chiral crystal habits. [22-28] Crystals like calcite and gypsum which crystallize in achiral space group can be transformed to chiral space group by treating calcite with chiral etchants or by crystallizing calcite in the presence of amino acids. Chiral morphologies of calcites can also be deposited electrochemically. [29]

In this chapter the terms and concepts employed in describing the enantiospecific electrodeposition are introduced. In addition, before embarking on a detailed consideration of methods of studying chiral deposition process and the various characterization techniques, we will try to understand the techniques involved in the preparation and characterization of these films.

2. Chirality

A chiral structure is non-superimposable on its mirror image. Pasteur reported in 1984 the concept of the molecular Chirality based on the distinction between the configurational isomers of molecules. Configurational isomers are compounds with the same molecular formula and same groups but different configurations. Enantiomers are pairs of configurational isomers that are mirror images of each other but are non-superimposable. Diastereomers are pairs of compounds that contain more than one chiral center, not all of which are superimposable. An equimolar mixture of opposite enantiomers is called racemic mixture or a racemate. Enantiomers when exposed to polarized light behave differently and have different catalyzing properties in a chiral medium. On the other hand, the racemic mixtures have completely different properties than enantiomers. The difference in the properties between the enantiomers and racemic mixtures arise due to different molecular interactions, and different crystal structures. [30, 31]

In an enantiomer, the molecular interactions are homochiral, which are the interactions between the assemblies of molecules with same Chirality. In a racemic compound the interactions are heterochiral, where the interactions are between opposite chiral molecules. The difference between the homochiral and heterochiral interactions leads to different physical properties. Particularly in a racemic compound, because the unit cell consists of enantiomeric molecules with opposite Chirality, the properties are completely different from enantiomers. Racemic compounds are the most common compounds that occur in nature. Such racemic compounds can exist in different forms based on the intermolecular interactions in their crystals. Analysis of the crystal structures facilitates enormously our understanding of the factors that determine the various physical and chemical properties, such as the thermodynamic stability of different types of racemates. [32] The details of such analysis are beyond the scope of this chapter.

Due to the presence of various chiral compounds, it is critical to have the right nomenclature for the differentiation. The internationally accepted nomenclature for chiral molecules uses the Cahn-Ingold-Prelog (CIP) rules for sp3 carbons. The four substituents are sorted by increasing mass of the first atom attached to the asymmetric center. If two atoms are identical, the next heaviest atom one bond further away is considered and so on. For example, in the case of 2-butanol with the order OH →ethyl → methyl rotating clockwise will be a R-enantiomer and the mirror image of that will be a S-enantiomer. These rules allow us for the absolute configuration of any chiral compounds. Another accepted form for nomenclature is Dextro (D-) and Levo (L-), based on the optical rotation of the compound. Setting glycine apart since it is nonchiral, it must be noted that all amino acids found in proteins are L-amino acids and also have the S-configuration at the exception of cysteine whose -CH2-SH substituent precedes the carboxylate -COOH in mass making L-cysteine the R-enantiomer. It is interesting to note that the electrodeposited chiral films also follow the CIP rules. It was shown that the CuO films grown from both R versions of tartaric acid and malic acid resulted in (1-1-1) orientation on Cu(111), while the S version of each resulted in a (-111) mirror image. [33] However, there are exceptions for films grown from amino acids which will be discussed in detail.

3. Chirality in crystals

The Chirality of a crystal depends on the symmetry operations present in the structure. Proper symmetry operations are those that do not change the handedness of an object. These operations include rotation axes, translations, and screw axes. If an object can be rotated about an axis and repeats itself after being rotated through either 360, 180, 120, 90 or 60° is said to have an axis of 1-fold, 2-fold, 3- fold, 4-fold, or 6-fold rotational symmetry. Although objects themselves may appear to have 5-fold, 7-fold, 8-fold or higher-fold rotation axes, these are not possible in crystals. The reason is that the external shape of a crystal is based on a geometric arrangement of atoms. In a translation operation, the object is translated up or down along an axes. Where as a screw axis also referred to as twist axis of an object are the axes that are simultaneously the axis of rotation and the axis along which a translation occurs. These symmetry operations do not change the original object because they are just movements of the same object. If only these operations are present, then the structure is chiral. However, improper symmetry operations such as rotoinversion operations or glide reflection produce the opposite "hand" of the object. [34] A rotoinversion operation is a combination of rotation and inversion where as the glide reflection is a combination of reflection and translation operation along a line. Any structure with one of these symmetery operators will produce an achiral structure.

The symmetry of the system determines what planes of a material are chiral. In the case of copper(II) oxide CuO (focus material in this chapter), the lattice parameters are a = 0.4685 nm, b = 0.3430 nm, c = 0.5139 nm, $\alpha = \Upsilon = 90°$, and $\beta = 99.08°$. The structure is centrosymmetric (i.e. it has an inversion center, i); therefore, the bulk crystal structure of CuO is achiral. However, crystallographic orientations/planes may be chiral. A monoclinic system has three axes of unequal length and an angle normally greater than 90o between two axes. In this arrangement, the b'axis is unique. The c and a axes do not intersect each other at right angles, but they are perpendicular to the b-axis. Based on the space group, a glide plane has a mirror or glide plane perpendicular to it. Therefore, achiral planes of CuO are planes that are parallel to the b-axis, planes k=0. Thus, planes such as (101), (001), and (102) are achiral. Chiral CuO planes lack glide plane symmetry; chiral planes are those with k ≠ 0, such as (111), (110), (022), and (020). For a chiral plane (hkl), its enantiomer is (-h-k-l).

Table 1 lists the symmetry of specific planes depending on the point group. [35] Screw axes are replaced by the highest possible rotation, and glides are replaced by mirrors. Planes that lack mirror symmetry (m) are chiral. For example, whereas all planes of triclinic structure are chiral, all planes of orthorhombic structure are achiral except where h≠k≠l≠0, the space group of CuO is C2/c; its point group is 2/m. According to table 1, chiral planes are {010}, {0kl}, {hk0}, and {hkl}, whereas achiral planes are {100}, {001}, and {h01}. From these select planes, a trend is visible; achiral planes are those that k=0. This trend is true for all monoclinic structures.

System	Space Group	Point Group	{100}	{010}	{001}	{0kl}	{h0l}	{hk0}	{hkl}
Triclinic	1	1	1	1	1	1	1	1	1
	2	1̄	2	2	2	2	2	2	2
Monoclinic (2nd setting)	3-5	2	m	2	m	1	m	1	1
	6-9	m	m	1	m	1	m	1	1
	10-15	2/m	2mm	2	2mm	2	2mm	2	2
Ortho-rhombic	16-24	222	2mm	2mm	2mm	m	m	m	1
	25-46	mm2	m	m	2mm	m	m	m	1
		m2m	m	2mm	m	m	m	m	1
		2mm	2mm	m	m	m	m	m	1
	47-74	mmm	2mm	2mm	2mm	2mm	2mm	2mm	2

System	Space Group	Point Group	{001}	{100}	{110}	{hk0}	{h0l}	{hhl}	{hkl}
Tetragonal	75-80	4	4	m	m	m	1	1	1
	81-82	4̄	4	m	m	m	1	1	1
	83-88	4/m	4	2mm	2mm	2mm	2	2	2
	89-98	422	4mm	2mm	2mm	m	m	m	1
	99-110	4mm	4mm	m	m	m	m	m	1
	111-122	4̄2m	4mm	m	2mm	m	m	m	1
	123-142	4̄m2	4mm	2mm	2mm	2mm	2mm	2mm	2

System	Space Group	Point Group	{001}	{100}	{110}	{hk0}	{h0l}	{hhl}	{hkl}
Trigonal (hexagonal axes)	143-146	3	3	1	1	1	1	1	1
	147-148	3̄	6	2	2	2	2	2	2
	149-155	321	31m	m	2	1	m	1	1
		312	3m1	2	m	1	1	m	1
	156-161	3m1	3m1	m	1	1	m	1	1
	162-167	3̄m1	6mm	2mm	2	2	2mm	2	2
		3̄1m	6mm	2	2mm	2	2	2mm	2
Hexagonal	168-173	6	6	m	m	m	1	1	1
	174	6̄	3	m	m	m	1	1	1
	175-176	6/m	6	2mm	2mm	2mm	2	2	2
	177-182	622	6mm	2mm	2mm	m	m	m	1
	183-186	6mm	6mm	m	m	m	m	m	1
	187-190	6̄m2	3m1	2mm	m	m	m	m	1
		6̄2m	31m	m	2mm	m	m	m	1
	191-194	6/mmm	6mm	2mm	2mm	2mm	2mm	2mm	2

System	Space Group	Point Group	{100}	{111}	{110}	{hk0}	{hhl} h>l	{hhl} h<l	{hkl}
Cubic	195-199	23	2mm	3	m	m	1	1	1
	200-206	m3	2mm	6	2mm	2mm	2	2	2
	207-214	432	4mm	3m	m	m	m	m	1
	215-220	4̄3m	4mm	3m	m	m	m	m	1
	221-230	m3m	4mm	6mm	2mm	2mm	2mm	2mm	2

Table 1. Symmetry of the planes of a point group. Orientations that lack mirror symmetry (i.e. having rotational symmetry) are chiral. The chiral orientations are highlighted in yellow.

Other methods to determine whether a surface is chiral or achiral are stereographic projections and interface models. Like a point group, stereographic projections is a two-dimensional plot that shows the angular relationships of the crystal's planes and directions based on its crystallographic symmetry. Simply, a stereographic projection is a way to represent a three dimensional crystal on a two dimensional page. If two orientations have a stereographic projections that are superimpossible mirror images, then the orientations are achiral. If the two orientations produce stereographic projections that are nonsuperimposable mirror images, then the orientations are chiral.

Calculated stereographic projections are shown in figure 1 for the achiral planes (001) and (00-1) of CuO and in figure 2 for the chiral planes (111) and (-1-1-1) of CuO. In figure 1 A and 1B, the stereographic projections of the (001) and (00-1) planes are superimposable mirror images of each other. In addition, each projection has mirror symmetry. This mirror symmetry in the stereographic projection indicates that there is an improper symmetry operator perpendicular to this surface, resulting in planes that are achiral. In figure 2A and 2B, although the stereographic projections of the (111) and (-1-1-1) planes are mirror images of each other, they are not superimposable. Also, neither projection has mirror symmetry. The absence of mirror symmetry in the stereographic projection indicates that only proper symmetry operators are perpendicular to this surface. The presence of only symmetry operators indicates that these planes are chiral.

Thus based on the structure and orientation of the thin films, one can determine whether or not the film is chiral.

4. X-ray Diffraction characterization of Films

X-ray diffraction (XRD) is a non-destructive technique that reveals crystallography of an unknown material using monochromatic X-rays. X-rays are generated by an X-ray tube that uses a high voltage to accelerate the electrons released by a cathode to a high velocity. The so generated electrons collide with a metal target, the anode, creating the X-rays. [36] Different X-ray sources are used based on the need. Tungsten or a crack-resistant alloy of rhenium (5%) and tungsten (95%) are generally used in the medical field. When soft X-rays are needed for special applications like mammography, a molybdenum source is used. In crystallography, a copper target is most common, with cobalt often being used when fluorescence from iron content in the sample might otherwise present a problem. For a copper target, X-ray emissions commonly contains a continuous white radiation and two characteristic x-rays, $K\alpha$ ($\lambda = 0.15418$ nm) and $K\beta$ ($\lambda = 0.13922$ nm) leading from $2p \rightarrow 1s$ and $3p \rightarrow 1s$ transitions, respectively. In general, the $K\alpha$ transition is more intense than $K\beta$, and is a combination of $K\alpha1$ and $K\alpha2$. This is because of the slight difference between two possible spin states of $2p$ electrons. Monochromatic $K\alpha$ X-rays can be obtained by using suitable filters that absorb the unneeded white radiation and $K\beta$. For example, a Ni foil is commonly used for radiation of copper target. [37]

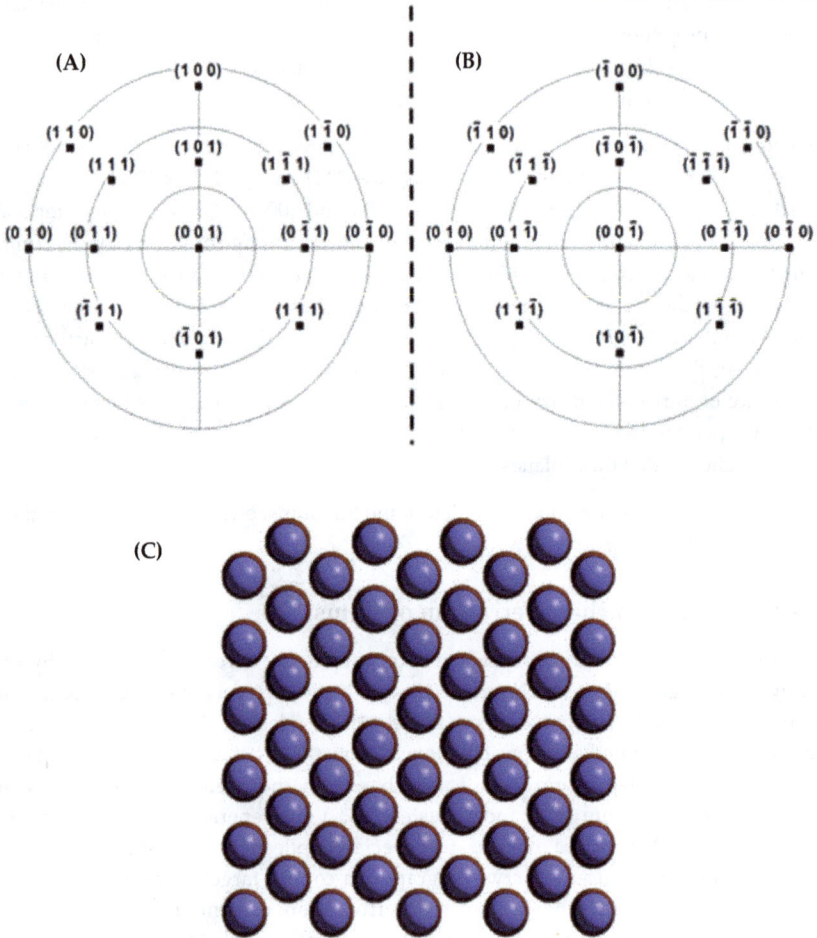

Figure 1. Stereographic projections of the A) (001) and B) (00-1) orientations of CuO. These two orientations are superimposable mirror images of each other; they are achiral. The radial grid lines on the stereographic projections correspond to 30° increments of the tilt angle, χ. C) the interface model of (00-1) CuO (front, blue Cu atoms) on (001) CuO (back, brown Cu atoms) with a common [010] directions. These two orientations are superimposed onto each other, indicating that they are achiral.

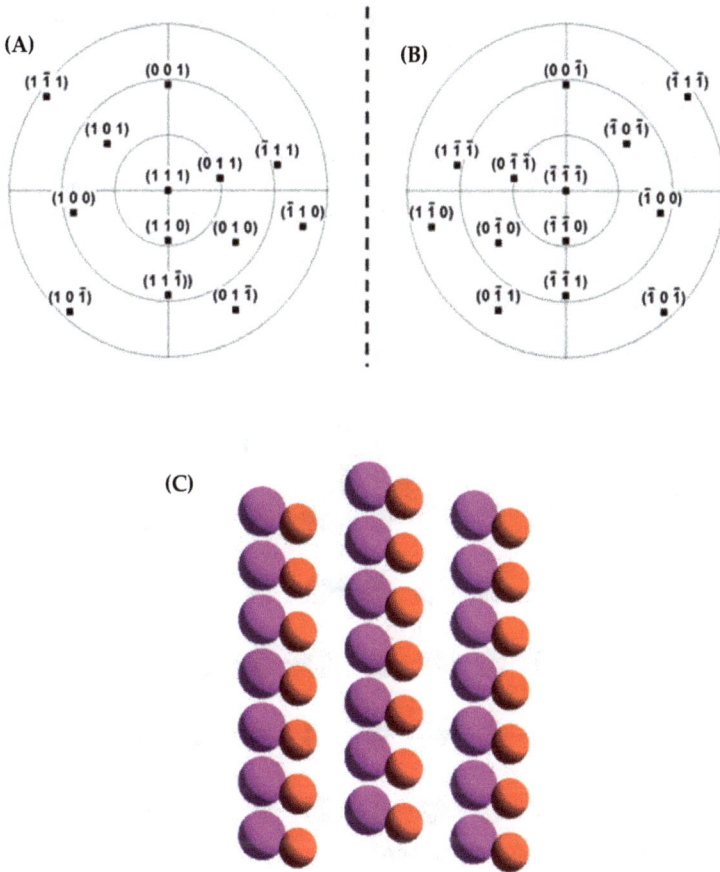

Figure 2. Stereographic projections of A) (111) and B) (-1-1-1) orientation of CuO. These two orientations are nonsuperimposable mirror images of each other; they are chiral. The radial grid lines on the stereographic projections correspond to 30° increments of the tilt angle, χ. C) the interface model of (-1-1-1) CuO (red oxygen atoms) on (111) CuO (violet oxygen atoms) with a common [1-10] direction. These two surfaces are nonsuperimposable; therefore, they are chiral.

X-ray diffraction works on the principle of Bragg's law, $n\lambda = 2d \sin \theta$, where λ is the x-ray wavelength, d is the lattice spacing and θ is the Bragg angle. The layers of a crystal act like weak reflecting mirrors for the X-rays. Only if the path difference of the reflected X-rays is a whole number of wavelengths does constructive interference occurs, as shown in Figure 3.

In general, X-ray diffraction patterns are plots of intensity versus 2θ angle. Different planes in a crystal diffract at different angles giving a pattern which is unique to a crystal. The intensities of the reflections are determined by the distribution of the electrons in the unit cell. X-rays going through areas with high electron density will reflect strongly and areas with low electron density will give weak intensities. Therefore, every crystalline material has a unique X-ray diffraction pattern that can be used to determine the crystallinity and phase of the deposited material.

Thin films deposited on polycrystalline substrates can be analyzed by running two types of experiments. In the first case, a symmetric scan (gonio scan) can be used to evaluate the out-of-plane texture of the film. It is a conventional θ-2θ scan for the Bragg-Brentano geometry. Figure 4 (A) shows a schematic representation of a gonio scan. Where, ω is the incident angle and θ is the diffracted angle.

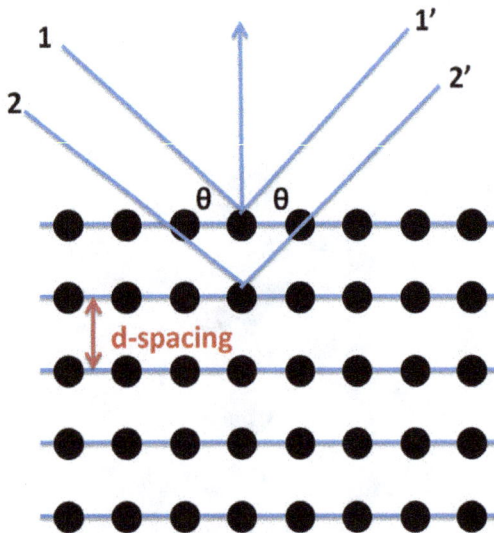

Figure 3. Schematic representation of Bragg's law.

In a gonio scan, $\omega = \theta$ and $\omega + \theta = 2\theta$. Another way to characterize thin films is through glancing angle measurements, as shown in Figure 4 (B). In this case, $\omega + \theta = 2\theta$ and ω is fixed but θ varies. Glancing angle measurement is a good technique to measure polycrystalline grains. As the w is fixed and the θ is varied, all the planes in the material are brought into the Bragg condition. However, this technique cannot be used for highly oriented films or epitaxial films.

Epitaxial films deposited on single crystals are oriented both out-of-plane as well as in-plane. In this case, X-ray characterization such as diffraction patterns, pole figures, azimuthal scans, and rocking curves are performed.

Unlike polycrystalline films, epitaxial films grow with one orientation and show only a family of planes in the pattern. In Figure 4, an epitaxial magnetite film deposited on a Ni(111) substrate shows only the {111} family peaks. [38] The experimental setup to determine out-of-plane orientation is similar to the gonio scan but instead a 2θ-omega scan is run. The difference between this and the gonio scan is that there can be an offset between 2θ and omega, so that omega = ½ 2θ + offset. This is useful when collecting a diffraction scan from an epitaxial film, when the tilt of the film is compensated by the offset. Identification of the pattern is done by comparing the pattern with the existing patterns in the database. For example, a magnetite film is identified by JCPDS#19-0629 pattern. However, this does not provide any information about in-plane orientation of the film. To determine the in-plane orientation of the film, X-ray pole figures and azimuthal scans are run. Pole figures can be used to probe planes which are not parallel with the geometric surface of the sample. The sample is moved through a series of tilt angles, χ, and at each tilt angle the sample is rotated through azimuthal angle, ϕ, of 0 to 360°. Peaks occur in the pole figure when the Bragg condition is satisfied. During the experiment 2θ is fixed, which is normally the highest intensity peak of a randomly oriented powder diffraction pattern of the material. Figure 6 is a schematic representation of a pole figure measurement. Azimuthal scans can be considered as a cross-section of a pole figure. They are obtained when the measurement is carried out at a specific 2θ for only specific tilt angle, χ, and rotated azimuthally, ϕ, from 0 to 360°. Comparing the azimuthal scans at specific tilt angle for a substrate and a film, we can obtain the epitaxial relationships.

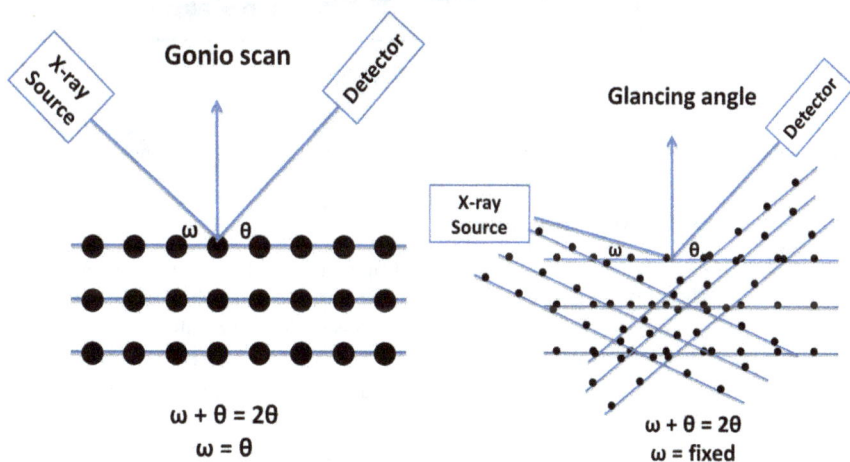

Figure 4. Schematic representation of (A) gonio scan used for textured films and (B) glancing angle used for polycrystalline grains.

To determine the quality of epitaxy, X-ray rocking curves are run. The rocking curves indicate the mosaic spread of the film relative to the substrate. The larger the full width at half maximum (FWHM), the larger the mosaic spread. In the experimental setup for a rocking curve, only the omega axis is scanned as data are collected. All other axes, such as 2θ, are fixed at specific angles. In a perfect single crystal, the FWHM is small indicating that all domains are aligned. The width of the rocking curve is a direct measurement of the range of orientation of the sample. In general, the rocking curves are performed for substrate as well as the film for comparison of mosaic spread. If the mosaic spread of the film is low and comparable to the substrate, the peaks in the pole figure become sharper and more intense. Rocking curves have been used to understand the in-plane misorientation of ZnO, AlN, and GaN on sapphire and MgO films grown on GaAs.

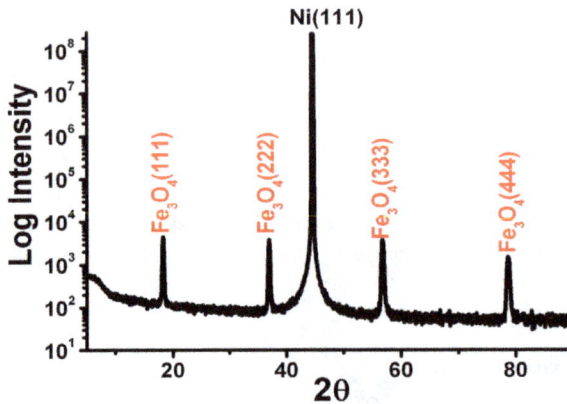

Figure 5. X-ray diffraction pattern of an epitaxial magnetite (Fe3O4) film on Ni(111) single crystal.

5. Electrodeposition of chiral films

Electrodeposition is a versatile method for depositing metals, metal oxides, semiconductors, and biomaterials. [20, 21] It is a simple and a low cost process which uses electrically conductive aqueous, organic, or fused salt solutions. Electrodeposition also offers the ability to tune the characteristics of the film by varying factors such as the electrolyte composition, the additives, the pH and temperature of the electrolyte, and the applied overpotential or current density. [40]

The most familiar method to electrodeposit a film involves electrochemically oxidizing or reducing the cation in the electrolyte at the electrode surface. In this process, the electrochemically produced species reacts with water or hydroxide ions to form a metal oxide on the electrode. [41-57] A second method to produce thin films is to electrochemically change the pH of the electrolyte at the electrode surface. Because the solubility of any material is dependent on pH, the pH can be electrochemically changed at the electrode surface, thus lowering the solubility of the material and resulting in the precipitation of the material only on the electrode surface. The pH can be decreased by electrochemically oxidizing water, ascorbates, hydroquinone, and other organics to produce metal oxides. Likewise, the pH can be electrochemically increased by reducing water, molecular oxygen, nitrate, peroxide, or organic molecules such as quinones. [58]

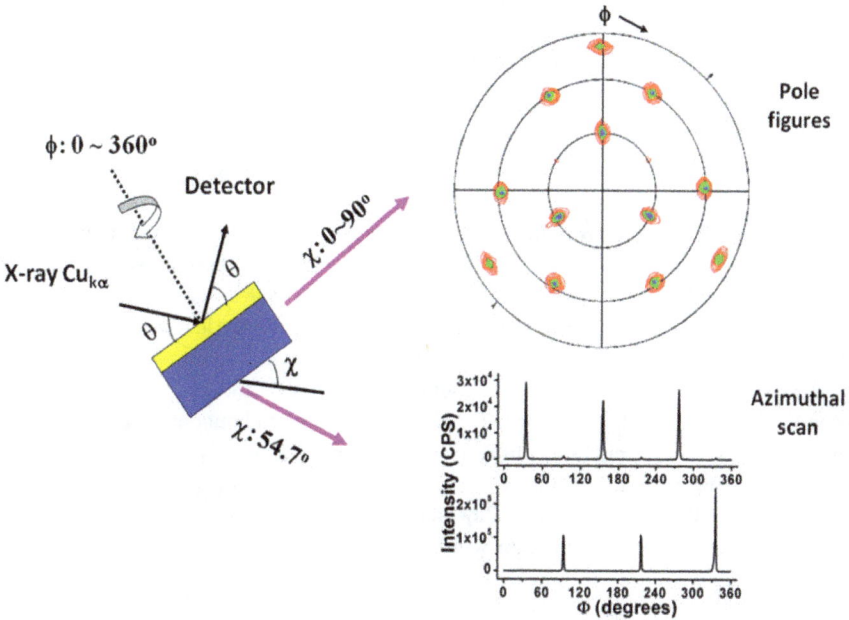

Figure 6. An illustration of the method to produce a (311) magnetite pole figure of a (111) oriented magnetite film on a Ni(111) substrate. The azimuthal scan on the bottom right is obtained by running the ϕ scan at a fixed χ.

Electrodeposited films orientation and morphology are determined by the electrochemical/chemical reaction, potential, current, pH, temperature, and additives. [40] For example, Switzer group has previously shown that varying the applied potential and pH changes the orientation and size of the electrodeposited Cu_2O films. [59, 60] Films follow the orientation of the substrate at low overpotentials and change to a kinetically controlled orientation at a critical thickness. As the overpotential increases, the critical thickness for the transition decreases. Similarly, Cu_2O films from a pH 9 solution have a [001] preferred orientation while films from a pH 12 solution have a [111] preferred orientation. Siegfried and Choi have examined the stability of facets of Cu_2O in the presence of additives where ammonium salts stabilize {111} planes, whereas sodium chloride stabilizes {100} planes. [61]

Thin films of metals and metal oxides can be electrodeposited on polycrystalline substrates or single crystalline substrates. Polycrystalline films grow on polycrystalline substrate and are usually used for identification and characterization. Different planes of atoms produce different diffraction peaks, which contain information about atomic arrangement within the crystal. Therefore, every crystalline material has a unique X-ray diffraction pattern. [62] This pattern can be used to determine the crystallinity and phase of the deposited polycrystalline material. The lattice parameters of a material can be calculated from the peaks by properly assigning Miller indices (hkl) and accurately measuring the d-spacing. However, if the films are grown on single crystals such as Au(001), Au(111), Si(111), Ni(110) and InP(110) we get epitaxial films. Single crystals eliminate the presence of grains of various orientations, which may have different film growth rates as discussed above. If a crystalline film grows with an out-of-plane and in-plane orientation that is dependent on the orientation of the substrate, the film is said to be epitaxial.

The properties of epitaxial films depend on the substrate. The film will have better crystalline quality and more uniform morphology when fewer defects are present on the substrate. Generally, an epitaxial film will deposit with an orientation providing the lowest lattice mismatch with the substrate. In some cases, the film will become strained or will rotate to obtain this low mismatch. Thus, different substrates will produce different orientations of the film. Since properties of a material can be dependent on the orientation, this control over selection of orientation permits optimization of the properties of a deposited material. Epitaxial films are useful in device applications, since the intrinsic properties of the material can be exploited rather than its grain boundaries. Epitaxial films can be characterized using X-ray diffraction, pole figures, azimuthal scans, and rocking curves. For example, the diffraction pattern of a polycrystalline film is different than that of an epitaxial film. Whereas a polycrystalline film has all possible planes, resulting in multiple peaks in the pattern, an epitaxial film will normally grow with one orientation and show family of planes in the pattern. The details of X-ray characterization for epitaxial films have been discussed in the above section. [40]

Enantiospecific epitaxial thin films of CuO can be electrodeposited onto Au(100), Cu(111), and Cu(110) single crystals from alkaline solutions. [18, 19, 33] The Chirality of the electrodeposited film in alkaline bath was determined by the enantiomer used to complex Cu(II). Chiral CuO films were deposited on Au(001), Cu(111) and Cu(110) single crystals from alkaline solutions of copper(II) complexed with tartaric acid, malic acid, and amino acids. The films were deposited either at constant current density or at constant potential. The CuO films deposited on Au(001) using tartaric acid as complexing agent in the deposition bath resulted in {1-1-1} set of peaks. [18, 19] Chirality of the film cannot be determined by the θ-2θ X-ray scan. The films grown from a solution of L-tartaric acid or D-tartaric acid can have either (1-1-1) or (-111) orientations. It is impossible to distinguish between the orientations based on the θ-2θ scans as they have identical d-spacing. The orientations can be assigned with the help of X-ray diffraction pole figure. The CuO(111) pole figures and stereographic projections indicated that the films grown from L-tartaric acid has (1-1-1) orientation whereas the film grown from D-tartaric acid has (-111) orientation. These two stereographic projections are non-superimposable mirror images of each other. The enantiomeric excess of one orientation over the other was determined from azimuthal scans probing the higher angle CuO (111)-type reflections. The percentage enantiomeric excess for the (1-1-1) orientation can be calculated quantitatively from the area under all the peaks obtained at higher angle using the formula.

$$ee = (A_{(1-1-1)} - A_{(-111)}/A_{(1-1-1)} + A_{(-111)}) *100$$

For a film deposited from L-tartaric acid has an enantiomeric excess of 95% while the films deposited from D-tartaric acid has an enantiomeric excess of 93%. The film from DL-tartaric acid has equal amounts of both the orientations and zero enantiomeric excess. For the CuO films grown from alkaline solutions using amino acids on Au(001) resulted in {1-1-1} and {-111} orientations. The films were grown with the alkaline amino acids solutions of alanine, valine, and glycine. The films grown from Cu(II) complexes of amino acids have two chiral orientations with a smaller enantiomeric excess. The films grown from L-alanine and L-valine solutions have a small excess of (-111) and (-1-1-1) orientations, while the films grown from D-alanine and D-valine have a small excess of the (1-1-1) and (111) orientations. The pole figures obtained from achiral glycine as a complexing agent have equal amounts of the chiral CuO(-111) and CuO(1-1-1) orientations.

For the films grown from Cu(II) complexes of malic acid on Cu(111) single crystals, two chiral orientations were obtained. The films produced from L-malate in the deposition bath have (-111) and (311) orientations and the ones produced from D-malate have (1-1-1) and (-3-1-1) orientations. CuO films grown on Cu(110) from L-malate had (110) and (31-1) orientations while the one grown from D-malate deposition bath had (-1-10) and (-3-11) orientations. The films grown from racemic malate showed equal amounts of only (-111) and (1-1-1) orientations on Cu(111) and CuO(110) and CuO(-1-10) orientations on Cu(110) single crystal. It is still not clear as to why the CuO films grown from malic acid in the deposition bath show 2 chiral orientations where as the films grown from tartaric acid show only one dominant chiral orientation. One argument can be that in a strong alkaline solution that the crystal structure of chiral Cu(II) malate has a different packing than the racemix Cu(II) malate. It is difficult to grow crystals of Cu(II) malate or Cu(II) tartarate from highly alkaline solutions. To understand the phenomenon of chiral electrodeposition, the crystals of copper(II) malate were grown at pH=1.5. Both L- and D- chiral complexes polymerize in a similar fashion where each copper is coordinated to two malate ligands of the same chirality via one hydroxyl oxygen atom and one carboxyl oxygen atom. Both the chiral complexes crystallized in $P2_1$ monoclinic space group. Whereas the complxes grown from racemic Cu(II)-malate crystallized in a centrosymmetric space group P21/c with copper atoms at the inversion center. Similarly, other copper(II) complexes have structures which are determined by the handedness of the ligand. It is believed that the chiral complexes formed in the solution are initially adsorbed on the surface and imprint chirality on the surface. However, further study has to be done to understand this phenomenon better.

To get a better understanding for the chiral electrodeposition further studies were done by growing CuO films on Cu(111) single crystal with different amino acids in the solution bath. Table 2 shows different chiral reagents used for depositing CuO films. The table also indicates their handedness in different conventions.1 Usually, there are three different types of conventions which are followed for chiral molecules. Optical activity (+/-) is based on the direction in which the enantiomer rotates the plane polarized light. The (+) and (-) enantiomer are also identified as dextrorotatory (d) and levorotatory (l). This notation can be easily confused with D and L labeling. The D/L labeling is unrelated to +/- notation; it does indicate which enantiomer is dextrorotatory or levorotatory. Rather, it says that the compound's stereochemistry is related to the stereochemistry of the dextrorotatory and levorotatory enantiomer of glyceraldehydes. Another convention which is generally used is the R/S configuration. In this convention, each chiral center in the molecule is either named R or S according to a system by which its substituent's are each assigned a priority. The priorities are based on Cahn Ingold Prelog rules, based on atomic number. The R/S system has no fixed relation with +/- or D/L systems. An R isomer can be either dextrorotatory or levorotatory, depending on its exact substituents. [63]

Amino Acids	(+) /(-) notation	d/l – notation	D/L-notation	R/S-Notation
Alanine	+	d	L	S
Arginine	+	d	L	S
Asparatic acid	-	l	L	S
Glutamic acid	+	d	L	S
Valine	-	l	L	S
Tartaric acid	+	d	L	R, R
Malic acid	-	l	L	S
Proline	-	l	L	S

Table 2. List of chiral precursors used for depositing CuO on Cu(111) single crystal. The table also lists different types of notations used for chiral molecules.

In this work, different chiral reagents were used to deposit CuO films on Cu(111) single crystals to understand the role of the precursor on chiral film formation. Figure 7 shows the (111) pole figures for CuO on Cu(111) deposited from different chiral reagents. The pole figures were obtained by setting 2θ equal to the angle of diffracted intensity for the (111) planes ($2\theta = 38.742^\circ$) and performing azimuthal scans at tilt angles, χ, from 0 to 90°. The pole figures in Figure 7 (A) - (E) show peaks at 57° and 63°. These peaks correspond to either (-111) orientation or (1-1-1) orientation. On the other hand, the pole figures in Figures 7 (F) – (H) have extra peaks at tilt angle of 27°, which correspond to either (311) or (-3-1-1) orientation. To interpret pole figures, we use stereographic projections. Figures 8(A) and 8(B) shows the (-111) and (1-1-1) stereographic projections of monoclinic CuO. The radial direction is the tilt angle, χ, while the azimuthal angle, ϕ, is the rotation of the sample about the axis. The positions of CuO{111} and CuO{200} reflections are specified. Figure 8(A) shows that for the (-111) orientation, reflections from the (-11-1) plane at a χ of 57°, and the (111) planes at a χ of 63°, are separated azimuthally by 115° rotated counterclockwise ($\Delta\phi$ = -115°). Figure 8(B) shows that for (1-1-1) orientation, reflections from the (1-11) plane at χ = 57° and the (-1-1-1) plane at χ = 63° are separated azimuthally by 115° rotated clockwise ($\Delta\phi$

= 115°). The calculated interplanar angles obtained from stereographic projections correspond with the measured interplanar angles of the pole figures. Figure 7 (F)-(H) also show reflections at χ = 27°. These reflections can be explained by figures 8(C) and 8(D). Figure 8(C) shows that for the (311) orientation, reflections from the (111) plane at a χ of 27.19°, and the (1-11) plane at a χ of 68.14°, are separated azimuthally by 125.18° rotated counterclockwise. Figure 8(D) shows that for the (-3-1-1) orientation, reflections from the (-1-1-1) plane at χ of 27.19°, and the (-11-1) plane at χ of 68.14°, are separated azimuthally by 125.18° rotated clockwise. Figures 8(A) and 8(C) clearly explains the pole figure of Figure 7 (F)-(H) and confirms the presence of (-111) and (311) orientations for the film grown from different amino acid deposition bath. Similarly, Figures 8(A) and 8(B) explain the pole figure of Figure 7 (A)-(E) and confirm the presence of (1-1-1) and (-111) orientations.

Figure 9 shows the SEM images of the CuO films deposited on Cu(111) single crystal from different chiral reagents. The chiral reagents influence the morphology of the deposited films. Figure 9 (D) shows the SEM image of the CuO film deposited using L-asparatic acid precursor. The image shows that CuO grows like spears and are approximately 50 nm in diameter. On the other hand, films deposited using L- arginine has cross hatch morphology with individual crystallites made up of multiple platelets. The morphology of the films deposited from other amino acids and simple carboxylic acids like malic acid and tartaric acid have simple cross hatch morphology with different crystallite sizes.

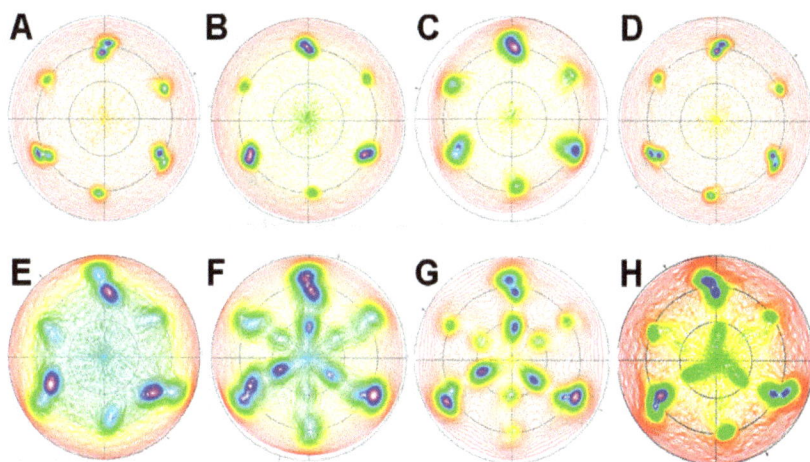

Figure 7. (111) CuO pole figure for CuO films deposited from different amino acids on Cu(111)single crystals. (A) tartaric acid, (B) asparatic acid, (C) glutamic acid, (D) proline, (E) arginine, (F) alanine, (G) valine and (H) malic acid.

The scanning electron microscopy (SEM) does not correlate the handedness of the electrodeposited films with their microstructure. However, the SEM images of the electrodeposited biominerals show very clearly the effect of additives in the solution. Different additives in the solutions are known to cause face-selective crystallization of inorganic crystals in biosystems. It was recently shown that in presence of chiral precursors, calcite and other biominerals with chiral morphology can be electrodeposited. [29] A combination of geometric matching, electrostatic interactions and stereochemistry is believed to be the cause of bimonineralization. [64] Researchers have shown that in the presence of chiral amino acids, asymmetric crystals of gypsum can be crystallized. [28] In nature, gypsum crystallizes in symmetric space group but by changing the solution precursors systematically one can control the handedness of the crystallites. Other researchers have studied the affect of step-edge free energies for the formation of chiral etch pits. [26]

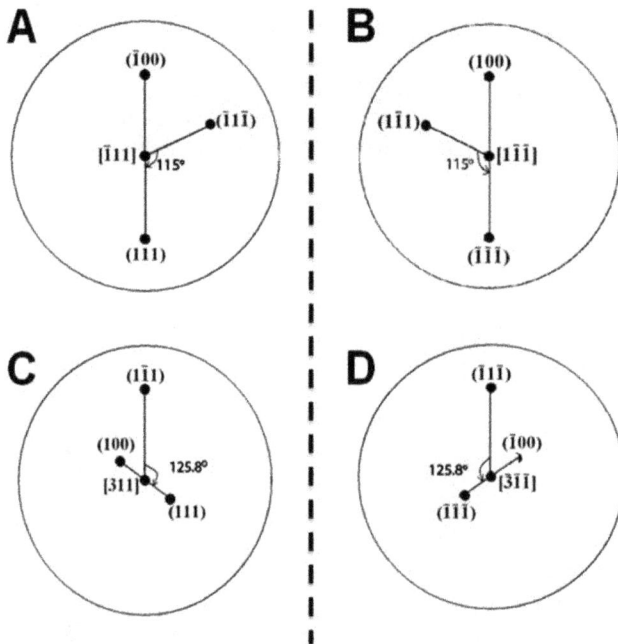

Figure 8. Stereographic projections for (A) (-1-11), (B) (11-1), (C) (311) and (D) (-3-1-1) orientations indicating the positions where the (111) type and (100) type reflections should be observed in the pole figures. For the   (-1-11) orientation, reflections from the (1-11) plane at $\chi = 57°$ and the   (-1-1-1) plane at $\chi = 63°$ are separated azimuthally by 115° rotated counterclockwise. For the (-3-1-1) orientation the (1-11) plane is at $\chi = 68.14°$, the (-1-1-1) planes at $\chi = 27.41°$ and (100) planes at $\chi = 27.19°$. For the (11-1) orientation the (-11-1) plane is at $\chi = 57°$ and the (111) plane is at $\chi = 63°$ are separated azimuthally by 115° rotated clockwise. For the (311) orientation the (1-11) plane at $\chi = 68.14°$, the (111) plane at $\chi = 27.41°$ are separated azimuthally by 125.8°.

Surface templating and imprinting can be the two main ways of obtaining chiral surfaces. Studies have shown that adsorption of chiral tartaric acid on low index surfaces of Cu and Ni can break the symmetry of the substrate causing chiral templates. On the other hand, adsorption of (R,R)-tartaric acid on Ni(110) surfaces results in chiral imprinting. [6] The difference between the two can be best described as following, in the case of chiral imprinting; the adsorbed molecule irreversibly reconstructs the substrate, whereas in the case of templating; the underlying substrate is not affected upon the removal of adsorbed molecule.

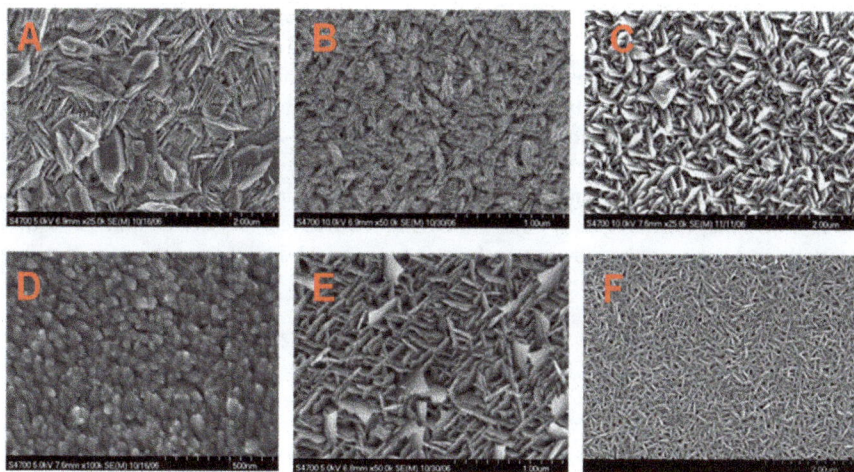

Figure 9. SEM micrographs for CuO films deposited on Cu(111) single crystal from different chiral precursors. (A) L(+)-Alanine, (B) L(-)-Arginine, (C) L(+)-Valine, (D) L(-)-Aspartic acid, (E) L(+)-Glutamic acid

In the case of electrodeposited chiral CuO films, it would not be unreasonable to expect chiral Cu(II) complexes to adsorb on the single crystal surfaces and break its symmetry. As discussed earlier, complexes of copper(II) with different chiral reagents have a dimeric structure with a symmetry that is determined by the handedness of the ligands. Surface reconstruction of the underlying single crystal substrate from adsorbed molecules can be another cause for the asymmetry. The precise mechanism, however, is not known at this time. Studies have shown that presence of chiral modifiers is necessary for the nucleation of chiral CuO domains. Once the nucleation layer is produced, the film maintains the orientation regardless of the precursor. [47] Studies on adsorption of organic molecules on surfaces have shown that local chiral imprinting can be obtained without the need of extended surface Chirality. [6] Adsorption of amino acids on Au [66] and Ag [67] surfaces are reported to be physical in nature, while amino acids chemisorb on Cu single crystals. Based on this argument the films grown on Cu single crystals should have more order as compared to other crystals. We have shown in this study that this is indeed true. To get more insight into the mechanism of chiral electrodeposition, further investigation is required. In-situ AFM during the chiral electrodeposition can be helpful in understanding the initial nucleation and whether or not it is local chiral imprinting which results in chiral films.

An interesting aspect of the electrodeposited chiral films is that they show chiral selectivity. [18, 19] In this process, the chiral film selectively oxidizes the chiral molecules in the solution. This property of the chiral films can be utilized post enantiomeric separation to detect the amount of enantiomeric excess in the mixture.

6. Electrochemical selectivity

Researchers have studied the electrochemical oxidation of sugars on chiral Pt surfaces. [15] In this study, enantiospecific electrochemical oxidation of glucose was demonstrated on Pt(643) and Pt(-6-4-3) surfaces. Recent studies show that chiral cyclodextrin based coatings applied to a three-transducer microsystem shows distinct chiral discrimination for different compounds. [68] CuO has been shown to be catalytic for the electrochemical oxidation of amino acids, carbohydrates and amines with very high sensitivities. [69-71] It has been reported that increased sensitivity for the electroxidation of amino acids with an increase in solution pH. [72] Electrodeposited chiral CuO films also show chiral selectivity in high pH chiral solutions. [18] It was shown that the film grown from L-tartaric acid selectively oxidizes L-tartaric acid over D-tartaric acid, whereas the films grown from D-tartaric acid does the opposite. For the films grown from a racemic tartaric acid, no selectivity was observed.

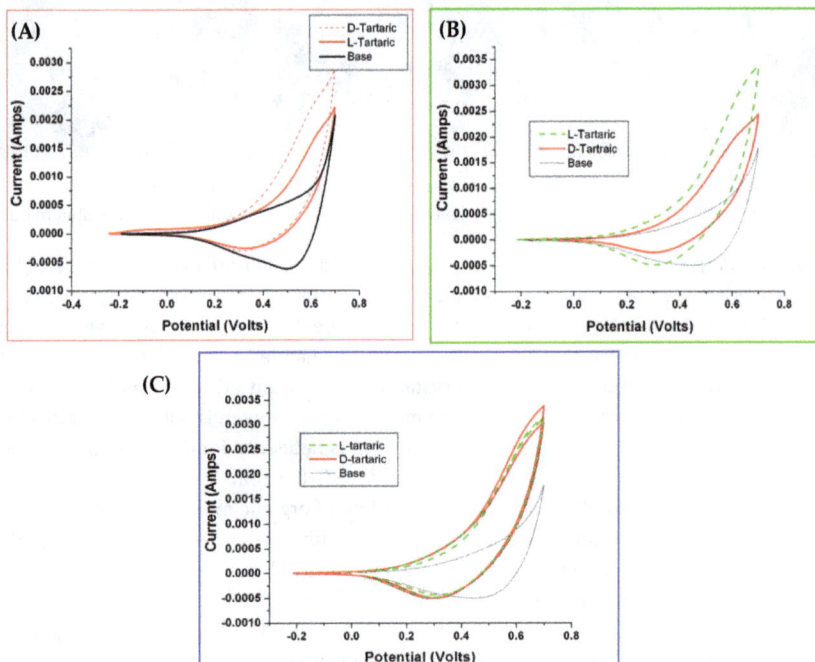

Figure 10. Cyclic voltammograms in unstirred solutions of alkaline solutions of L-, D-, and DL- tartaric acid by scanning from open circuit potentials to 0.75 V vs. SCE at a scan rate of 10 mV/sec for the films grown (A) L-malic acid, (B) D-malic and (C) DL-malic acid

The study was extended to understand the selectivity behavior of a chiral film deposited with malic acid precursor and oxidizing tartaric acid films. The cyclic voltammograms (CV) were obtained in unstirred solutions by scanning from open circuit potentials to 0.75 V vs. SCE at a scan rate of 10 mV/sec. Before switching solutions the electrode was cleaned by scanning in 0.1 M NaOH from the OCP to 0.75 vs. SCE to remove adsorbed remnants from the electrode. Figure 10(A) shows the film grown from L-malic acid selectively oxidizes for D-tartaric acid, whereas the Figure 10 (B) film grown from D-malic acid does the opposite. Figure 10 (C) shows for the film grown from racemic malate has a very slight selectivity for D-tartaric acid but it is insiginificant difference compared to the selectivity shown by the chiral films. This study is in agreement with the pole figures for the CuO films grown on Cu(111) single crystals from the solutions of malic and tartaric acid precursors. The CuO films grown from D-malic acid and L-tartaric acid have similar dominant orientations and hence the similar selectivity. Another study has shown that by selectively etching the chiral CuO films the selectivity can be enhanced considerably. However, a detail study on more complex molecules still needs to be studied to validate this model. The final goal should be to have a sensor grown from one chiral molecule which can selectively distinguish any chiral molecule with the same handedness.

In summary, enantiospecific epitaxial CuO films can be electrodeposited onto single crystal substrates from copper (II) complexes of chiral molecules. Chirality of a crystal is dependent on the symmetry operations in the structure. Chirality can be imparted to non chiral crystals by depositing the films in the presence of chiral precursors. The chiral orientation of the films can be determined by X-ray pole figures and stereographic projections. The chiral CuO films show selectivity in the alkaline chiral solutions. At this point, the mechanism of the chiral electrodeposition is not very well understood and needs further study.

Author details

Rakesh Gudavarthy
Missouri University of Science and Technology, Rolla, Missouri, USA Current address: Intel Corporation, Hillsboro, OR, USA

Elizabeth A. Kulp
Missouri University of Science and Technology, Rolla, MO, USA

Acknowledgement

The authors acknowledge Prof. Jay A. Switzer and his group members for technical contributions and providing research facilities. The support and encouragement from Anand Sadekar and Gayatri Angara is greatly acknowledged. The authors wish to thank the reviewers for their critical comments and suggestions.

7. References

[1]   Rouhi, A.M. Chem. Eng. News 2004; 82, 47.

[2]   Rentsch K. M. The importance of stereoselective determination of drugs in the clinical laboratory. Journal of Biochemical and Biophysical Methods 2002; 54(1-3), 1-9.

[3]   Walther W, Netscher T. Design and Development of Chiral Reagents for the Chromatographic Determination of Chiral Alcohols. Chirality, 1996; 8, 397-401.

[4]   Katzung B. G. The Nature of Drugs. In: Basic and Clinical Pharmacology. 9th Ed. New York, Lange Medical Books/McGraw Hill, 2004; 3-5.

[5]   Stinson S. C. Chiral Pharmaceuticals. Chem. Eng. News 2001; 79, 79–97

[6]   Lorenzo M. O, Baddeley C. J, Muryn C, Raval R Nature 2000; 404, 376–379.

[7]   Humblot V, Haq S, Muryn C, Hofer W. A, Raval R. J. Am. Chem. Soc. 2002; 124, 503–510.

[8]   Kühnle A, Linderoth T. R, Hammer B, Besenbacher F. Nature 2002; 415, 891–893.

[9]   Izumi Y. Advances in Catalysis 1983; 32, 215–271.

[10] LeBlond C, Wang J, Liu J, Andrews A. T, Sun Y. K. J. Am. Chem. Soc.1999; 121, 4920–4921.

[11] McFadden C. F, Cremer P. S, Gellman A. J. Langmuir 1996; 12, 2483–2487.

[12] Horvath J. D, Gellman A. J. J. Am. Chem. Soc. 2001; 123, 7953–7954.

[13] Horvath J. D, Gellman A. J. J. Am. Chem. Soc. 2002; 124, 2384–2392.

[14] Ahmadi A, Attard G, Feliu J, Rodes A. Langmuir 1999; 15, 2420–2424.

[15] Attard G. A. J. Phys. Chem. B 1999; 103, 1381–1385.

[16] Attard G. A. J. Phys. Chem. B 2001; 105, 3158–3167.

[17] Sholl D. S, Asthagiri A, Power T. D. J. Phys. Chem. B 2001; 105, 4771–4782.

[18] Switzer J. A, Kothari H. M, Poizot P, Nakanishi S, Bohannan E. W. Nature 2003; 425, 490.

[19] Kothari H. M, Kulp E. A, Boonsalee S, Nikiforov M. P, Bohannan E. W, Poizot P, Nakanishi S, Switzer J. A. Chem. Mater. 2004; 16, 4232.

[20] Switzer J. A, Electrochemistry of Nanomaterials, G. Hodes, Ed. Wiley-VCH, Weinheim 2001; chap 3.

[21] Switzer J. A. Handbook of Nanophase Materials, A. Goldstein, Ed. Marcel Dekkar, New York 1996; chap 4.

[22] Hazen R. M, Sholl D. S. Nat. Mater 2003; 2, 367.

[23] Orme C. A, Noy A, Wierbicki A, McBride M. T, Grantham M, Teng H. H, Dove P, De Yoreo J. J. Nature 2001; 411, 775.

[24] Addadi L, Weiner W. Nature 2001; 411, 753.

[25] Teng H.H, Dove P. M, Orne C. A, De Yoreo J. J. Science 1998; 282, 724.

[26] Cody A. M, Cody R. D. J. Cryst. Growth 1991; 113, 508.

[27] Aizenberg J, Black A. J, Whitesides G. M. Nature 1999; 398, 495.

[28] Hazen R. M, Filley T. R, Goodfriend G. A. Proc. Natl. Acad. Sci. USA 2001; 98, 5487.

[29] Kulp E. A, Switzer J. A. J. Am. Chem. Soc. 2007; 129, 15120.

[30] Challener C.A. Overview of Chirality. In: Chiral Drugs. 1st Ed. Aldershot (England), Ashgate Publisher 2001; 3-14.

[31] Drayer D.E. The Early History of Stereochemistry. In: Drug Stereochemistry. Analytical Methods and Pharmacology. 2nd Ed., Wainer I. W, Editor. New York, Marcel Dekker Publisher 1993. P 1-24.

[32] Reddy I. K, Mehvar R, Chirality in Drug Design and Development, Marcel Dekker, Inc; 2004.

[33] Gudavarthy R. V, Burla N, Kulp E. A, Limmer S. J, Sinn E, Switzer J.A. J. Mater. Chem. 2011; 21, 6209-6216.

[34] Glusker J. P, Lewis M, Rossi M. Crystal Structure Analysis for Chemists and Biologists, Wiley, New York; 1994.

[35] Henry N. F, Londsdale K. International Tables for X-ray Crystallography Vol. 1, Kynoch, Birmingham; 1952. p37-38.

[36] Whaites E, Roderick C. Essentials of Dental Radiography and Radiology 2002; 19-22.

[37] Culity B. D, Stock S. R. Elements of X-ray Diffraction, 3rd Ed. Upper Saddle River, New Jersey: Prentice Hall; 2001.

[38] Gudavarthy R. V, Gorantla S, Mu G, Kulp E. A, Gemming T, Eckert J, Switzer J. A. Chem. Mater. 2011; 23 (8), 2017–2019.

[39] Srikanth V, Speck J. S, Clarke D. R. J. Appl. Phys. 1997; 82, 4286.

[40] Switzer J. A, Hodes G, MRS Bulletin 2010; 35, 743.

[41] Liu R, Bohannan E. W, Switzer J. A, Oba F, Ernst F. Appl. Phys. Lett. 2003; 83, 1944.

[42] Liu R, Oba F, Bohannan E. W, Ernst F, Switzer J. A. Chem. Mater. 2003; 15, 4882.

[43] Liu R, Kulp E. A, Oba F, Bohannan E. W, Ernst F, Switzer J. A. Chem. Mater. 2005; 17, 725.

[44] Sorenson T. A, Morton S. A, Waddill D. G, Switzer J. A. J. Am. Chem. Soc. 2002; 124, 7604.

[45] Kothari H. M, Kulp E. A, Limmer S. J, Poizot P, Bohannan E. W, Switzer J. A. J. Mater. Res. 2006; 21, 293.

[46] Mitra, S, Poizot P, Finke A, Tarascon J. M. Adv. Funct. Mater. 2006; 16, 2281.

[47] Kulp E. A, Kothari H. M, Limmer S. J, Yang J, Gudavarthy R. V, Bohannan E. W, Switzer J. A. Chem. Mater. 2009; 21, 5022.

[48] Switzer J. A, Shumsky M. G, Bohannan E. W, Science 1999; 284, 293.

[49] Bohannan E. W, Jaynes C. C, Shumsky M. G, Barton J. K, Switzer J. A, Solid State Ionics 2000; 131, 97.

[50] Kulp E. A, Limmer S. J, Bohannan E. W, Switzer J. A. Solid State Ionics 2007; 178, 749.

[51] Breyfogle B. E, Hung C. J, Shumsky M. G, Switzer J. A. J. Electrochem. Soc. 1996; 143, 2741.

[52] Switzer J. A. J. Electrochem. Soc.1986; 133, 722.

[53] Van Leeuwen R. A, Hung C. J, Kammler D. R, Switzer J. A. J. Phys. Chem. 1995; 99, 15247.

[54] Vertegel A. A, Bohannan E. W, Shumsky M. G, Switzer J. A. J. Electrochem. Soc.2001; 148, C253.

[55] Mu G, Gudavarthy R. V, Kulp E. A, Switzer J. A. Chem. Mater. 2009; 21, 3960.

[56] Boonsalee S, Gudavarthy R. V, Bohannan E. W, Switzer J. A, Chem. Mater.2008; 20, 5737.

[57] Switzer J. A, Gudavarthy R. V, Kulp E. A, Mu G, He Z, Wessel A. J. Am. Chem. Soc. 2010; 132, 1258.

[58] Limmer S. J, Kulp E. A, Bohannan E. W, Switzer J. A. Langmuir 2006; 22, 10535.

[59] Liu R, Oba F, Bohannan E. W, Ernst F, Switzer J. A. Chem. Mater. 2003; 15, 4882.

[60] Golden T. D, Shumsky M. G, Zhou Y, VanderWerf R. A, Van Leeuwen R. A, Switzer J. A, Chem. Mater. 1996; 8, 2499-2504.

[61] Siegfried M, Choi K. S, Angew. Chem. Int. Ed. 2005; 44, 3218-3223.

[62] Cullity B. D. Elements of X-ray Diffraction, Addison-Wesley, Reading, 2nd Ed; 1978.

[63] Eliel E. L, Wilen S. H. The Stereochemistry of Organic Compounds, Wiley-Interscience; 1994.

[64] Mann S. Nature 1998; 332, 119-124.

[65] Gellman A. J, Hovarth J. D, Buelow M. T. J. Mol. Catal. A 2001; 167, 3.

[66] Zhao X, Yan H, Zhao R. G, Yang W. S. Langmuir 2002; 18, 3901.

[67] Zhao X, Yan H, Tu X, Zhao R.G, Yang W.S. Langmuir 2003; 19, 5542.

[68] Kurzawski P, Schurig V,   Hierlemann A, Anal. Chem.2009; 81, 9353-9364.

[69] Wels B, Johnson D.C. J. Electrochem. Soc. 1990, 137, 2785.

[70] Xie Y., Huber C.O., *Anal. Chem* 1991, 63, 1714.

[71] Kano K., Torimura M., Esaka Y, Goto M, J. Electroanal. Chem. 1994; 372, 137.

[72] Labuda J, Meister A, Glaser P, Werner G. Fresenius J. Anal. Chem. 1998; 360, 654.

Chapter 8

Halogen Bonding in Crystal Engineering

Xin Ding, Matti Tuikka and Matti Haukka

Additional information is available at the end of the chapter

1. Introduction

The structural features of a molecule are determined by the covalent bonds within the molecule. Modification of the structure requires the breaking and creation of covalent bonds. Similarly, the intrinsic reactivity of a molecule arises from the covalently bonded functional groups and active sites on a molecule. Again, changing the properties requires modifications of the covalent bonds. Even if the ability of a molecule to react is dependent on the covalent molecular structure, the reaction itself is typically launched by non-covalent intermolecular contacts. The reacting molecules need to "see each other" and they need to be brought close enough to each other for the reaction to take place. In other words, the non-covalent interactions provide the glue that brings and holds molecules together, thus making the intermolecular interplay possible. Supramolecular chemistry and processes such as self-assembly and crystallization are strongly guided by non-covalent interactions. These interactions cover a range of different types of intermolecular forces including Columbic interactions, hydrogen bonds, π-interactions, metallophilic interactions, agostic interactions, and halogen bonds. Building predesigned supramolecular entities and molecular assemblies imposes some requirements on intermolecular contacts. They must be sufficiently strong and they must have directional preferences. If these conditions are met it is possible to design molecular building blocks with suitable acceptor/and donor sites for different types of contacts and also to build molecular assemblies in a controlled way.

2. Crystal engineering

When the properties of a molecule are tailored in a conventional way by creating or removing covalently bound functional groups or active sites on a molecule, the physical or chemical properties obtained are solely the intrinsic properties of the molecule designed. An alternative approach to modifying the functionality of molecular material is to link molecular units together to create coordination polymers or extended molecular systems. In such systems the interactions between the building units give rise to new properties that do not exist in the building block molecules. This is the very essence of crystal engineering. Desiraju has stated that "crystal engineering is the rational design of functional molecular solids" and defines crystal engineering as "the understanding of intermolecular interactions in the context of crystal packing and in the utilization of such understanding in the design of new solids".[1] In other words, the goal is to create functional systems by assembling molecular units into extended molecular structures. Over the past few decades vast numbers of papers (Fig. 1) and textbooks have been published on this topic.[2–4]

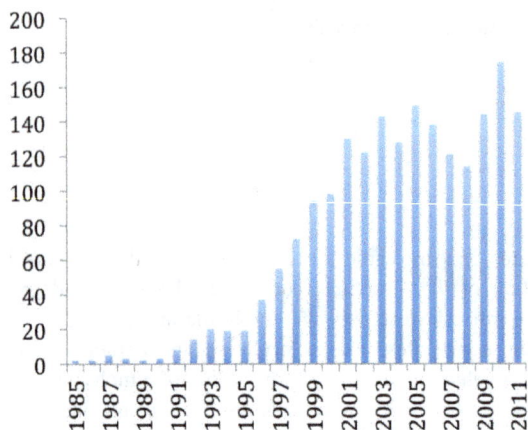

Figure 1. The number of crystal engineering publications since 1985 (ISI WoK, March 2012, topic ="crystal engineering"). These figures include only those publications whose topic includes "crystal engineering". The true number in this field including all related publications is much larger.

As mentioned above, bringing molecules together in a predictable way requires that the intermolecular forces are directional and strong enough to maintain a certain molecular architecture. Non-covalent interactions such as hydrogen bonds, halogen bonds, $\pi\cdots\pi$ interactions, metallophilic interactions, and agostic interactions all have directionality to at least some extent. Especially hydrogen bonds, halogen bonds and $\pi\cdots\pi$ interactions are relatively strong electrostatic forces with strong directionality. The bond energies of very strong hydrogen bonds range between approximately 65 and 170 kJ/mol, strong bonds between 15 and 65 kJ/mol, and weak hydrogen bonds around 15 kJ/mol or less.[5] The $\pi\cdots\pi$ interactions are somewhat weaker with interaction energies up to 50 kJ/mol.[6,7] The strength of the halogen bonds is comparable with the hydrogen bonds ranging between weak (ca. 5 kJ/mol) to strong (180 kJ/mol) contacts.[8] In addition to strength and directionality, the third requirement is that the intermolecular interactions should be selective. If the molecular building blocks contain different types of active sites the contacts must be predictable. Aakeröy et al have shown that even closely related interactions such as hydrogen bonds and halogen bonds can be used side-by-side in the same structure in a hierarcial way to build predictable molecular assemblies.[9,10] The challenge in this kind of combination is that the halogen bond donor (typically iodine or bromine) can interact not only with the halogen bond acceptor (electron-pair donor) but also with the hydrogen bond donor. An example of the coexistence of halogen bond and hydrogen bond is shown Fig. 2. In this example two-point N-H···O hydrogen bond contacts and one point I···N halogen bond contacts between the 2-aminopyrazin-1-ium and 2,3,5,6-tetrafluoro-4-iodobenzoate are used to build linear chain structure.[10]

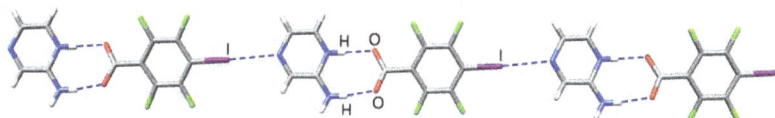

Figure 2. Halogen bond and hydrogen bond contacts in the linear chain assembly of 2-aminopyrazin-1-ium and 2,3,5,6-tetrafluoro-4-iodobenzoate.[10]

Building predictable assemblies and extended molecular systems is possible only after the very natures of the different types of interactions are understood. When the potential and limitations of these contacts are recognized, they serve as a versatile toolbox for crystal engineering. The following sections will focus on one of the "new" intermolecular contacts i.e. halogen bonds. In fact this is not a new discovery. The first observations about halogen bonds were published as early as 1863. This intermolecular force was, however, almost forgotten for years. But because of the interest in crystal engineering it was "rediscovered" and for the past decade it has become topic of growing interest.

3. Halogen bonds (XB)

The definition of halogen bond is not as well established as the definition of hydrogen bond although these interactions have a lot of similarities. Both contacts are electrostatic intermolecular interactions involving an electron donor and an electron acceptor. In hydrogen bonds D-H acts as a hydrogen bond donor, i.e., the electron acceptor. The hydrogen bond acceptors are, then, electron donors such as oxygen or nitrogen atoms (Fig. 3).

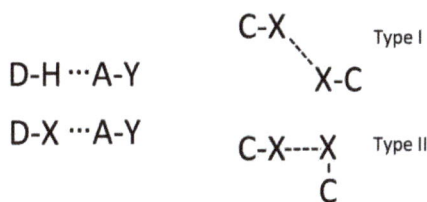

Figure 3. Comparison of the hydrogen bond (top left) and the halogen bond (bottom left). (D = donor, A = acceptor). Classification of the halogen bonds based on the geometry (right).

Because of the similarities involved, the same terminology has also been adapted in halogen bonds. The halogen in D-X acts as the halogen bond donor (electron acceptor). While electron donors such as nitrogen, oxygen, sulfur etc. act as the halogen bond acceptors (Fig.

3). The key to the halogen bonds is the polarizability of the halogen atom. Therefore, the strongest halogen bonds are formed by the most easily polarizable halogens, and the strength of the halogen bonds typically decreases in the order I > Br > Cl > F.

Halogen bonds are commonly defined as electrostatic interactions between Lewis acids (the halogen atom) and neutral or anionic Lewis bases and abbreviated as XB, where X refers to the halogen and B the Lewis base.[11] The strong directional preferences of a halogen bond arise from the tendency to maximize the main two directional attractive contributions to the interaction energy i.e. electrostatics and charge transfer. These, in turn, minimize the exchange repulsion that is also strongly directional. Optimizing the electrostatic and charge transfer aspects have been successfully used in designing of drugs, liquid crystals, organic semiconductors, magnetic materials, nonlinear optical materials, and templates for solid synthesis.[12–16] Conventionally, halogen bonds have been divided into two classes, TypeI and Type II (Fig. 3), based solely on the bonding geometry.[4] A few theories and concepts have been proposed for rationalizing the XB in greater detail. The most familiar one is the σ-hole theory. Other theories such as the lump-and-hole theory and the concept of amphoteric halogen bonds have been used to cover the "blind spots" in the σ-hole theory.

4. σ-hole theory

In most cases, the σ-hole theory has successfully explained the contradictory nature of halogen bonding. Conventionally covalently bonded halogens are seen as negatively charged entities. How, then, is it possible that they can participate in inter-atomic interactions as electron acceptors? In the σ-hole theory the σ-holes are defined as regions of positive electrostatic potential on the outer sides of halogen atoms, centered close to the extension of the halogen atoms' covalent bonds (Fig.4).[17]In general three factors determine the σ-hole's presence or absence and their magnitudes: a) the polarizability of the halogen atom, b) its electronegativity, and c) the electron-withdrawing power of the remainder D of the D-X molecule.[17] When the halogen is more polarizable and has lower electronegativity, the potential of the σ-hole can become more strongly positive. So the positivity of the σ-hole increases in the order F ‹ Cl ‹ Br ‹ I.

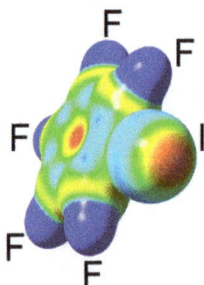

Figure 4. Regions of concentrated negative electrostatic potential (blue) and regions of depleted potential (red) on pentafluoroiodobenzene.

Apparently, the σ-hole determines the existence and, the strength of the halogen bonding. Since the σ-hole is located on the extension of the covalent bond along the D-X axis, it generates the directional preferences of the halogen bonding. When the halogen atom in D-X⋯A acts as the halogen bond donor, the D-X⋯A angle is close to 180° (Fig. 3 and Fig. 4). If the halogen acts as the halogen bond acceptor (electron donor), the angle is close to 90° because the electron density around the halogen atom is concentrated at an angle of 90° from the D-X bond (Fig. 3).

5. Lump-hole theory

The σ-Hole theory has satisfactorily explained most halogen bonding interactions, but it fails in some cases. For example, CH_3Cl can form a halogen bond with OCH_2, which is impossible according to the σ-hole theory because of the lack of a positive potential region around the Cl.[17,18] In lump-hole theory there are no true positive regions of the halogen bond donor. The charge density is, however, polarized and there are regions of charge depletion and charge concentration on the donor and acceptor. When these two interact, the areas with charge concentration on the halogen bond acceptors are interacting with the charge depleted areas of the halogen bond donor. Based on lump-hole theory, the participation of fluorine in halogen bonds can be also explained, since a true positive σ-hole is not needed.

6. Amphoteric character of halogen bonds

The amphoteric character of halogen bonding was first proposed by Nelyubina et al. in 2010.[19] They were unable to find a true σ-hole in halogen bonds between I_2 and I^-. Rather, the I in both I_2 and I^- had regions of electron density accumulation and depletion. When the I^- and I-I were interacting, both of them acted as donors and acceptors of electron density simultaneously. Nelyubina et al. defined this type of halogen bond as an 'amphoteric' interaction.

7. The effect of the halogen bond acceptor

When fine-tuning halogen bonding interactions it is, at least in principle, possible to modify the properties of both the halogen bond acceptor and the halogen bond donor. According to the σ-hole theory, the XB acceptor should be rich in negative electrostatic potential, or the acceptor is at least expected to be charge concentrated, which is required by lump-hole theory. Even according to the concept of amphoteric halogen bonds, the acceptor should be able to act as an electron donor. Probably the most commonly used halogen bonding acceptors are covalently bonded halogens and nitrogen atoms.[8,11,20–24] Metal halides, oxygen, sulfur, selenium and even silicon have, however, been reported to have the capacity to act as halogen bond acceptors with suitable donors.[25–31]

8. C-X/N/S/O…1, 2-diiodoterfluorobenzene systems

One of the most commonly used XB donor is 1,4-diiodoterfluorobenzene due to its strong positive σ-hole on the iodine atom.[28,32,33] Similarly, 1,3-diiodoterfluorobenzene has been widely used as the Lewis acid in halogen bonds.[22,34,35] In the present study, however, we focus first on the XB systems with the less commonly used isomer of diiodotetrafluorobenzene, i.e., 1,2-diiodoterfluorobenzene (1,2-TFIB).[33,34,36–39] We will use 1,2-TFIB as the "probe donor" to study the halogen bonding interactions with different type of XB acceptors and to elucidate how the halogen bonding acceptor affects the properties of halogen bonding.

Cauliez. et al, used thiocyanate anion as the XB donor to construct co-crystals with 1,2-TFIB (Fig. 5).[34] Halogen bonds were observed between the neutral iodinated species (XB donor) and both the S and the N end of the thiocyanate anions, demonstrating the bidentate nature of SCN⁻. Both C-I··N and C-I··S presented strong linearity, and relatively strong halogen bonding interactions. On the other hand, the C-S··I and C-N··I angles follow roughly the directions of the free electron pairs on the acceptor atoms, providing an example of how the electronic structure of the XB acceptor affects the overall geometry of the molecular assembly. In this particular structure, weaker C-I··F interactions were also observed. In these contacts the fluorine atoms of the halogenated benzene acted primarily as XB acceptors. In addition to the XBs mentioned above, series of F··π, I··π, π··π, and F··H hydrogen bond contacts contributed to the crystal structure of the co-crystals of thiocyanate and 1,2-TFIB.

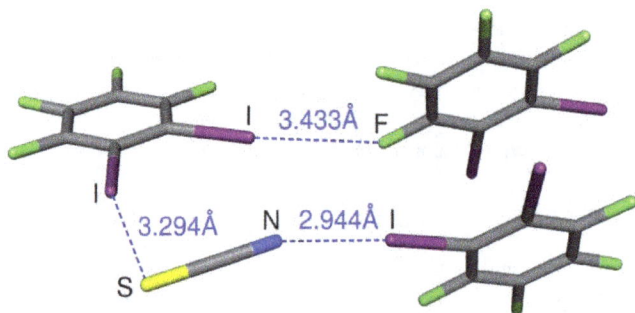

Figure 5. The halogen bond contacts in the co-crystal of the thiocyanate anion and 1,2-TFIB.[34]

A zigzag chain structure have also been obtained through C-I··N, and C-I··S halogen bonds in co-crystals of thiomorpholine (TMO) and 1,2-TFIB (Fig. 6).[33] Like thiosyanate, the thiomorpholine is able to act as a bidentate N, S halogen bond acceptor.[33,34] The co-crystal of TMO and 1,2-TFIB have been obtained by a simple mechanocemical synthesis i.e. by grinding the components together. The interesting feature of this stepwise co-crystallization process is that it is proposed to be guided by the competition of the strong and weak halogen bonds. The initially formed finite molecular assemblies are held together mainly by the stronger N··I bonds. These intermediates are then polymerized into infinite chains by cross-linking through weaker S··I interactions.

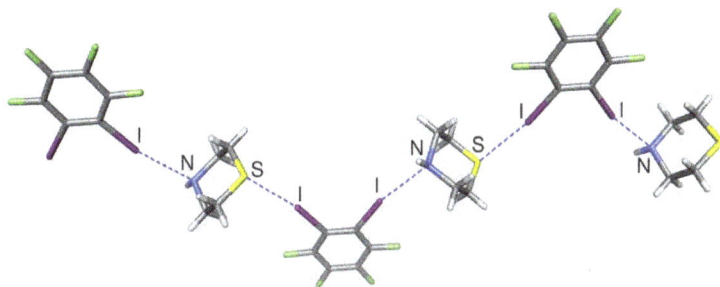

Figure 6. The zigzag chain of TMO/1,2-TFIB co-crystals.[33]

Yet another zigzag chain structure has been obtained by co-crystallizing 1,2-TFIB with phenazine through C-I···N halogen bonds (Fig. 7).[36] Here the C-I···N halogen bonding is slightly weaker than that of the example shown in Fig. 6. This was most probably caused by the steric effect of the XB donor. It demonstrates that the overall geometry of the XB acceptor has an effect on the halogen bonding. The directionality of XB is clearly shown in the example on Fig. 7. The angle of the C-I···N is 169°, which is almost linear.

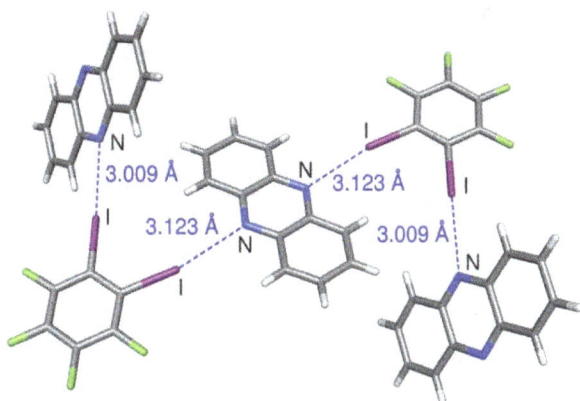

Figure 7. C-I···N halogen bonding in the zigzag chain structure of the co-crystal of 1, 2-TFIB with phenazine.[36]

The co-crystals of 4,4'-bipyridine and (1,2-TFIB) (Fig. 8) provide another example of C-I···N contacts. In this structure the C-I···N bonds linked the bipyridine (XB acceptor) and 1,2-TFIB together into two independent and almost perpendicular wave-like chains.[39]

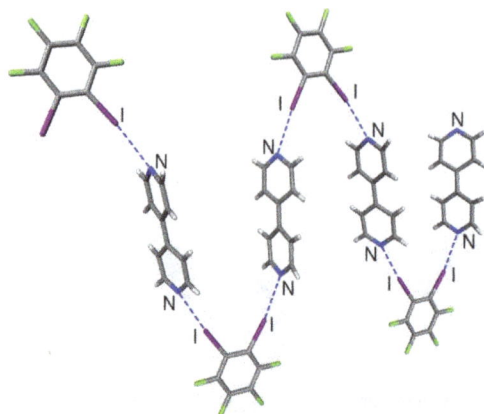

Figure 8. Co-crystals 4,4'-bipyridine and (1,2-TFIB). The I⋯N distances range from 2.909 to 2.964 Å.[39]

The co-crystals of 2-mercapto-1-methylimidazole (mmim) and 1,2-TFIB show nicely the bonding preferences of XB (Fig. 9).[38] The mmim molecule and 1,2-TFIB form a complex, in which N-H⋯S bound imidazole dimers are connected through C-I⋯S interactions to a pair of 1,2-TFIB molecules, forming infinite chains. The C-I⋯S bonds are different in strength (the C-I⋯S distance of the weaker one was 3.843 Å, while the stronger one was 3.291 Å) and the bonds involving sulphur can be defined as trifurcated bonds. They consisted of two halogen bonds and one hydrogen bonds. The sulfur acts as the electron donor for both bonding types. In principle, the iodine could be also a hydrogen bond acceptor. However, it is solely devoted to the halogen bond, while the hydrogen bonds are formed only between the sulfur and NH of the imidazole ring.

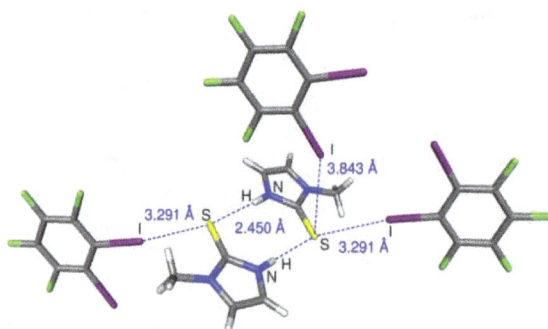

Figure 9. The halogen and hydrogen bonds in the structure of (mmim)·(1,2-TFIB).[38]

As mentioned previously, oxygen can also be used as an XB acceptor to construct supramolecular structures. Co-crystals of the nitroxide 1,1,3,3-tetramethylisoindolin-2-

yloxyl (TMIO) and 1,2-TFIB are formed under standard sublimation conditions.[37] The formed 2:2 cyclic tetramer structure (Fig. 10), (TMIO)$_2$·(1,2-TFIB)$_2$, showed that each nitroxide oxygen atom, when serving as the XB acceptor, set up bifurcated halogen bonding with two iodine atoms from two 1,2-TFIB molecules, respectively. Again, the N-O···I angle follow the direction of the free electron pairs on the oxygen atoms, thus encouraging the tetrameric assembly of molecules. The O···I contacts in this motif were clearly shorter than the van der Waals contacts (down to 81.2%-83.1%), with strong directionality (C-I···O angles range from 170.30°-179.2°).

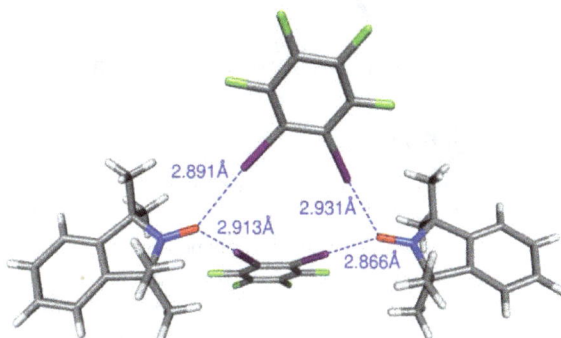

Figure 10. The tetrameric unit (TMIO)$_2$·(1,2-TFIB)$_2$ (red = oxygen, purple iodine).[37]

As the examples above show, the 1,2-TFIB can be used as the XB donor with various acceptors. If, however, there are no other acceptors available, the "amphoteric" nature of 1,2-TFIB is revealed. When the 1,2-TFIB is crystallized from methanol, a structure with series of weak XB bonds can be obtained (Fig. 11).[†] The iodines form XB contacts, functioning as both the donors and the acceptors. The I···I distances are relatively long, ranging from 3.258 Å to 3.740 Å. Nevertheless, the distances are less than the sum of the van der Waals radii, and the directionality support the existence of the halogen bonds. The I···I contacts resulted in a zigzag structure that is further expanded through the F···F, F···I, and F···π halogen bonds. The F···F and F···I contacts are weak with long distances consisting of 2.783-2.924 Å and 3.258 Å for F···F and F···I, respectively. It should be noted that, to judge from the C-F···F angle (147.3°), the F···F contacts showed some amphoteric character. In the case of C-I···F, the fluorine atom behaves more clearly as the halogen bond acceptor, due to the existence of the negative lateral sides of the fluorine atom caused by the aspherical charge density distribution.[40]

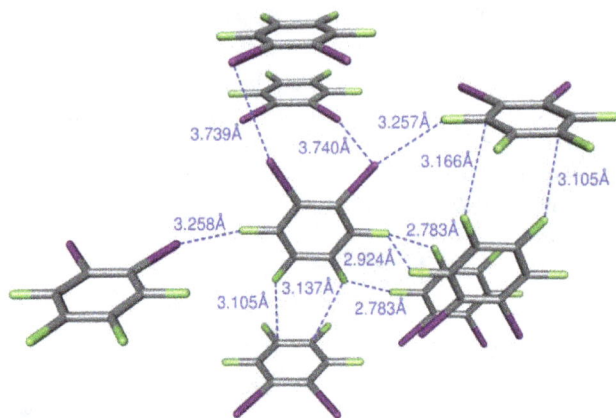

Figure 11. Halogen bonding interactions in the crystal structure of 1,2-TFIB.[†]

9. X/Se···X system (X=Cl, Br, I)

As early as in 1984, a study of Cl···Cl halogen bond in crystal structures of six dichlorophenols was carried out by Thomas and Desiraju.[41] This study was extended in 2011 by Mukherjee and Desiraju to 3,4,5-trichlorophenol and 2,3,4-trichlorophenol.[42] In the crystal structure of 3,4,5-trichlorophenol (Fig. 12), one Cl atom forms bifurcated halogen bonds with another chlorine and oxygen, respectively. This Cl, however, functions as XB acceptor with the other Cl, while at the same time, it serves as the XB donor for the oxygen atom, thus showing its dual nature.

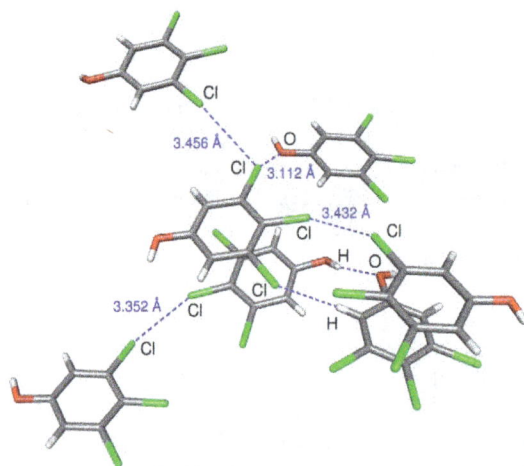

Figure 12. Halogen and hydrogen bonding in the crystal structure of 3,4,5-trichlorophenol.[42]

Crystals with a zigzag sheet packing structure have been obtained by crystallizing 1-butyl-4, 5-dibromo-3-methylimidazolium iodide (Fig. 13).[43] The two bromine atoms of the cation act as the XB donors, while the iodine anion is the XB acceptor. The crystal structure shows that the iodide anions in the c-axis direction are positioned either at the top or the bottom of the zigzag structure, suggesting that the size of the halide anion has a strong effect on the zigzag sheet formation.[43]

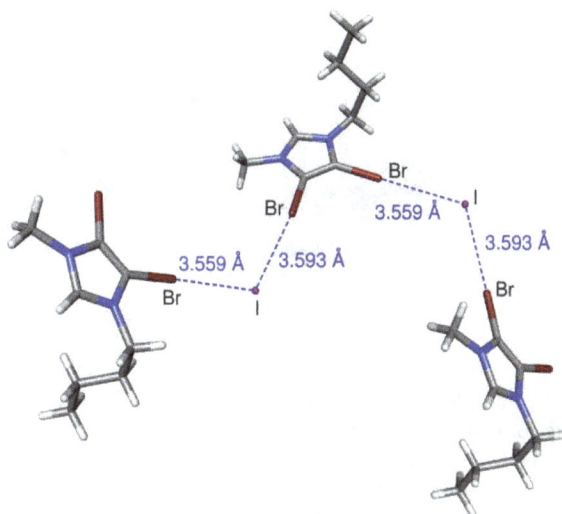

Figure 13. Halogen bonding interactions in crystal structure of 1-butyl-4,5-dibromo-3-methylimidazolium iodide.[43]

The structure of co-crystal of 1,2-diiodoimidazole with 1,3,4-triiodoimidazole (Fig. 14) was studied by our group.‡ In this structure the two 1,2-diiodoimidzol molecules are linked by a I···I halogen bonding interaction (3.916Å), forming a dimeric unit. This unit is further connected with other 1,3,4-triiodoimidzole molecules, expanding the structure along crystallographic directions a, and b. The N-H···N hydrogen bonds expand the structure further in direction of c axis, constructing a 3D network. The bonding preferences of different interactions can be clearly seen in this structure. Halogen bonds are formed only between the iodine atoms, and each iodine atom of diiodoimidzol is trifurcated. In addition, it demonstrated the dual character of iodines, serving both as XB donors and as acceptors.

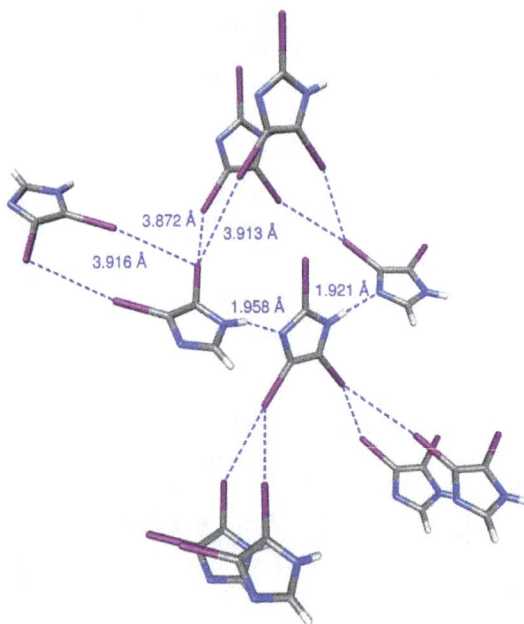

Figure 14. Hydrogen and halogen bonds in the co-crystal of 1,2-diiodoimidazol and 1,3,4-triiodoimidazol.‡

Halogen bonds involving selenium as the XB acceptor have not been widely studied.[29] The chain structure formed by connecting di-ter-butyliodophosphane selenide molecules through Se⋯I halogen bonding is one example of such a system.[29] In this structure the Se⋯I distance was found to be only slightly shorter than the sum of the van der Waals radii of selenium and iodine. Another example of Se⋯I interaction can be found in the crystal structure of iodoisopropylphosphane selenide.[29] In both structures, selenium is also involved in Se⋯H hydrogen bonds, providing another example of the interplay between the closely related electrostatic interactions.

10. M-X/N/NCS/CN…X systems

Simple metal bound ligands, capable of donating electrons, can also be used as XB acceptors in the construction of supramolecular structures.[44] Metal complexes are of particular interest because of the possibilities of using halogen bonds as a tool in the modification of the redox, magnetic, optical, and chemical reactivity of the metal complexes.[44–46]

The palladium pincer complex {2,6-bis[(di-t-butylphosphino)methyl]-phenyl}palladium (PCPPd) halides, PCPPdX (X=Cl, Br, or I) have been studied by Johnson and Rissanen as the XB acceptor in systems with I₂ as XB donor.[47] The all three crystal structures have similar basic features (Fig. 15). However, the halogen bond strength was found to increase in the order Cl ‹ Br ‹ I, suggesting that the XB interactions are mainly electrostatic as expected.[47] The other intermolecular contacts were relatively weak. It is, however, worth mentioning that in PCPPdI·I₂ the Pd···I-I···π interaction also appeared to result in the formation of a chain-like structure. This was not observed in the other two structures.

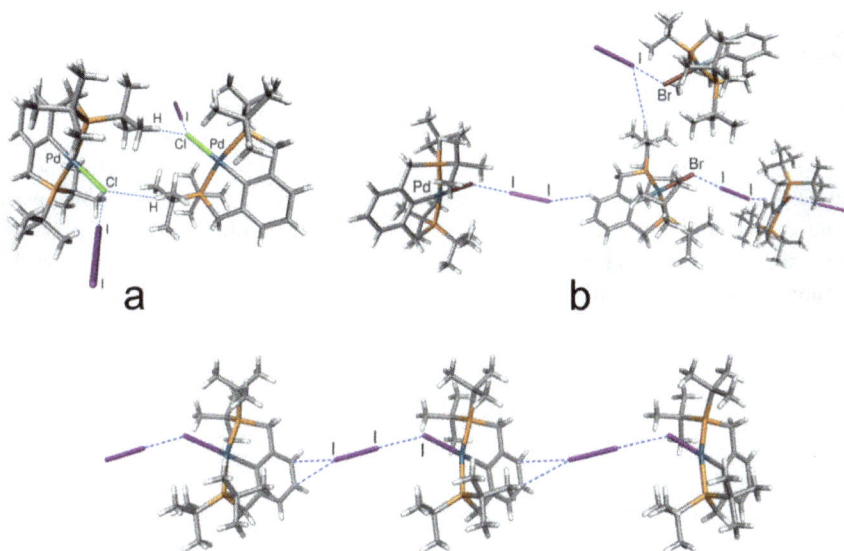

Figure 15. Crystal structures with halogen bonds in PCPPdX···I₂ (X=Cl, Br, or I). (a) PCPPdCl···I interactions, (b) PCPPdBr···I interactions, (c) PCPPdI···I interactions.[47]

The bonding preferences of halogen and hydrogen bonds can also be found among the assemblies of metal complexes. The co-crystals of [RuI₂(H₂dcbpy)(CO)₂] (H₂dcbpy = 4,4′-dicarboxylic acid-2,2′-bipyridine), I₂ and methanol is an example of such system (Fig. 16).[48] The two [RuI₂(H₂dcbpy)(CO)₂] complexes are held together strongly by the hydrogen bonds between the carboxylic acid groups. The iodide ligands bonded to the ruthenium centers are involved only in halogen bond, thus extending the structure into a chain of metal complexes. The halide ligands are linked by halogen bonds through two I₂ molecules. It is worth noticing that the halogen bonds between the I₂ molecules are bifurcated, and the solvent molecule is supporting the structure via hydrogen bond. The I-I···I bond angle was nearly linear (167°) for the first I₂, but due to the bifurcated nature only 137.8° for the second I···I₂ contact, which differs from the conventional XB bond angles.

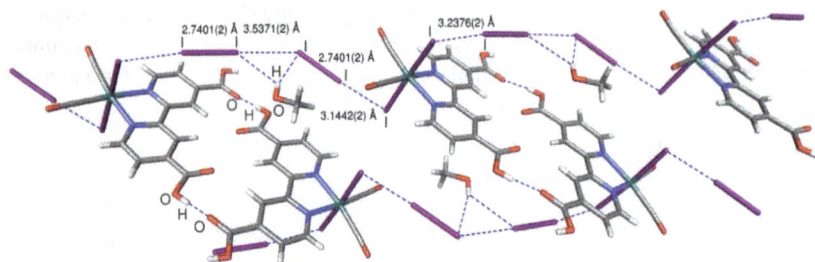

Figure 16. Halogen and hydrogen bonds in the structure of [RuI₂(H₂dcbpy)(CO)₂]·I₂.[48]

Any ligand possessing a free electron pair can be seen as a potential halogen bond acceptor. The N-bound thiocyanate in *cis*-diisothiocyanato-bis(2,2′-bipyridyl-4,4′-dicarboxylato)ruthenium(II) provides an example of such a ligand. The structure of the [RuI₂(H₂dcbpy)(CO)₂]·2I₂(Fig. 17) adduct have been obtained at room temperature by mixing I₂ and the complex in methanol.[49] In this structure, the sulfur atom of one of the thiocyanate ligands forms bifurcated halogen bonds with two I₂ molecules. Based on the distances, these bifurcated bonds are weaker than the non-bifurcated one.

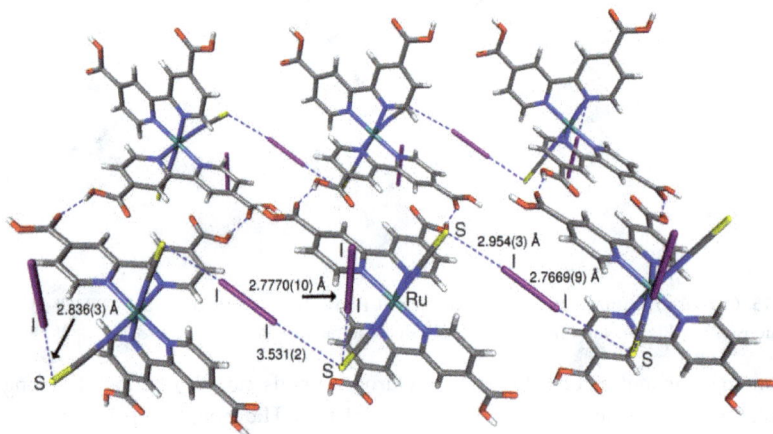

Figure 17. Halogen bonding interactions in [RuI₂(H₂dcbpy)(CO)₂]·2I₂.[49]

Ormond-Prout, Smart, and Brammer proposed that halogen bonds can be used to predict and control the process of self-assembly and to fine-tune the electronic properties of cyanometallates.[50] To confirm this assumption they synthesized two types of halopyridium hexacyanometallate salts, (3-XpyMe)₃[M(CN)₆] and (3, 5-X₂pyMe)₃[M(CN)₆] (X=Cr, Fe, Co).[50] The authors harvested a total of ten crystals, and found out that five out of each family of compounds were isostructural, while other structures were the solvates, (3-IpyMe)₃[Fe(CN)₆]·2MeCN(2·2MeCN) and (3,5-Br₂pyMe)₃[Cr(Cr(CN)₆]·(10·4H₂O). The halogen bonding distances in these structures were shorter than the sum of the van der Waals radii. In the case of (3,5-Br$_2$pyMe)$_3$[Cr(Cr(CN)$_6$]·(10·4H$_2$O), a weak additional C-Br···O halogen bond was found, which can be attributed to the competition between the halogen bonding and the O-H···N hydrogen bonding. The close-to-linear geometry of the CN···X halogen bonds found in the all structures suggests that these interactions predominantly involved the exo lone pair of nitrogen atom. However, the structures contained another type of halogen bonds with less linear CN···X contacts (CN···X < 105°). Such angles indicate that the triple bond between the C and N contributes to the halogen bonding interaction leading to X···π contact (Fig. 18). In this series the strength of the halogen bonds was found to be dependent on the metal center (Cr < Fe < Co). This is a good example of how the metal center can be used for modification of halogen bonds. In this particular example, the primary reason for the different behavior of the different metals has been attributed to the metal-cyanide π-back-donation.[50]

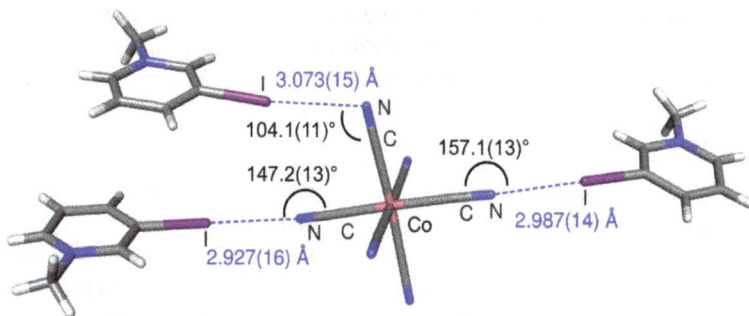

Figure 18. Halogen bonding contacts in (3-IpyMe)$_3$[Fe(CN)$_6$]·2MeCN.[50]

11. Hydride-halogen systems

In addition to the more conventional electron donors, the hydride, R-H$^{\delta-}$, has also been proposed as potential XB acceptors.[51] This type of halogen bonds, R-H$^{\delta-}$···XR$_x$, has been investigated computationally by analyzing a series of model systems. The results indicate that H$^{\delta-}$ is a potential electron donor for halogen bonds.[51–53] The halogen bonding interaction between LiH or HBeH and either XCF$_3$ or XCCH (X= F, Cl, Br, I) has been studied with high-level quantum mechanical calculations, quantum theory of atoms in molecules (QTAIM), and natural bond orbital (NBO) methods. The most important finding of these studies has been that the hydride-halogen bond formation causes the elongation of R$^{\delta+}$-H$^{\delta-}$ bond due to the involvement of hydride in the halogen bond. It has been suggested that the interaction is inductive in nature, and the formation of hydride-halogen bond results in the charge transfer from the hydride to a halogen-donor molecule.[51]

12. Modification of the halogen bond donor

The modification of the halogen bonds acceptor molecule for halogen bonding has been reviewed in many papers. As the above examples show a variety of acceptors can be used for halogen bonds. In general, it can be said that a good halogen bond acceptor is a strong electron donor. Much less have been written about the modification of the XB donor. [21,54]

Some of the strongest halogen bond donors are dihalogen molecules, which form strong halogen bonds.[55] This is of course due to the polarizing effect of the other halogen atom. All dihalogen molecules can act as halogen bond donors and the order of the XB bond strength follows the order I_2> Br_2> Cl_2> F_2. This is again the result of the polarizability of the halogens increasing in the same series.[20] The bond strength can be increased further by substituting the second halogen atom in dihalogens with fluorine, which polarizes the other halogen even more strongly. [56,57] The lighter dihalogens are volatile and so the crystal structures of such systems are rather rare. However, some of these structures have been characterized in gas phase by rotational spectroscopy.[58] The results indicate that for Cl_2, ClBr, ClF and ICl the covalent halogen-halogen bond strength increase in the order Cl_2 < BrCl < ClF < ICl. When combined with the known crystallographic data the increasing strength of the halogen bond donors for dihalogens can be given: F_2< Cl_2< Br_2< I_2< IBr < ICl. This order seems to be independent on the halogen bond acceptor.[54]

Polyhalides is another well known group of XB donors.[59] In these systems the halogen bonding typically occurs solely between the polyhalogenides and do not include other molecules. This is especially true with the polyiodides.[59] The polyiodide networks are often complicated three-dimensional networks, layers, or chains. The properties of these compounds have been intensively studied. There are even examples of systems where some of the iodines can be released into solution without breaking the crystal structure.[60] The removal of iodines have then been used to change the nonlinear optical properties of the compound.[9]

In the polyhalides iodine and bromine can often act both as halogen bond donors and as acceptors, and occasionally, the same atom can act both as acceptor and donor.[59] Amphoteric halogen bonds have also been found within polyhalide networks, though they are not very common.[8] In addition to the homonuclear polyhalides, mixed polyhalides are also known. The most common type is the mixed trihalide.[61,62]

Fig. 19 shows the typical features of the polyiodides. In this example, there are two crystallographically different I_3^- units, one of which has two nearly identical I-I bonds. The second I_3^- has unequal I-I bonds and it is closer to a I_2-I^- motif. The two I_3^- units are then linked via I_2 molecule. This is a typical example of a polyiodide structure. In this particular network the I_2 molecule acts as a halogen bond donor and the I_3^- units act as acceptors.[12]

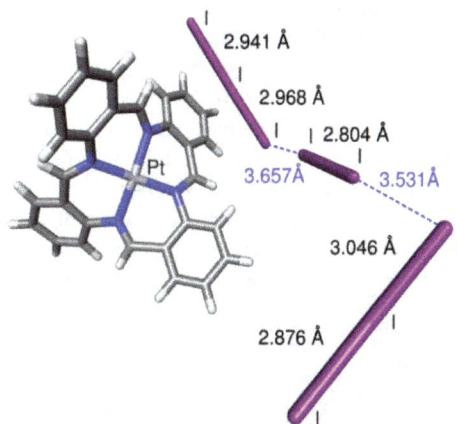

Figure 19. An example of a polyiodide network of $(C_{28}H_{20}N_4\,Pt)^{2+}\,(I_8)^{2-}$.[63]

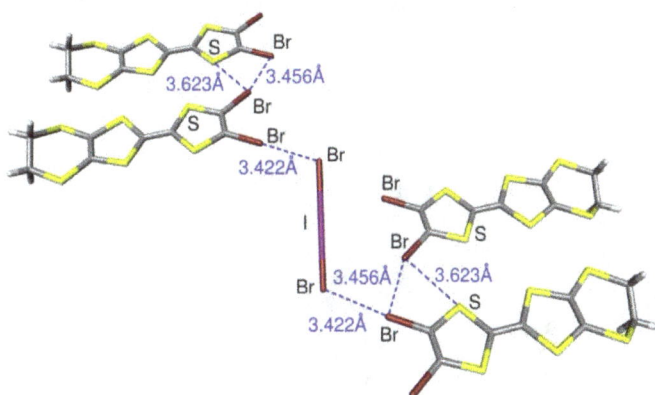

Figure 20. The halogen bonding in $(C_8H_4Br_2S_6)_2\,IBr_2$.[64]

Fig. 20 provides an example of a structure containing a mixed trihalide. The structure contains halogen bonds between the trihalide and two dibromo tetrathiofulvalene molecules. Additionally, there are also bifurcated halogen bonds between the bromine and sulfur atoms of the tetrathiofulvalenes. The very similar halogen bond distances between the trihalide (Br-I-Br) and Br-atoms of tetrathiofulvalene indicate that the bond strengths are nearly identical for both of the Br···Br contacts. In this case the Br-atoms dibromo tetrathiofulvalene act as the halogen bond donors and the trihalide as the acceptor.[64]

Organic compounds containing a C-X bond are often relatively easy to modify, which makes them attractive halogen bond donors. As discussed earlier, among the most commonly used

XB donors are fluorinated iodobenzenes.[22,47,65] There is obviously a large number of halogen containing organic compounds that could be used as halogen bonding donors. For the purpose of crystal engineering, the interesting parameters of these compounds are the geometry and expected strength of the halogen bonds. If the halogen atom is only singly bonded to a carbon atom, the formed sigma hole will be pointing in the opposite direction of that bond.[22] Hence, the direction of the halogen bond is clear and easy to predict. However, controlling the strength of the halogen bond donor requires further information. Numerous studies of this topic have been published, especially on aromatic halogen bond donors.[24,55–57,66] On the basis of both the existing experimental and theoretical studies, it can be stated that electron withdrawing substituents increase the strength of the halogen bond donor, while the electron-donating groups reduce it.[24,55,66]

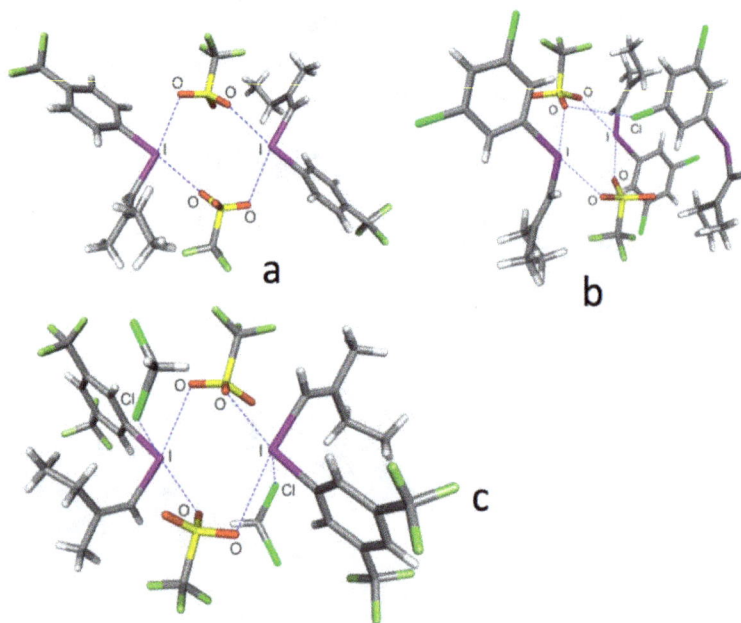

Figure 21. Halogen bonding of the iodine atoms for $(C_{12}H_{13}F_3\ I^+)(CF_3O_3S^-)$ (a); $(C_{11}H_{12}Cl_2I^+)(CF_3O_3\ S^-)$ (b) and $2(C_{13}H_{12}F_6I^+)\ 2(CF_3O_3S^-)\ CH_2Cl_2$ (c) [67]

In Fig. 21 there are three structures each of which contains the same basic building blocks of the alkenyl(aryl)iodonium trifluoromethanesulfonate salts, while one of them also contains dichloromethane. From the point of view of halogen bonds, these provide a useful illustration of the effects that the electron withdrawing groups have on the halogen bonding donor. In the first compound (a) there is one –CF$_3$ substituent on the aromatic ring. The halogen bonds are formed between the iodine of alkenyl(aryl)iodonium cation and oxygen atoms of the triflate. The I-O distances are 2.910 Å and 2.991Å. In the second structure (b) there are two chlorine substituents on the aromatic ring, which makes it more electron-deficient than the previous one.

There are now two crystallographically independent dimers (only one is shown in Fig. 21.) that are involved in similar type of halogen bonding between the iodines and triflates. However, the iodine-triflate distances are different. The I···O distances are 2.848 Å and 2.802 Å for the first dimer and 2.832 Å and 2.850 Å for the second. The halogen bond distances for the structure (a) are clearly shorter compared to the distances of the (b) structure as one might expect based on the electron withdrawing substituents. The third compound (c) in Fig. 21 contains two –CF₃ substituents on the aromatic ring, making it the most electron deficient of the three. Again there are two crystallographically independent dimers (only one is shown if Fig. 21) that have slightly different geometries. Despite of this the basic halogen bonding geometry involving the iodine atoms is similar with slightly different I···O distances (2.767Å and 2.985Å for the first dimer and 2.881Å and 2.893Å for the second). Now the message obtained from the halogen bond distances is not so obvious. Although the shortest distance in (c) is clearly the shortest of them all, the variation of the distances is large. This is a useful reminder that the final solid state structure is a result of several competing interactions and conclusions based only on distances is often an oversimplification and may be misleading.[67]

In most cases metal bound halogens act as halogen bond acceptors.[44,68] There are some examples where the interaction seems to be more amphoteric, but these are relatively rare cases.[69–71] In general, using metal centers to form synthons for halogen bonding networks can be beneficial, because they readily permit the formation of well defined geometries. In addition, by changing the oxidation state of the metal, the geometry and chemical properties of the system can also be changed.[72] Metal compounds can also possess interesting magnetic and luminescent properties.[46,73]

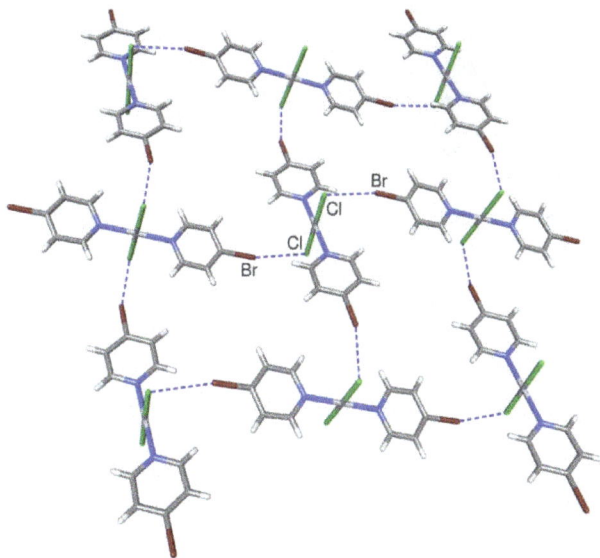

Figure 22. The halogen bonding network of PtCl₂(C₅NBrH₄)₂.[70]

In Fig. 22 there is an example of a network consisting of $PtCl_2(NC_5H_4Br)_2$ linked together via halogen bonds. In addition to the halogen bonds, the two dimensional layers consist of weak hydrogen bonds and π-π stacking interactions. The structure shown is a good example of a network structure of a metal complex formed by halogen bonding.[70]

13. Future perspectives

Even if the very essence of crystal engineering is to produce functional materials, a large number of studies and papers in this field are still devoted solely to the structural aspects of the molecular assemblies and frameworks. The same is true with halogen bonding. The latter is understandable since establishing the whole concept of halogen bonding has required (and still requires) a considerable amount of work. Nevertheless, examples already exist of the utilization of halogen bonding in the production of functional materials. The role of halogen bonding has been investigated in the context of the inhibition of the human protein kinase CK2α.[74]It has also been used for selective recognition of halide anions and employed in host-guest systems.[23] There are examples of the use of halogen bonding for controlling the luminescent properties of Au_2-Ag_2 clusters and the birefringence properties of chains of square planar Au complexes[45,75]. There are also examples of the utilization of halogen bonding in catalysis.[76] In the future the number of these types of applications is expected to grow rapidly. All of this means that halogen bonding is in the process of being transformed from a strange solid-state phenomenon to a versatile tool in the hands of crystal engineers.

Author details

Xin Ding, Matti Tuikka and Matti Haukka
Department of Chemistry, University of Eastern Finland, Joensuu Campus, Joensuu, Finland

Acknowledgement

Financial support provided by the Academy of Finland (project no.139571) is gratefully acknowledged. Molecular graphics was created with the UCSF Chimera package. Chimera is developed by the Resource for Biocomputing, Visualization, and Informatics at the University of California, San Francisco, with support from the National Institutes of Health (National Center for Research Resources grant 2P41RR001081, National Institute of General Medical Sciences grant 9P41GM103311).

14. Footnotes

†Crystal data for 1,2-TFIB: $C_6F_4I_2$, M = 401.86, brown block, $0.43 \times 0.32 \times 0.20$ mm^3, monoclinic, space group $P2_1/n$ (No. 14), a = 10.606(4), b = 5.770(2), c = 14.548(5) Å, β = 110.344(4)°, V = 834.7(5) Å^3, Z = 4, D_c = 3.198 g/cm^3, F_{000} = 712, Bruker SMART APEX II CCD, MoKα radiation, λ = 0.71073 Å, T = 100(2)K, $2\theta_{max}$ = 70.0°, 18326 reflections collected, 3673 unique (R_{int} = 0.0353). Final $GooF$ = 1.105, $R1$ = 0.0184, $wR2$ = 0.0375, R indices based on 3154

reflections with I >2sigma(I) (refinement on F^2), 109 parameters, 0 restraints. Lp and absorption corrections applied, μ = 7.540 mm^{-1}. The crystal was obtained in methanol by slow evaporation. CCDC-875313 contain the supplementary crystallographic data for this structure. It can be obtained free of charge from The Cambridge Crystallographic Data Center via www.ccdc.cam.ac.uk/data_request/cif.

‡ Crystal data for co-crystals of 1,2-diiodoimidazole and 1,3,4.-diidoimidazole: $C_9H_5I_7N_6$, M = 1085.49, colourless plate, 0.19 × 0.15 × 0.08 mm^3, monoclinic, space group $P2_1/m$ (No. 11), a = 4.27080(10), b = 27.9241(6), c = 8.8926(2) Å, β = 101.6110(10)°, V = 1038.81(4) Å^3, Z = 2, D_c = 3.470 g/cm^3, F_{000} = 944, Bruker SMART APEX II CCD, MoKα radiation, λ = 0.71073 Å, T = 100(2)K, $2\theta_{max}$ = 71.2°, 17788 reflections collected, 4803 unique (R$_{int}$ = 0.0353). Final $GooF$ = 1.124, $R1$ = 0.0433, $wR2$ = 0.0785, R indices based on 3855 reflections with I >2sigma(I). (refinement on F^2), 103 parameters, 0 restraints. Lp and absorption corrections applied, μ = 10.461 mm^{-1}. The crystal was obtained in methanol by slow evaporation. CCDC-875314 contain the supplementary crystallographic data for this structure. It can be obtained free of charge from The Cambridge Crystallographic Data Center via www.ccdc.cam.ac.uk/data_request/cif.

15. References

[1]   Desiraju G.R. (2010) Crystal engineering: A brief overview. Journal of chemical sciences. 122(5):667–75.

[2]   Weber E., editor (1998) Design of Organic Solids, Topics in Current Chemistry. 1st ed. Springer;

[3]   Tiekink E.R.T., Vittal J., editors (2006) Frontiers in Crystal Engineering. 1st ed. Wiley;

[4]   Desiraju G.R. (1989) Crystal Engineering. 1st ed. Elsevier Science;

[5]   Desiraju G.R., Steiner T. (2001) The Weak Hydrogen Bond: In Structural Chemistry and Biology, IUCr Monographs on Crystallography. Oxford University Press;

[6]   Steed J.W., Atwood J.L. (2009) Supramolecular Chemistry. John Wiley & Sons;

[7]   Chen X., Tong M. Supramolecular Interactions in Directing and Sustaining Coordination Molecular Architectures. In: Tiekink ERT, Vittal JJ, editors. Frontiers in Crystal Engineering. John Wiley & Sons, Ltd; p. 219–63.

[8]   Metrangolo P., Resnati G., Pilati T., Liantonio R., Meyer F. (2007) Engineering functional materials by halogen bonding. Journal of Polymer Science Part A: Polymer Chemistry. 45(1):1–15.

[9]   Aakeröy C.B., Fasulo M., Schultheiss N., Desper J., Moore C. (2007) Structural Competition between Hydrogen Bonds and Halogen Bonds. J. Am. Chem. Soc. 129(45):13772–3.

[10] Aakeröy C.B., Chopade P.D., Ganser C., Desper J. (2011) Facile synthesis and supramolecular chemistry of hydrogen bond/halogen bond-driven multi-tasking tectons. Chem. Commun. 47(16):4688–90.

[11] Metrangolo P., Neukirch H., Pilati T., Resnati G. (2005) Halogen Bonding Based Recognition Processes: A World Parallel to Hydrogen Bonding†. Acc. Chem. Res. 38(5):386–95.

[12] Bruce D.W., Metrangolo P., Meyer F., Pilati T., Präsang C., Resnati G., et al. (2010) Structure–Function Relationships in Liquid-Crystalline Halogen-Bonded Complexes. Chemistry - A European Journal. 16(31):9511–24.

[13] Fourmigué M., Batail P. (2004) Activation of Hydrogen- and Halogen-Bonding Interactions in Tetrathiafulvalene-Based Crystalline Molecular Conductors. Chem. Rev. 104(11):5379–418.

[14] Hanson G.R., Jensen P., McMurtrie J., Rintoul L., Micallef A.S. (2009) Halogen Bonding between an Isoindoline Nitroxide and 1,4-Diiodotetrafluorobenzene: New Tools and Tectons for Self-Assembling Organic Spin Systems. Chemistry - A European Journal. 15(16):4156–64.

[15] Cariati E., Cavallo G., Forni A., Leem G., Metrangolo P., Meyer F., et al. (2011) Self-Complementary Nonlinear Optical-Phores Targeted to Halogen Bond-Driven Self-Assembly of Electro-Optic Materials. Crystal Growth & Design. 11(12):5642–8.

[16] Sun A., Lauher J.W., Goroff N.S. (2006) Preparation of Poly(diiododiacetylene), an Ordered Conjugated Polymer of Carbon and Iodine. Science. 312(5776):1030–4.

[17] Politzer P., Murray J.S., Clark T. (2010) Halogen bonding: an electrostatically-driven highly directional noncovalent interaction. Phys. Chem. Chem. Phys. 12(28):7748–57.

[18] Eskandari K., Zariny H. (2010) Halogen Bonding: A Lum-Hole Interaction. Chemical Physics Letters. 492:9–13.

[19] Nelyubina Y.V., Antipin M.Y., Dunin D.S., Kotov V.Y., Lyssenko K.A. (2010) Unexpected "amphoteric" character of the halogen bond: the charge density study of the co-crystal of N-methylpyrazine iodide with I_2. Chem. Commun. 46(29):5325.

[20] Legon A.C. (2010) The halogen bond: an interim perspective. Phys. Chem. Chem. Phys. 12(28):7736.

[21] Politzer P., Lane P., Concha M.C., Ma Y., Murray J.S. (2006) An overview of halogen bonding. J Mol Model. 13(2):305–11.

[22] Roper L.C., Präsang C., Kozhevnikov V.N., Whitwood A.C., Karadakov P.B., Bruce D.W. (2010) Experimental and Theoretical Study of Halogen-Bonded Complexes of DMAP with Di- and Triiodofluorobenzenes. A Complex with a Very Short N⋯I Halogen Bond. Crystal Growth & Design. 10(8):3710–20.

[23] Sarwar M.G., Dragisic B., Sagoo S., Taylor M.S. (2010) A Tridentate Halogen-Bonding Receptor for Tight Binding of Halide Anions. Angewandte Chemie International Edition. 49(9):1674–7.

[24] Tsuzuki S., Wakisaka A., Ono T., Sonoda T. (2012) Magnitude and Origin of the Attraction and Directionality of the Halogen Bonds of the Complexes of C_6F_5X and C_6H_5X (X=I, Br, Cl and F) with Pyridine. Chemistry - A European Journal. 18(3):951–60.

[25] Alkorta I., Sanchez-Sanz G., Elguero J., Del Bene J.E. (2012) FCl:PCX Complexes: Old and New Types of Halogen Bonds. J. Phys. Chem. A. 116(9):2300–8.

[26] Brezgunova M., Shin K.-S., Auban-Senzier P., Jeannin O., Fourmigué M. (2010) Combining halogen bonding and chirality in a two-dimensional organic metal (EDT-TTF-I2)2(D-camphorsulfonate)·H2O. Chem. Commun. 46(22):3926.

[27] Cametti M., Raatikainen K., Metrangolo P., Pilati T., Terraneo G., Resnati G. 2-Iodo-imidazolium receptor binds oxoanions via charge-assisted halogen bonding. Org. Biomol. Chem. 10(7):1329–33.

[28] Cinčić D., Friščić T., Jones W. (2008) Isostructural Materials Achieved by Using Structurally Equivalent Donors and Acceptors in Halogen-Bonded Cocrystals. Chemistry - A European Journal. 14(2):747–53.

[29] Jeske J., du Mont W., Jones P.G. (1999) Iodophosphane Selenides: Building Blocks for Supramolecular Soft–Soft Chain, Helix, and Base-Pair Arrays. Chemistry - A European Journal. 5(1):385–9.

[30] Madzhidov T.I., Chmutova G.A., Martín Pendás A. (2011) The Nature of the Interaction of Organoselenium Molecules with Diiodine. J. Phys. Chem. A. 115(35):10069–77.

[31] Politzer P., Murray J. (2012) Halogen bonding and beyond: factors influencing the nature of CN–R and SiN–R complexes with F–Cl and Cl2 Theoretical Chemistry Accounts: Theory, Computation, and Modeling (Theoretica Chimica Acta). 131(2):1–10.

[32] Aakeröy C.B., Desper J., Helfrich B.A., Metrangolo P., Pilati T., Resnati G., et al. (2007) Combining halogen bonds and hydrogen bonds in the modular assembly of heteromeric infinite 1-D chains. Chem. Commun. (41):4236–8.

[33] Cinčić D., Friščić T., Jones W. (2008) A Stepwise Mechanism for the Mechanochemical Synthesis of Halogen-Bonded Cocrystal Architectures. J. Am. Chem. Soc. 130(24):7524–5.

[34] Cauliez P., Polo V., Roisnel T., Llusar R., Fourmigué M. (2010) The thiocyanate anion as a polydentate halogen bond acceptor. CrystEngComm. 12(2):558–66.

[35] Präsang C., Whitwood A.C., Bruce D.W. (2008) Spontaneous symmetry-breaking in halogen-bonded, bent-core liquid crystals: observation of a chemically driven Iso–N–N* phase sequence. Chemical Communications. (18):2137.

[36] Cinčić D., Friščić T., Jones W. (2011) Experimental and database studies of three-centered halogen bonds with bifurcated acceptors present in molecular crystals, cocrystals and salts. CrystEngComm. 13(9):3224–31.

[37] Davy K.J.P., McMurtrie J., Rintoul L., Bernhardt P.V., Micallef A.S. (2011) Vapour phase assembly of a halogen bonded complex of an isoindoline nitroxide and 1,2-diiodotetrafluorobenzene. CrystEngComm. 13(16):5062–70.

[38] Jay J.I., Padgett C.W., Walsh R.D.B., Hanks T.W., Pennington W.T. (2001) Noncovalent Interactions in 2-Mercapto-1-methylimidazole Complexes with Organic Iodides. Crystal Growth & Design. 1(6):501–7.

[39] Liantonio R., Metrangolo P., Pilati T., Resnati G. (2002) 4,4'-Bipyridine 1,2-diiodo-3,4,5,6-tetrafluorobenzene. Acta Crystallographica Section E Structure Reports Online. 58(5):o575–o577.

[40] Metrangolo P., Murray J.S., Pilati T., Politzer P., Resnati G., Terraneo G. (2011) Fluorine-Centered Halogen Bonding: A Factor in Recognition Phenomena and Reactivity. Crystal Growth & Design. 11(9):4238–46.

[41] Thomas N.W., Desiraju C.R. (1984) An investigation into the role of chloro-substituents in hydrogen-bonded crystals: The crystal structures of the dichlorophenols. Chemical Physics Letters. 110(1):99–102.

[42] Mukherjee A., Desiraju G.R. (2011) Halogen Bonding and Structural Modularity in 2,3,4- and 3,4,5-Trichlorophenol. Crystal Growth & Design. 11(9):3735–9.

[43] Mukai T., Nishikawa K. (2010) Zigzag Sheet Crystal Packing in a Halogen-bonding Imidazolium Salt: 1-Butyl-4,5-dibromo-3-methylimidazolium Iodide. X-ray Structure Analysis Online. 26:31–2.

[44] Bertani R., Sgarbossa P., Venzo A., Lelj F., Amati M., Resnati G., et al. (2010) Halogen bonding in metal–organic–supramolecular networks. Coordination Chemistry Reviews. 254(5–6):677–95.

[45] Ovens J.S., Geisheimer A.R., Bokov A.A., Ye Z.-G., Leznoff D.B. (2010) The Use of Polarizable $[AuX_2(CN)_2]^-$ (X = Br, I) Building Blocks Toward the Formation of Birefringent Coordination Polymers. Inorg. Chem. 49(20):9609–16.

[46] Coronado E., Day P. (2004) Magnetic Molecular Conductors. Chem. Rev. 104(11):5419–48.

[47] Johnson M.T., Džolić Z., Cetina M., Wendt O.F., Öhrström L., Rissanen K. (2011) Neutral Organometallic Halogen Bond Acceptors: Halogen Bonding in Complexes of PCPPdX (X = Cl, Br, I) with Iodine (I_2), 1,4-Diiodotetrafluorobenzene (F_4DIBz), and 1,4-Diiodooctafluorobutane (F_8DIBu). Crystal Growth & Design. 12(1):362–8.

[48] Tuikka M., Niskanen M., Hirva P., Rissanen K., Valkonen A., Haukka M. (2011) Concerted halogen and hydrogen bonding in $[RuI_2(H_2dcbpy)(CO)_2]I2(CH_3OH)I_2[RuI_2(H_2dcbpy)(CO)_2]$. Chem. Commun. 47(12):3427–9.

[49] Tuikka M., Hirva P., Rissanen K., Korppi-Tommola J., Haukka M. (2011) Halogen bonding—a key step in charge recombination of the dye-sensitized solar cell. Chem. Commun. 47(15):4499–501.

[50] Ormond-Prout J.E., Smart P., Brammer L. (2011) Cyanometallates as Halogen Bond Acceptors. Crystal Growth & Design. 12(1):205–16.

[51] Jabłoński M., Palusiak M. (2012) Nature of a Hydride–Halogen Bond. A SAPT-, QTAIM-, and NBO-Based Study. J. Phys. Chem. A. 116(9):2322–32.

[52] Lipkowski P., Grabowski S.J., Leszczynski J. (2006) Properties of the Halogen–Hydride Interaction: An ab Initio and "Atoms in Molecules" Analysis. J. Phys. Chem. A. 110(34):10296–302.

[53] Li Q., Dong X., Jing B., Li W., Cheng J., Gong B., et al. (2010) A new unconventional halogen bond C-X···H-M between HCCX (X = Cl and Br) and HMH (M = Be and Mg): An ab initio study. Journal of Computational Chemistry. 31(8):1662–9.

[54] Ouvrard C., Le Questel J.-Y., Berthelot M., Laurence C. (2003) Halogen-bond geometry: a crystallographic database investigation of dihalogen complexes. Acta Crystallographica Section B Structural Science. 59(4):512–26.

[55] Chudzinski M.G., Taylor M.S. (2012) Correlations between Computation and Experimental Thermodynamics of Halogen Bonding. J. Org. Chem. Available from: http://dx.doi.org/10.1021/jo300279m

[56] Lu Y.-X., Zou J.-W., Wang Y.-H., Yu Q.-S. (2006) Ab initio and atoms in molecules analyses of halogen bonding with a continuum of strength. Journal of Molecular Structure: THEOCHEM. 776(1–3):83–7.

[57] Zou J.-W., Jiang Y.-J., Guo M., Hu G.-X., Zhang B., Liu H.-C., et al. (2005) Ab Initio Study of the Complexes of Halogen-Containing Molecules RX (X=Cl, Br, and I) and NH3: Towards Understanding the Nature of Halogen Bonding and the Electron-Accepting Propensities of Covalently Bonded Halogen Atoms. Chem. Eur. J. 11(2):740–51.

[58] Legon A.C. (1999) Angular and radial geometries, charge transfer and binding strength in isolated complexes B···ICl: some generalisations. Chemical Physics Letters. 314(5–6):472–80.

[59] Svensson P.H., Kloo L. (2003) Synthesis, Structure, and Bonding in Polyiodide and Metal Iodide–Iodine Systems. Chem. Rev. 103(5):1649–84.

[60] Yin Z., Wang Q.-X., Zeng M.-H. (2012) Iodine Release and Recovery, Influence of Polyiodide Anions on Electrical Conductivity and Nonlinear Optical Activity in an Interdigitated and Interpenetrated Bipillared-Bilayer Metal–Organic Framework. J. Am. Chem. Soc. 134(10):4857–63.

[61] Robertson K.N., Cameron T.S., Knop O. (1996) Polyhalide anions in crystals. Part 2. I3-asymmetry and N—H...I bonding: triiodides of the Me2NH2+, Ph2I+, tropanium, N,N,N',N'-Me4-1,2-ethanediammonium, N,N,N',N'-Me4-1,3-propanediammonium, N-Me-piperazinium(2+), and N,N'-Me2-piperazinium(2+) cations, and Me2NH2I. Canadian Journal of Chemistry. 74(8):1572–91.

[62] McIndoe J.S., Tuck D.G. (2003) Studies of polyhalide ions in aqueous and non-aqueous solution by electrospray mass spectrometry. Dalton Trans. (2):244–8.

[63] Jircitano A.J., Colton M.C., Mertes K.B. (1981) Unique octaiodide configuration in congested macrocyclic ligand complexes. Inorg. Chem. 20(3):890–6.

[64] Domercq B., Devic T., Fourmigué M., Auban-Senzier P., Canadell E. (2001) Hal···Hal interactions in a series of three isostructural salts of halogenated tetrathiafulvalenes. Contribution of the halogen atoms to the HOMO–HOMO overlap interactions. Journal of Materials Chemistry. 11(6):1570–5.

[65] Walsh R.B., Padgett C.W., Metrangolo P., Resnati G., Hanks T.W., Pennington W.T. (2001) Crystal Engineering through Halogen Bonding: Complexes of Nitrogen Heterocycles with Organic Iodides. Crystal Growth & Design. 1(2):165–75.

[66] Bauzá A., Quiñonero D., Frontera A., Deyà P.M. Substituent effects in halogen bonding complexes between aromatic donors and acceptors: a comprehensive ab initio study. Phys. Chem. Chem. Phys. 13(45):20371–9.

[67] Hinkle R.J., McDonald R. (2002) (Z)-2-Methylbuten-1-yl(aryl)iodonium trifluoromethanesulfonates containing electron-withdrawing groups on the aryl moiety. Acta Crystallographica Section C. 58(2):o117–o121.

[68] Libri S., Jasim N.A., Perutz R.N., Brammer L. (2008) Metal Fluorides Form Strong Hydrogen Bonds and Halogen Bonds: Measuring Interaction Enthalpies and Entropies in Solution. J. Am. Chem. Soc. 130(25):7842–4.

[69] Ovens J.S., Truong K.N., Leznoff D.B. (2012) Structural organization and dimensionality at the hands of weak intermolecular AuAu, AuX and XX (X = Cl, Br, I) interactions. Dalton Trans. 41(4):1345–51.

[70] Zordan F., Brammer L. (2006) M–X···X′–C Halogen-Bonded Network Formation in MX_2(4-halopyridine)2 Complexes (M = Pd, Pt; X = Cl, I; X′ = Cl, Br, I). Crystal Growth & Design. 6(6):1374–9.

[71] Hathwar V.R., Row T.N.G. (2010) Nature of Cl···Cl Intermolecular Interactions via Experimental and Theoretical Charge Density Analysis: Correlation of Polar Flattening Effects with Geometry. The Journal of Physical Chemistry A. 114(51):13434–41.

[72] Friese V.A., Kurth D.G. (2008) Soluble dynamic coordination polymers as a paradigm for materials science. Coordination Chemistry Reviews. 252(1–2):199–211.

[73] Cooke M.W., Chartrand D., Hanan G.S. (2008) Self-assembly of discrete metallosupramolecular luminophores. Coordination Chemistry Reviews. 252(8–9):903–21.

[74] Wąsik R., Łebska M., Felczak K., Poznański J., Shugar D. (2010) Relative Role of Halogen Bonds and Hydrophobic Interactions in Inhibition of Human Protein Kinase $CK2\alpha$ by Tetrabromobenzotriazole and Some C(5)-Substituted Analogues. J. Phys. Chem. B. 114(32):10601–11.

[75] Laguna A., Lasanta T., López-de-Luzuriaga J.M., Monge M., Naumov P., Olmos M.E. (2010) Combining Aurophilic Interactions and Halogen Bonding To Control the Luminescence from Bimetallic Gold–Silver Clusters. Journal of the American Chemical Society. 132(2):456–7.

[76] Dordonne S., Crousse B., Bonnet-Delpon D., Legros J. Fluorous tagging of DABCO through halogen bonding: recyclable catalyst for the Morita–Baylis–Hillman reaction. Chem. Commun. 47(20):5855–7.

Chapter 9

Crystal Structures of Organic Compounds

Nader Noroozi Pesyan

Additional information is available at the end of the chapter

1. Introduction

X-ray crystallography is an important method for determination of the arrangement of atoms within a crystal of compound in which a beam of X-rays strikes a crystal and causes the beam of light to spread into many specific directions. From electron density, in the molecule the mean positions of the atoms in the crystal can be determined, as well as their chemical bonds and various other information.

The hexamethylenetetramine as an organic compound was solved in 1923 "(Dickinson & Raymond, 1923)". Several studies of long-chain fatty acids were followed which are an important component of biological membranes "(Bragg, 1925; de Broglie & Trillat, 1925; Caspari, 1928; Müller, 1923, 1928, 1929; Piper, 1929; Saville & Shearer, 1925; Trillat, 1926)". In the 1930s, the structures of larger molecules with two-dimensional complexity began to be solved. An important advance was the structure of phthalocyanine "(Robertson, 1936)", that is closely related to porphyrin molecules important in biology, such as heme, corrin, chlorophyll and etc.

2. Crystal structures of organic compounds

In this chapter, crystal structures of some organic compounds such as organic torsion helicoids, organic compounds consists of intra- and intermolecular hydrogen bond and their some metal complexes and crystal structures of some organic spiro compounds were described.

2.1. Crystal structure of organic torsion helicoids

In recent years, several different interesting organic compounds structures have been found by X-ray crystallography. Helicenes are an extremely attractive and interesting class of conjugated molecules currently investigated for optoelectronic applications "(Groen et al., 1971; Katz, 2000; Rajca et al., 2007; Schmuck, 2003; Urbano, 2003)". They combine the electronic properties afforded by their conjugated system with the chiroptical properties "(Bossi et al., 2009; Collins

& Vachon, 2006; Larsen et al., 1996)" afforded by their interesting and peculiar helix-like structure, resulting from the condensation of aromatic (and/or heteroaromatic) rings, all of them in *ortho* position. For example, the formula and crystal structures of tetrathia-[7]-helicene **1** are shown in Figures 1 and 2, respectively. The compound **1** has been synthesized in three step by starting material of benzo[1,2-*b*:4,3-*b*']dithiophene and is shown in Scheme 1 "(Maiorana et al., 2003)" and this compound showed second-order non-linear optical (NLO) properties and has been investigated (Clays et al., 2003). In particular, carbohelicenes only include benzene rings, and also in heterohelicenes one or more aromatic rings are heterocyclic (pyridine, thiophene, pyrrole and etc.) "(Miyasaka et al., 2005; Rajca et al., 2004)". With increasing number of condensed rings (typically, $n > 4$), the steric interference of the terminal rings forces the molecule to be a helicoidal form. For $n > 4$ the energetic barrier is such that the two enantiomers can be separated and stored "(Martin, 1974; Newman, et al., 1955, 1967; Newman & Lednicer, 1956; Newman & Chen, 1972)". Of course, the conjugation of π system decreases with decreasing of planarity; however, in longer helicenes π-stack interactions can also take place between overlapping rings "(Caronna et al., 2001; Liberko et al., 1993)". All helicenes (generally, $n > 4$) are chiral molecules and exhibit huge specific optical rotations "(Nuckolls et al., 1996, 1998)" since the chromophore itself, in this case the entire aromatic molecule, is inherently dissymmetric (right-hand or left-hand helix), having a twofold symmetry axis, C_2, perpendicular to its cylindrical helix (in carbohelicenes), or inherently asymmetric (in heterohelicenes) "(Wynberg, 1971)".

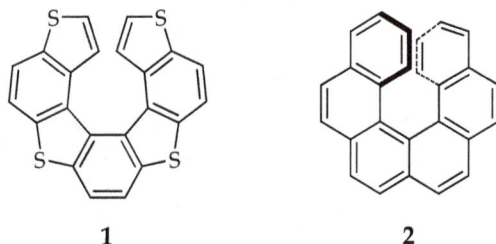

1   **2**

Figure 1. Formula structures of **1** and **2**.

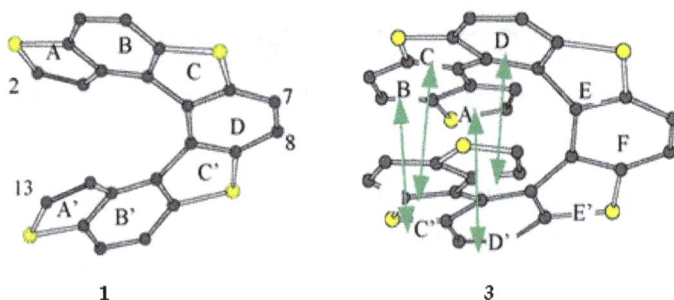

1   **3**

Figure 2. The helicoid structures of unsubstituted tetrathia-[7]-helicene **1** and unsubstituted hexathia-[11]-helicene **3** "(Caronna et al., 2001)" with the labelling scheme adopted for structural discussion "(Bossi et al., 2009)".

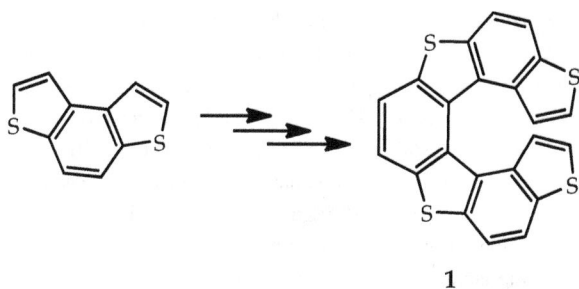

Scheme 1. The synthesis of **1** from benzo[1,2-*b*:4,3-*b*']dithiophene as a starting material.

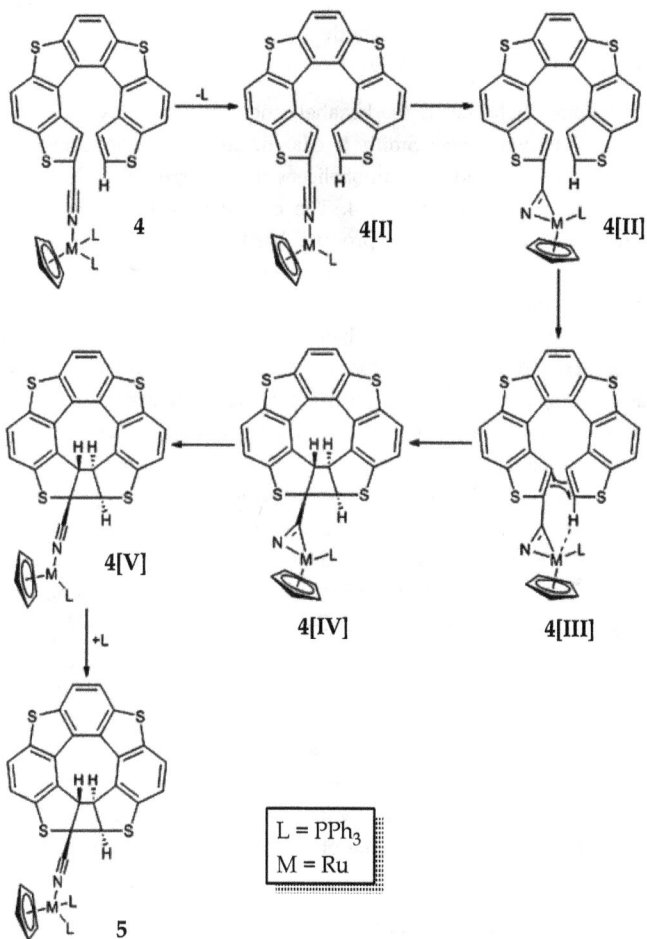

Scheme 2. Reaction mechanism for formation of **5** "Garcia et al., 2009)".

Tetrathia-[7]-helicene **1** have been used for the synthesis of organometallic complexes "(Garcia et al., 2009)". A series of organometallic complexes possessing tetrathia-[7]-helicene nitrile derivative ligands **5** as chromophores, has been synthesized and fully characterized by Garcia et al. "(Garcia et al., 2009)". This compound was analyzed by means of ^{1}H NMR, FT-IR, UV–Vis and X-ray crystallography techniques. The spectroscopic data of this compound was shown with in order to evaluate the existence of electronic delocalization from the metal centre to the coordinated ligand to have some insight on the potentiality of this compound as non-linear optical molecular materials. Slow crystallization of compound **4** revealed an interesting isomerization of the helical ligand with formation of two carbon-carbon bonds between the two terminal thiophenes, leading to the total closure of the helix **5**. The reaction mechanism for the formation of **5** is shown in Scheme 2. Crystal structure of **5** is shown in Figure 3. A selected bond length, angles and torsion angles for compound **5** is summarized in Table 1 "Garcia et al., 2009)".

Another example about helicenes is the hexahelicene **2** and its derivatives that is a chiral molecule "(Noroozi Pesyan, 2006; Smith & March, 2001)". A convenient route for the synthesis of [7]-helicene (**6a**) and [7]-bromohelicene (**6b**) is reported "(Liu et al., 1991)". The crystal structure of 6b is shown in Fig. 4. The crystal structure of **6b** and its unusual oxidation reaction product **7** (as a major product) has been reported "(Fuchter et al., 2012)" (Figure 4 and Scheme 3). Alternatively, compound **6** may be an option for a neutral helicene-derived metallocene complex, since the seven-membered benzenoid rings give rise to a scaffold that completes one full turn of the helix with the two terminal rings being co-facial. It has been theoretically predicted and reported that the **6** has potential to bind some metal cation such as Cr, Mo, W, and Pt in a sandwich model "(Johansson & Patzschke, 2009)". Fuchter and co-workers "(Fuchter et al., 2012)" also reported the crystal structure of **7** that obtained via unusual oxidation rearrangement of **6**. In this structure, The bonds within the pyrenyl unit range between 1.3726(19) and 1.4388(14) Å with the exception of one outlier at 1.3512(18) Å for the C(26)–C(27) bond. The C=C double bonds in rings D and E are 1.3603(15) and 1.3417(16) Å respectively, and the C=O bond is 1.2419(14) Å. The structure of **7** revealed the dominant canonical form to have a pyrenyl group consisting of rings A, B, C and H linked by single bonds to a C–C=C–C=C–C=O unit to form rings D and E (Scheme 3). The pyrenyl unit is flat, the sixteen carbon atoms being coplanar. Ring I has four single bonds and two aromatic bonds, and has a folded conformation with the methylene carbon lying ca. 0.87 Åout of the plane of the other five carbons which are coplanar. Aryl ring G forming the five-membered ring J, links to ring I. The planes of the five coplanar atoms of ring I and the four coplanar atoms of ring J are inclined by ca. 108° to each other. The ring of E is slightly distorted in a boat-like fashion with the carbon shared just with ring D and that shared with rings I and J, out of the plane of the other four atoms which are coplanar to within ca. 0.01 Å.

The formula structure of Katz's helical ferrocene **8** is shown in Figure 5 "(Katz & Pesti, 1982; Sudhakar & Katz, 1986)".

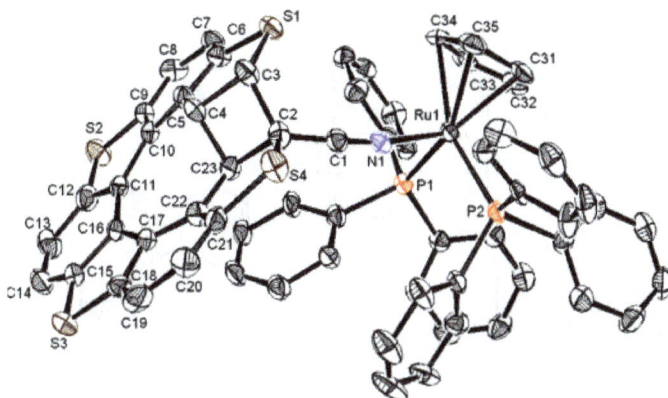

Figure 3. Crystal structure of **5**.

Bond distances (Å)			
Ru(1)–N(1)	2.030(5)	P(2)–C(221)	1.850(7)
Ru(1)–Cp[a]	1.8595(6)	P(2)–C(231)	1.829(6)
Ru(1)–P(1)	2.350(2)	N(1)–C(1)	1.147(8)
Ru(1)–P(2)	2.353(2)	C(1)–C(2)	1.433(9)
P(1)–C(111)	1.843(6)	C(2)–C(3)	1.598(9)
P(1)–C(121)	1.831(7)	C(3)–C(4)	1.535(9)
P(1)–C(131)	1.817(7)	C(4)–C(23)	1.519(9)
P(2)–C(211)	1.846(7)	C(23)–C(2)	1.565(9)
Bond angles (°)			
N(1)–Ru(1)–Cpa	121.79(14)	C(1)–C(2)–C(3)	115.3(6)
N(1)–Ru(1)–P(1)	87.93(14)	C(2)–C(3)–C(4)	88.8(5)
N(1)–Ru(1)–P(2)	90.83(15)	C(2)–C(23)–C(4)	90.6(5)
P(1)–Ru(1)–P(2)	99.84(6)	C(3)–C(4)–C(23)	91.2(5)
P(1)–Ru(1)–Cpa	124.15(5)	C(3)–C(2)–C(23)	87.2(5)
P(2)–Ru(1)–Cpa	122.98(5)	C(5)–C(4)–C(23)	106.9(6)
Ru(1)–N(1)–C(1)	169.9(5)	C(4)–C(23)–C(22)	107.7(5)
N(1)–C(1)–C(2)	175.7(7)	C(2)–C(3)–S(1)	119.3(5)
C(1)–C(2)–C(23)	116.4(6)	C(3)–C(2)–S(4)	116.5(5)
Torsion angles (°)			
Ru(1)–N(1)–C(1)–C(2)	29(11)	N(1)-C(1)–C(2)–C(23)	62(9)
N(1)–C(1)–C(2)–C(3)	-38(9)	C(2)–C(3)–C(4)–C(23)	-11.0(5)
N(1)–C(1)–C(2)–S(4)	-175(9)		

[a] Cp ring centroid.

Table 1. Selected bond distances and bond and torsion angles for compound **5** "(Garcia et al., 2009)".

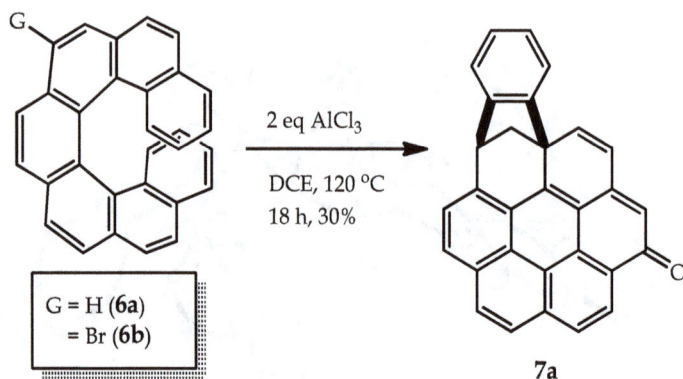

Scheme 3. The formula structures of **6a** and **6b** and its unusual reaction for synthesis of **7a** (and also its structure).

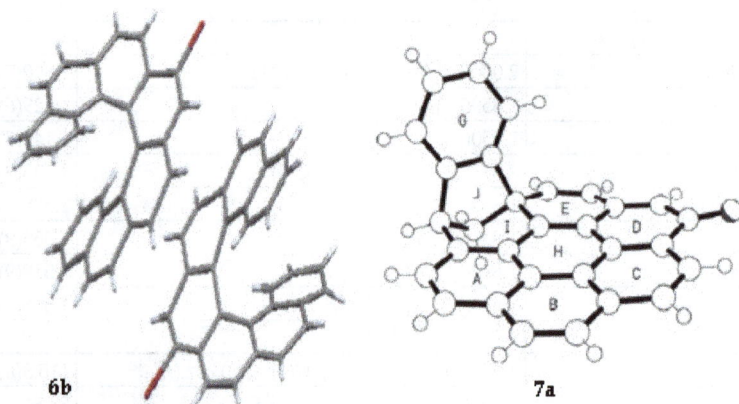

Figure 4. Crystal structures of **6b** and **7a**.

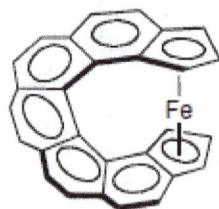

8

Katz's helical
ferrocene

Figure 5. The formula structure of Katz's helical ferrocene **8**.

Diazepinone dervatives are of pharmaceutical compounds. Another interesting helical diazepinone compound that is discussed in this section, is 1,9-dimethyl-4,5-dihydro-6*H*-pyrido[3',2':4,5]thieno[2,3-*f*]pyrrolo[1,2-*a*][1,4]diazepin-6-one (**9**). This molecule show two crystallographically independent molecules that form the asymmetric unit of the structure are shown in Figure 6. The X-ray crystallographic analysis shows the molecular structure of the compound **9** and reveals an interesting fact that this structure features two stereochemically different molecules (**9A** and **9B**) that can be understood as different torsion helicoids (Figure 6). The compound has two stereoisomers (*R* and *S* conformers). In each structure the seven-membered diazepinone ring exhibits a boat conformation.

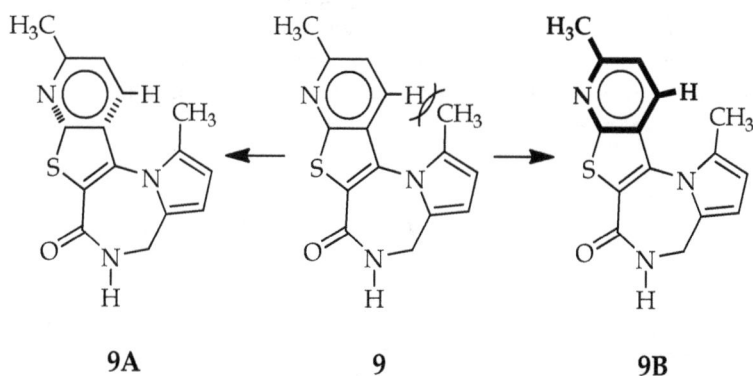

Scheme 4. Two possible different torsion helicoids of **9**.

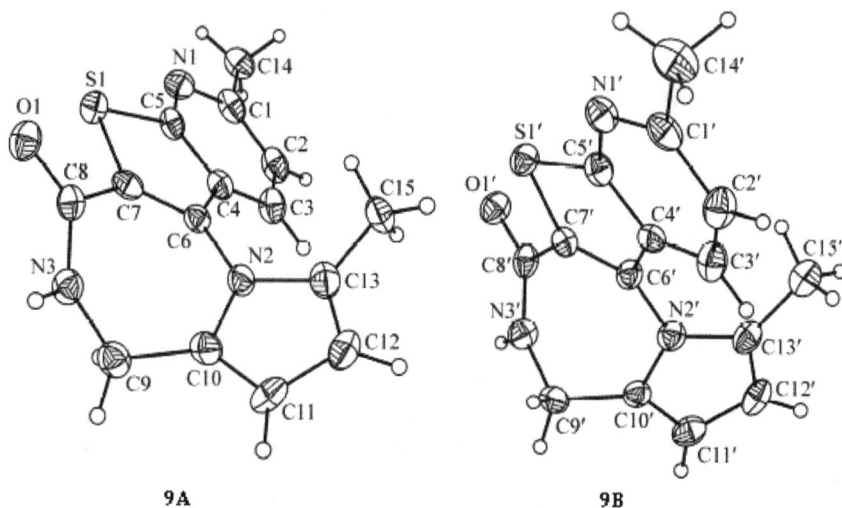

Figure 6. Two independent molecules of **9** in the crystal studied.

The fused pyrido[3',2':4,5]thieno ring moiety has planar geometry. The C3–H3 bond is slightly off the fused pyrido[3',2':4,5]thieno ring plane. The hindrance repulsion between the hydrogen atom at C3 on pyridine ring and methyl group on pyrrole ring makes the molecule of **9** essentially non-planar (repulsion of C3–H3A with C15 and C3'–H3'B with C15' of methyl groups) (Scheme 4). The torsion angles between the pyrrole and thiophene rings in **9A** and **9B** are 45.7(6)° and –49.3(6)°, respectively "(Noroozi Pesyan, 2010)".

The –NH– group of each molecule (e.g. molecule **9A**) makes an intermolecular hydrogen bond to the C=O functional group of the molecule of another kind (molecule **9B**), and *vice versa*. For example, the intermolecular hydrogen bond N3–H3····O1' involves the N3 atom from molecule **9A** and O1' atom from the carbonyl group of molecule **9B**, and *vice versa* for N3'–H3'····O1 (Figure 7). The crystal packing diagram indicates zigzag hydrogen-bonded chains along the crystallographic axes with two distinct hydrogen bonds (Figure 7). The intermolecular hydrogen bonds play a principal and important role in the crystal packing diagram of **9** "(Noroozi Pesyan, 2010)".

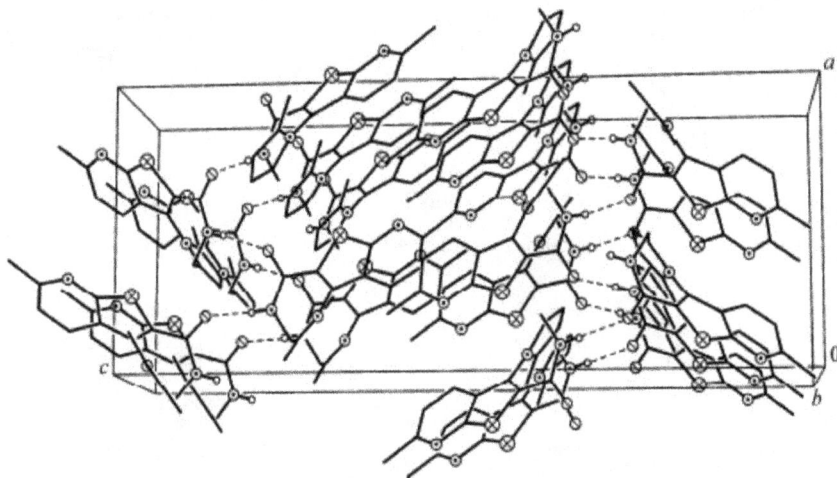

Figure 7. Crystal packing diagram of **9** showing zigzag H-bonds (shown by dashed lines).

One of the most interesting helical primary structure is sown in Figure 8 has been reported by Fitjer et al. "(Fitjer et al., 2003)". Helical primary structures of spiro annelated rings are unknown in nature but have been artificially produced, both in racemic and enantiomerically pure form. The formula structure of 1-cyclobutylidenespiro[3.3]heptane (**10**) as a starting material is shown in Scheme 5. The compound **10** yielded enantiomeric mixture of **11** and **12** in the presence of zinc and 2,2,2-trichloroacetyl chloride. Reductive dehalogenation of **11** and **12** then Wolff–Kishner reduction yielded the desired trispiro[3.0.0.3.2.2]tridecane [rac-(**15**), (symmetry, C_2)]. The crystal structure of the camphanic acid derivative of **15** ((1S,5'S,10'S)-**16**) is shown in Figure 8 "(Fitjer et al., 2003)".

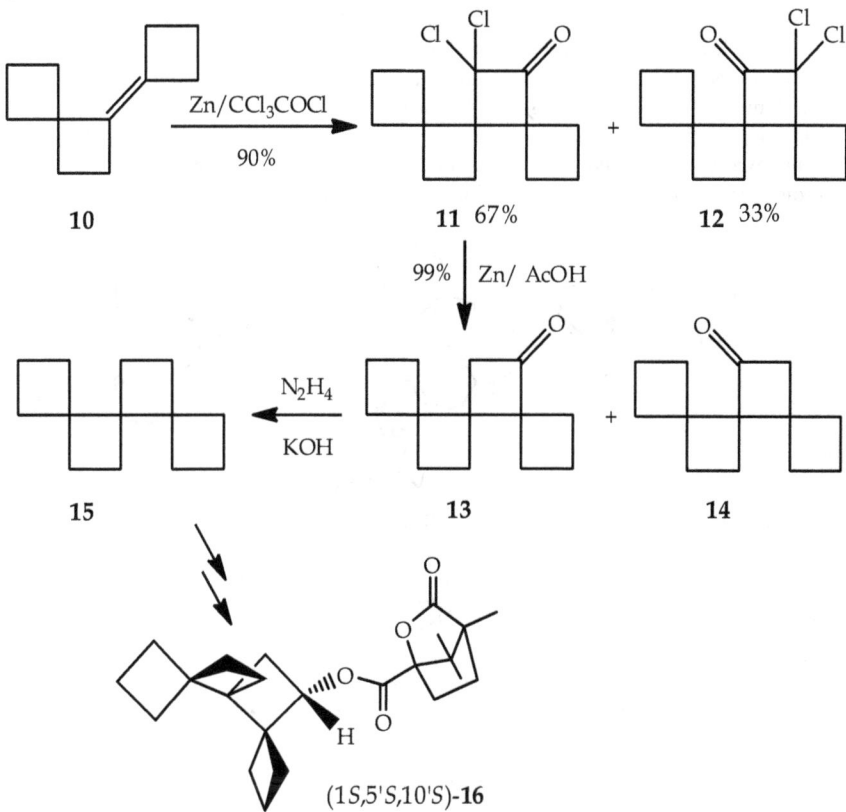

Scheme 5. Synthesis of the compounds trispiro[3.0.0.3.2.2]tridecane (**15**) and the formula structure of its derivative (1S,5'S,10'S)-**16** "(Fitjer et al., 2003)".

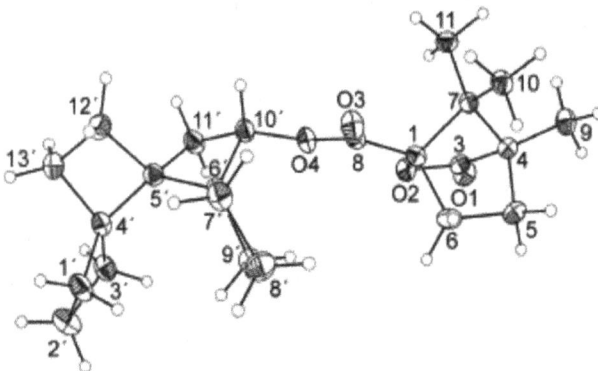

Figure 8. Crystal structure of (1S,5'S,10'S)-16.

Helquats, the family of *N*-heteroaromatic cations "(Arai & Hida, 1992)", recently were introduced helical dications that represent a missing structural link between helicenes and viologens"(Casado et al., 2008)". Specifically, basic [7]-helquat (**17**) "(Severa et al., 2010)" is a structural hybrid between [7]-helicene and a well-known herbicide diquat (Scheme 6). Synthesis of [7]-helquat (**17**) starts with bisquaternization of bis-isoquinoline precursor (**18**) with an excess of 3-butynyltriflate followed by the key metal catalyzed [2+2+2] cycloisomerization of the resulting triyne, formed **17** (Scheme 7).

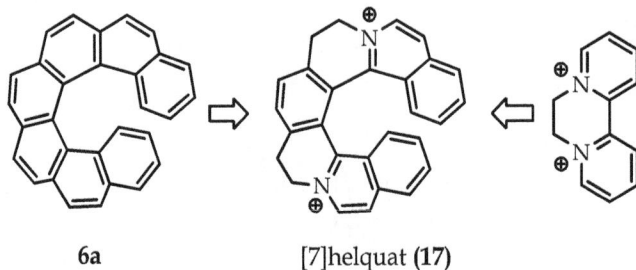

6a   [7]helquat (**17**)

Scheme 6. Structural relation of [7]-helquat (**17**) to [7]-helicene **6a** and herbicide diquat.

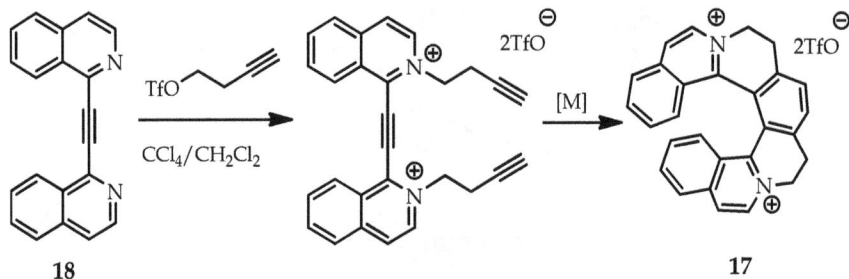

18   17

Scheme 7. Synthesis of **17** via one-pot bis-quaternization of **18**.

Recently, Nakano et al. have been reported the helical structure, λ^5-phospha [7]-helicenes 9-phenyl-9*H*-naphtho[1,2-*e*]phenanthro[3,4-*b*]phosphindole-9-oxide (**21**) and its thio analogue 9-phenyl-9*H*-naphtho[1,2-*e*]phenanthro[3,4-*b*]phosphindole-9-sulfide (**22**) "(Nakano et al., 2012)". The formula structure of **21** and **22** and the crystal structure of **21** are shown in Fig. 10. Phospha [7]-helicenes **21** and **22** have more distorted structures than the other heterohelicenes. In the structure of **21**, the sums of the five dihedral angles that are derived from the seven C–C bonds [C(17)-C(17a)-C(17b)-C(17c), C(17a)-C(17b)-C(17c)-C-(17d), C(17b)-C(17c)-C(17d)-C(17e), C(17c)-C(17d)-C(17e)-C(17f), and C(17d)-C(17e)-C(17f)-C(1)] are 95.28 for **21** and 99.68 for **22**. These angles are larger than those of hetero[7]-helicenes **23–25** (79–88°). This case can be attributed to the large angles between the two double bonds of phosphole oxide (50°) and phosphole sulfide (50°) relative to furan (32°), pyrrole (35°), and thiophene (45°). Owing to the larger angle, a larger overlap of the two terminal benzene rings was occurred in the λ^5-phospha[7]-helicenes, therefore, a stronger steric repulsion. These larger distortions in **21** and **22** explain the higher tolerance of **21** and **22** towards racemization.

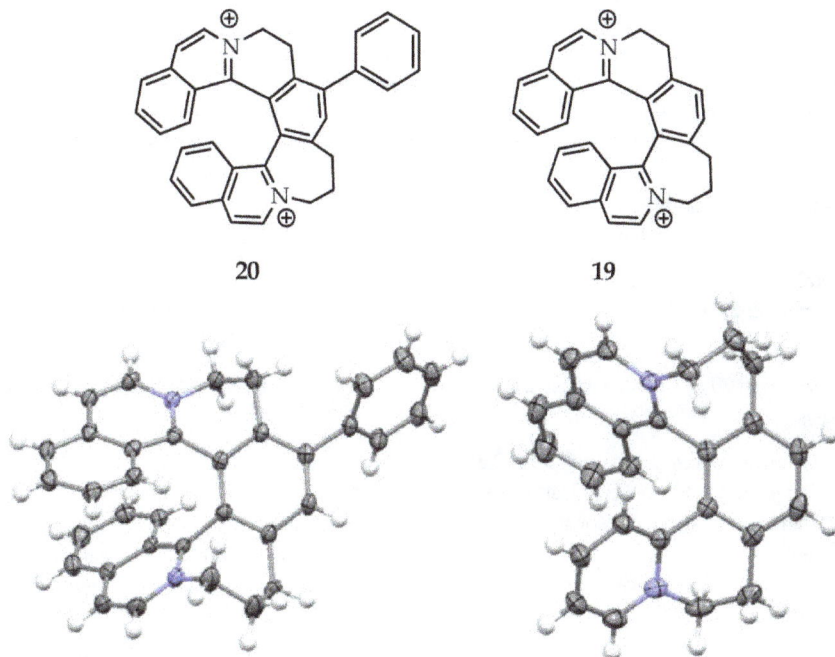

Figure 9. Formula and X-ray single crystal structure of compounds **19** and **20** (Triflate counterions are omitted for clarity) "(Severa et al., 2010)".

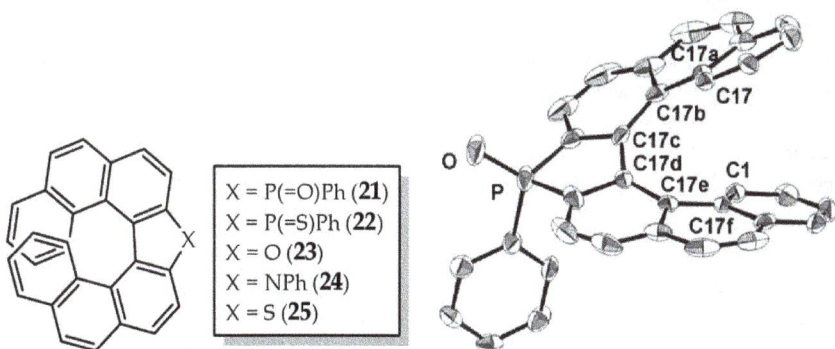

Figure 10. Formula structures of λ^5-Phospha[7]-helicenes **21** and **22** and crystal structure of **21** as representative.

2.2. Inter- and intramolecular hydrogen bonds in the crystal structure of organic compounds

Hydrogen bond plays a key and major role in the biological and pharmaceutical systems and remains a topic of intense current interest. Few selected recent articles exemplify the general scope of the topic, ranging from the role of H-bonding such as in: weak interaction in gas phase "(Nishio, 2005; Wang et al., 2005)", supramolecular assemblies "(McKinlay et al., 2005)", helical structures "(Azumaya et al., 2004; Noroozi Pesyan, 2010)". Important consequences of both inter- and intra-molecular H-bonding have long been recognized in the physicochemical behavior of DNA and RNA "(Jeffery & Saenger, 1991)".

Several kinds of hydrogen bond have been reported. If the donor-acceptor distance to be in the range of; $2.50 \leq d \ (O \cdots O) \leq 2.65$, this kind of hydrogen bond is strong and when shorter than 2.50 Å $(d(O \cdots O) \leq 2.50)$, to be very strong hydrogen bond "(Gilli et al., 1994)".

In very short $O \cdots H \cdots O$ bonds (2.40-2.45 Å) the major distribution of the proton are as follows:

i.   The proton is closer to one of the O atoms (asymmetric hydrogen bond).
ii.   The proton is located precisely at the centre (symmetric or centred hydrogen bond).
iii. There is statistically disorder of the proton between two positions on either side of the centre (the proton is closer to one or the other side in different domains of the crystal).
iv.   There is a dynamical disorder between two positions as in (iii); the proton jumps between the two positions in the same hydrogen bond "(Gilli et al., 1994; P. Gilli & G. Gilli, 2000; Olovsson et al., 2001; Steiner, 2002)".

For instance, the structure of the potassium hydrogen dichloromaleate (26) has been studied by neutron diffraction at 30 and 295 K, with the emphasis on the location of the protons. There are two crystallographically independent hydrogen atoms in two very short hydrogen bonds, 2.437(2) and 2.442(2) Å at 30 K. For the centrosymmetric space group P1, with the hydrogen atoms located at the centres of symmetry, the structure could be refined successfully. Olovsson et al. have then been applied several different types of refinements on this structure, including unconventional models; with all atoms except hydrogen constrained in P1, but with hydrogen allowed to refine without any constraints in P1, anisotropic refinement of all atoms resulted in clearly off-centred hydrogen positions. The shifts of the two hydrogen atoms from the centres of symmetry are 0.15(1) and 0.12(1) Å, respectively, at 30 K, and 0.15(1) Å for both hydrogen atoms at room temperature. At 30 K: R(F) = 0.036 for 1485 reflections; at 295 K: R(F) = 0.035 for 1349 reflections (Olovsson et al., 2001)" (Fig. 11).

One of the most interesting example about intermolecular hydrogen bond is the heptan-4-yl (2'-hydroxy-[1,1'-binaphthalen]-2-yl) phosphonate (27a) "(Dabbagh et al., 2007)". The phosphonate 27a was existed in dimmer form via two strong intermolecular hydrogen bonds with centrocymmetric (C_i) 18-membered dimmer form consisting of two monomers strongly hydrogen-bonded between the oxygen of P=O units and hydroxyl hydrogen atoms (Fig. 12). The crystal structure of 27a was determined by X-ray crystallography and is shown

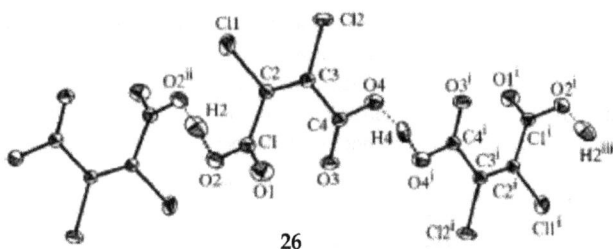

Figure 11. Crystal structure of potassium hydrogen dichloromaleate (**26**).

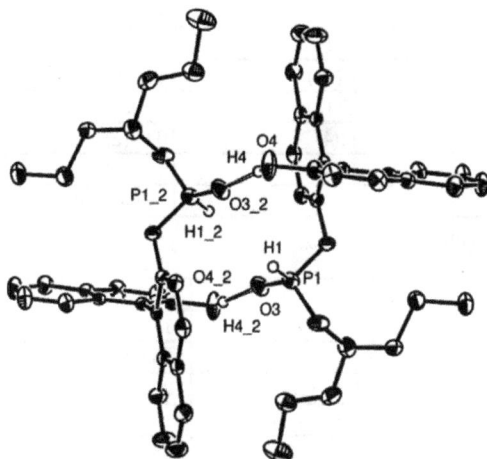

Figure 12. Crystal structure of **27a**.

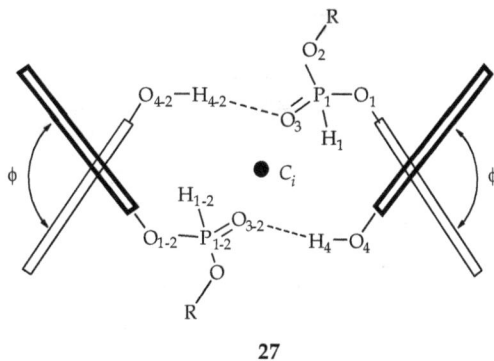

R: 4-Heptyl (**a**), cyclohexyl (**b**)

Figure 13. Representatively, two strong intermolecular hydrogen bonds with centrocymmetric 18-membered dimmer form in **27a** and **27b**.

in Fig. 12. The selected bond lengthes, angles and torsion angles of **27a** are summarized in Tables 2-4, respectively. Crystal data indicated the torsion angles (ϕ) between two naphthalenic rings moieties in BINOL species are 95.28(16)° and are transoid forms (Fig. 13). The intermolecular hydrogen bond distance in the structure of **27a** was obtained 2.70 Å (strong hydrogen bond) and comparised with other hydrogen bonds P-containing systems (Table 5).

Entry	Bond	length (Å)
1	P(1) – O(3)	1.4578(11)
2	P(1) – O(2)	1.5544(12)
3	P(1) – O(1)	1.5865(11)
4	P(1) – H(1)	1.295(16)
5	O(1) – C(1)	1.4073(17)
6	O(2) – C(21)	1.5006(19)
7	O(4) – C(12)	1.3634(18)
8	O(4) – H(4)	0.89(2)
9	O(3) – H(4)	1.81(2)
10	C(10) – C(11)	1.4942(19)

Table 2. Selected bond length (Å) of dimmer **27a**.

Entry	Bond	Angle (θ, °)
1	O(3) – P(1) – O(2)	118.32(7)
2	O(3) – P(1) – O(1)	113.50(7)
3	O(2) – P(1) – O(1)	102.16(6)
4	O(3) – P(1) – H(1)	112.7(7)
5	O(2) – P(1) – H(1)	103.9(7)
6	O(1) – P(1) – H(1)	104.7(8)
7	C(1) – O(1) – P(1)	122.24(9)
8	C(21) – O(2) – P(1)	122.27(10)
9	C(12) – O(4) – H(4)	114.4(14)
10	C(10) – C(1) – C(2)	123.47(14)
11	C(10) – C(1) – O(1)	117.78(13)
12	C(2) – C(1) – O(1)	118.66(13)
13	C(9) – C(10) – C(11)	120.98(12)
14	O(4) – C(12) – C(11)	124.08(14)
15	O(4) – C(12) – C(13)	114.68(13)
16	O(2) – C(21) – C(22)	107.92(12)
17	O(2) – C(21) – C(25)	111.86(18)
18	C(22) – C(21) – C(25)	108.84(18)
19	O(2) – C(21) – H(21)	109.4
20	C(22) – C(21) – H(21)	109.4
21	C(25) – C(21) – H(21)	109.4

Table 3. Selected bond angle of dimmer **27a**.

Entry	Bond	Torsion angles (Φ, °)
1	O(3) – P(1) – O(1) – C(1)	51.11(13)
2	O(2) – P(1) – O(1) – C(1)	179.64(11)
3	O(3) – P(1) – O(2) – C(21)	58.93(13)
4	O(1) – P(1) – O(2) – C(21)	-66.49(12)
5	P(1) – O(1) – C(1) – C(10)	110.03(13)
6	P(1) – O(1) – C(1) – C(2)	-73.13(15)
7	O(1) – C(1) – C(2) – C(3)	-177.12(12)
8	C(7) – C(8) – C(9) – C(10)	178.21(14)
9	C(2) – C(1) – C(10) – C(11)	-178.59(12)
10	O(1) – C(1) – C(10) – C(11)	-2.11(19)
11	C(1) – C(10) – C(11) – C(12)	-97.30(17)
12	C(9) – C(10) – C(11) – C(12)	83.69(18)
13	C(1) – C(10) – C(11) – C(20)	83.73(17)
14	C(9) – C(10) – C(11) – C(20)	-95.28(16)
15	C(20) – C(11) – C(12) – O(4)	179.63(140
16	C(10) – C(11) – C(12) – O(4)	0.6(2)
17	P(1) – O(2) – C(21) – C(22)	-119.33(13)
18	P(1) – O(2) – C(21) – C(25)	120.97(19)
19	O(2) – C(21) – C(22) – C(23)	62.75(18)
20	C(25) – C(21) – C(22) – C(23)	-175.66(19)
21	C(21) – C(22) – C(23) – C(24)	166.78(15)
22	O(2) – C(21) – C(25) – C(26)	-68.8(3)
23	C(22) – C(21) – C(25) – C(26)	172.0(2)
24	C(21) – C(25) – C(26) – C(27)	-173.4(3)

Table 4. Selected torsion angles of dimmer **27a**.

Linkages	Bond distance [Range, donor....H....acceptor] (Å)	Strength
P–O–H....O–P	2.39-2.50	Very strong
P–O–H....O–P	2.50-2.65	Strong
P–O–H....O–C	2.41-2.82	Strong
P–H....OH₂	2.56-3.15	Moderate
P–O....H–N	2.65-3.10	Moderate
Dimmer **27**	2.70	Strong

Data taken from references "(Corbridge, 1990; Gilli et al., 1994; Gilli & & Gilli, 2000; Steiner, 2002)".

Table 5. Classification of hydrogen bonds within P-containing systems.

Dimeric centrosymmetric ring structures are quite common within phosphorous chemistry: for example; the structures of **28** and **29** are of 12- and 8-membered structures, respectively "(Corbridge, 1990)". According to spectroscopic evidence, esters of (trichloroacetyl) amidophosphoric acid (**29**) exist as **29I** rather than **29II**, which suggests that the hydrogen bond in N-H⋯O=P is stable than that of in N-H⋯O=C "(Corbridge, 1990)".

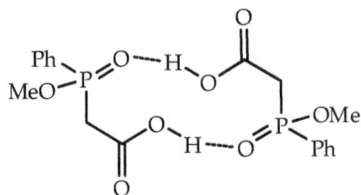

28

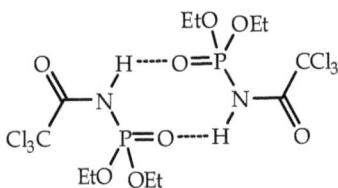

29I

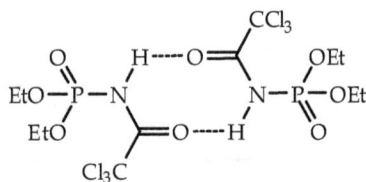

29II

The formula structures of (E)-2-benzamido-3-(pyridin-2-yl)acrylic acid (**30a**) and (E)-2-benzamido-3-(pyridin-4-yl)acrylic acid (**30b**) are shown in Figure 14. The isomer **30a** possesses a strong seven-membered ring intramolecular hydrogen bonding and shows quite different physicochemical properties, such as solubility and pKa, comparing with its isomer **30b**. The p-conjugation between pyridyl and acrylate moieties is extended by intramolecular hydrogen bonding leading to a strong absorption at about 340 nm. Intramolecular proton transfer facilitates in the excited state, resulting in dual emission at around 420 nm and 490 nm in acetonitrile "(Guo et al., 2011)". Crystal structure of **30a** show a strong seven-membered ring intramolecular hydrogen bonding (Figure 15). The intramolecular proton transfer is facilitated by intramolecular hydrogen bond of O–H⋯N. Tautomeric forms of **30a** is shown in Scheme 8.

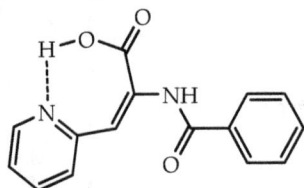

30a

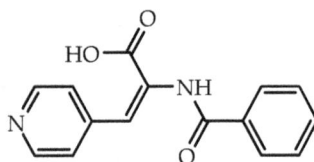

30b

Figure 14. Formula structures of **30a** and **30b**.

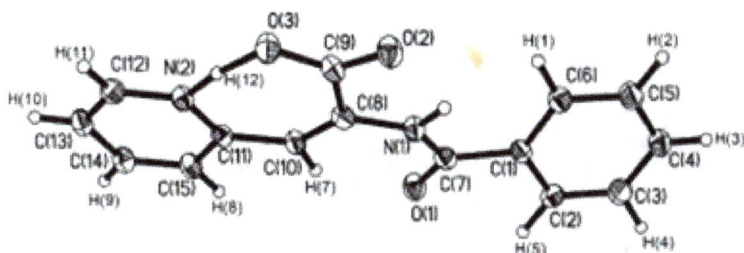

Figure 15. Crystal structure of **30a**.

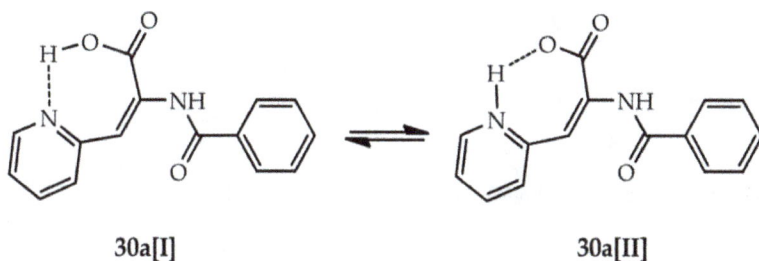

30a[I]   **30a[II]**

Scheme 8. Possible tautomeric forms of **30a**.

There is hydrogen bonding between the acrylate O(3) and the pyridine N(2) atoms; the distance between these two atoms is 2.483 Å, and the O(3)-H(12)-N(2) angle is 171.3°. The O(3)-H(12) distance is 1.345 Å (the theoretical distance is 0.920 Å for general carboxyl O-H bond), which is longer than the N(2)-H(12) distance of 1.145 Å (the general distance is 0.960 Å). The distance difference revealed that H(12) is closer to the pyridine N(2) than it is to the acrylate O(3). The O(2)-C(9) and O(3)-C(9) distances are 1.233 Å and 1.272 Å, respectively. These results show that H(12) is involved in a strong intramolecular hydrogen bonding. N(2)-H(12)···O(3), in which the H(12) interaction with the pyridine N(2) is stronger than that with O(3) atom. The carboxylic acid proton moves to the pyridine N atom, while an electron delocalizes across O(2), O(3), and C(9) to form two almost equivalent carbonyl groups. These results provide further evidence that compound **30a** exists mainly as a tautomeric form **30a** (NH) in the solid state (**30a[II]**) form "(Guo, et al. (2011)".

Resorcarene derivatives are used as units in self-assembled capsules via hydrogen bonds. Like to calixarenes, resorcarenes are the core to which specific functional groups are attached. These groups are responsible for the hydrogen bonds while the resorcarenes offer the right spatial arrangement of them. McGillivray and Atwood found that **31** forms in the crystalline state a hexameric capsule with the internal volume of about 1375Å^3. There are 60 hydrogen bonds in hexameric with the help of eight molecules of water (Fig. 16) "(McGillivray & Atwood, 1997)".

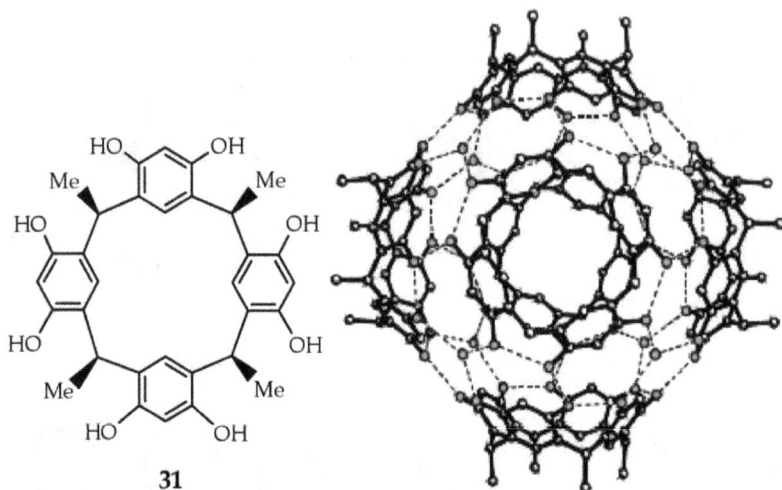

Figure 16. Formula structure of **31** unit and crystal structure of **(31)₆·8H₂O**.

Yoshida et al. have also been reported the formation of a three-dimensional hydrogen bonding network by self-assembly of the Cu(II) complex of a semi-bidentate Schiff base "(Yoshida et al., 1997)". The crystal structure of the Cu(II) complex of Shiff base **32** is shown in Fig. 17. The infinite overall structure of **32** is found to be organized by a three-dimensional hydrogen-bonding network in which the $-NH_2\cdots O_2S-$ type intermolecular hydrogen bonds play an important role, as shown in Fig. 18. One complex molecule is surrounded by four adjacent complexed molecules through four $-NH_2\cdots O_2S-$ hydrogen bonds. These hydrogen bonds would be strong judging from the $NH\cdots O$ distances in the range 2.032–2.941 Å. From the neutron diffraction study of sulfamic acid ($NH_3^+SO_3^-$), a comparably strong hydrogen bond has been observed ($-N^+H\cdots O-S-$ distances in the range 1.95–2.56 Å) "(Jeffrey & Saenger, 1991)". Similar hydrogen bonds between sulfone and hydroxyl groups [2.898(6) Å] have been found in a supramolecular carpet formed *via* self-assembly of bis(4,4'-dihydroxyphenyl) sulfone "(Davies et al., 1997)". Furthermore, four weak $Br\cdots H$ hydrogen bonds may participate in the hydrogen-bonding arrays "(Yoshida et al., 1997)".

Yang et al. reported the crystal structure of Bis(barbiturato)triwater complex of copper(II). The neutral $Cu(H_2O)_3(barb)_2$ molecules are held together to form an extensive three-dimensional network *via* $-OH\cdots O-$ and $-NH\cdots O-$ hydrogen-bonded contacts "(Yang et al., 2003)". Hydrogen bonding motifs in fullerene chemistry have been reported by Martín et al. as a minireviewe. The combination of fullerenes and hydrogen bonding motifs is a new interdisciplinary field in which weak intermolecular forces allow modulation of one-, two-, and three-dimensional fullerene-based architectures and control of their function "(Martín et al., 2005)".

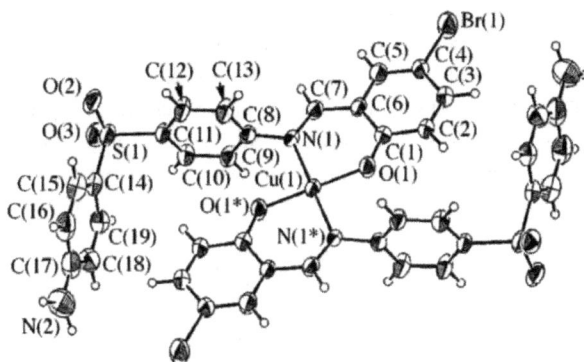

Figure 17. Crystal structure of **32** unit.

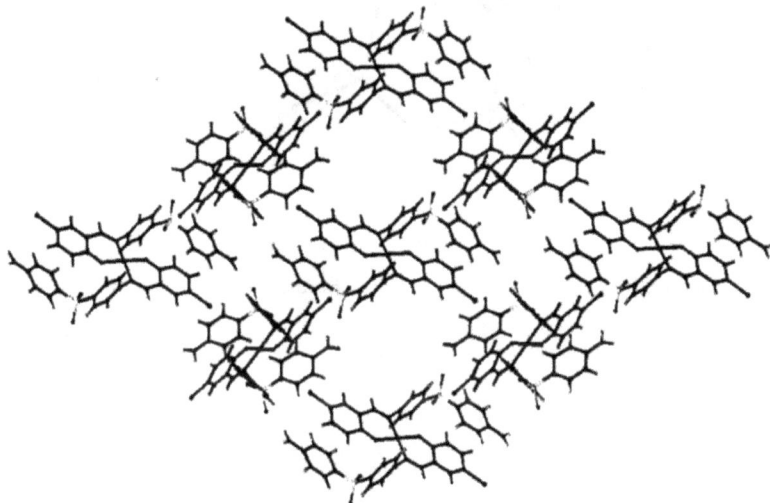

Figure 18. Crystal packing diagram of **32**.

Methyl 2,4-dimethoxy salicylate (**33**) as potential antitumor activity, was synthesized from the reaction of 1,3,5-trimethoxybenzene (the most electron-rich aromatic ring) with 2-methoxycarbonyl-5-(4-nitrophenoxy) tetrazole, under solvent-free conditions, a low yield product was obtained (< 2%), while in the presence of a Lewis acid (AlCl₃), the yield was increased to 30% (a kind of trans esterification reaction) "(Dabbagh et al., 2003)".

Crystal structure of **33** is shown in Fig. 19. The carbon-oxygen framework of the molecular structure of **33** is essentially planar; bond lengths and angles are summarized in Table 6, while a structural diagram is shown also in Fig. 20. Planarity is maintained by a strong intramolecular hydrogen bonding interaction between the carbonyl-oxygen and phenolic-H

210

Innovations in Crystallography

atom [H(1)⋯O(1) = 1.68(4) Å; O(5) – H(1) = 1.00(4) Å], and a much weaker intramolecular hydrogen bond of distance 2.535 Å between Me hydrogen's [H(8)] and the C=O group (in what we label a "bisected" conformation with Cs symmetry, Figs. 19 and 20). The orientations of the o-OMe and ester-OMe are such to minimize steric interactions. The structure of **33** was also calculated by semi-empirical *ab-initio*, PM3 and AM1 methods, and data for bond lengths, angles and torsion angles are in good agreement together with the experimental ones (Tables 6 and 7), while the corresponding calculated H(1)⋯O(1) bond lengths were 1.57, 1.78 and 1.97 Å,

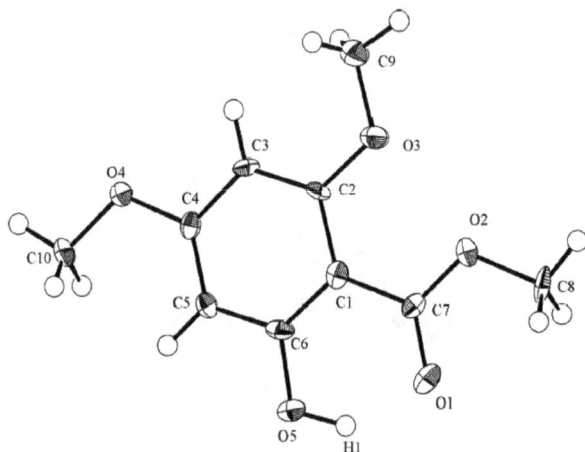

Figure 19. Crystal structure of **33** with 50% probability ellipsoids.

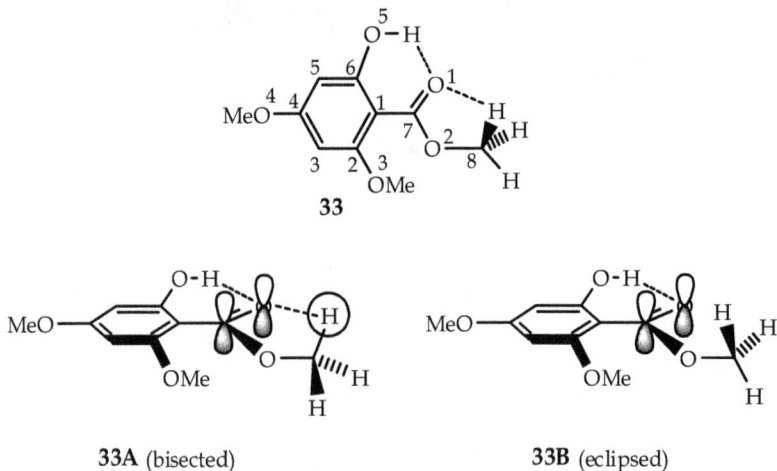

33

33A (bisected)   **33B** (eclipsed)

Figure 20. Diagrams showing the favored so-called "bisected" (left) and "eclipsed" (right) conformations of **33**.

and the calculated O(5) – H(1) values were 1.00, 0.980, 0.970 Å, respectively. The *ab-initio* value for the weaker hydrogen bonding interaction was 2.574 Å. The *ab-initio* calculation also revealed a 1.40 Kcal higher energy, eclipsed conformation with C1 symmetry (Fig. 21 c and d, Fig. 20, Table 7) with an H(8) – carbonyl bond length of 2.14 Å "(Dabbagh et al., 2004)".

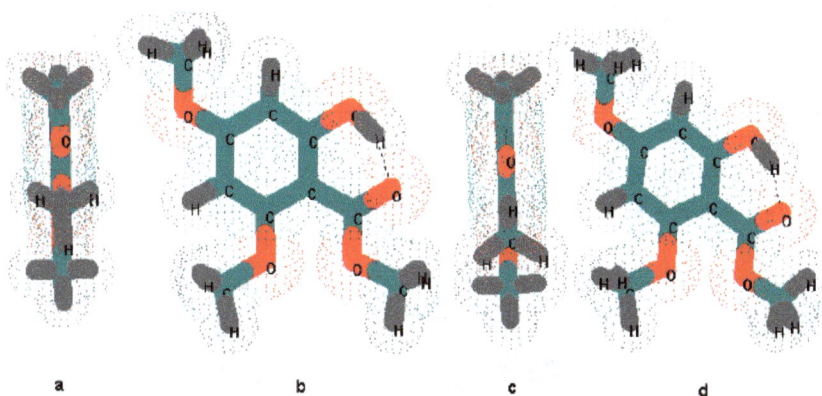

      a   **b**   **c**   **d**

Figure 21. Molecular structures for **33** from *ab-initio* analysis [side-view: **a** (bisected); **c** (eclipsed), and front-view: **b** (bisected0; **d** (eclipsed)].

Atoms	Bond lengths (Å)	Atoms	Bond angles(º)
O(1) – C(7)	1.247(4)	C(7) – O(2) – C(8)	116.1(3)
O(2) – C(7)	1.325(4)	C(2) – O(3) – C(9)	117.1(3)
O(2) – C(8)	1.465(4)	C(4) – O(4) – C(10)	115.8(2)
O(3) – C(2)	1.363(4)	C(2) – C(1) – C(6)	116.9(3)
O(3) – C(9)	1.438(4)	C(2) – C(1) – C(7)	124.7(3)
O(4) – C(4)	1.362(4)	C(6) – C(1) – C(7)	118.3(3)
O(4) – C(10)	1.443(4)	O(3) – C(2) – C(1)	117.1(3)
C(1) – C(2)	1.432(4)	O(3) – C(2) – C(3)	121.9(3)
C(1) – C(6)	1.393(4)	C(1) – C(2) – C(3)	121.03
C(1) – C(7)	1.465(4)	C(2) – C(3) – C(4)	119.8(3)
C(2) – C(3)	1.378(4)	O(4) – C(4) – C(3)	114.4(3)
C(3) – C(4)	1.406(4)	O(4) – C(4) – C(5)	125.1(3)
C(4) – C(5)	1.364(4)	C(3) – C(4) – C(5)	120.6(3)
C(5) – C(6)	1.402(5)	C(4) – C(5) – C(6)	119.7(3)
O(5) – C(6)	1.349(4)	O(5) – C(6) – C(1)	122.3(3)
-	-	O(5) – C(6) – C(5)	115.7(3)
-	-	C(1) – C(6) – C(5)	122.0(3)
-	-	O(1) – C(7) – O(2)	120.7(3)
-	-	O(1) – C(7) – C(1)	122.0(3)
-	-	O(2) – C(7) – C(1)	117.3(3)

[a] Atom numbering is as in Fig. 19.

Table 6. Bond lengths (Å) and angles (o) of **33**.

	Bisected			Eclipsed			Relative Energy
Method	E_{total}	[C=O--H-C]	[C=O--H-O]	E_{total}	[C=O--H-C]	[C=O--H-O]	
	(kcal/mol)	Å	Å	(kcal/mol)	Å	Å	(E_{eclip} -E_{bist})
X-Ray	-	2.535	1.68(4)	-	-	-	-
Ab-intio	- 470792.80	2.574	1.565	-470791.40	2.139	1.557	1.40
PM3	-2812.50	2.647	1.780	-2811.10	2.309	1.780	1.40
AM1	-2813.50	2.554	1.974	-2812.50	2.186	1.969	1.0

[a] Average of three calculations.

Table 7. Experimental and calculated[a] hydrogen bond lengths and energies (kcal/mol) for bisected and eclipsed structure of **33**.

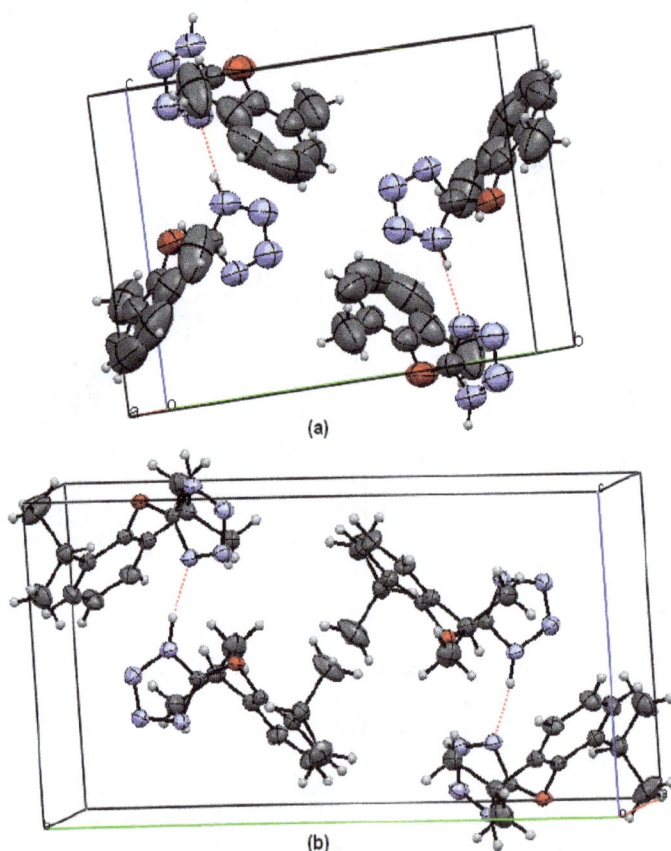

(a)

(b)

Figure 22. Crystal packing diagram of **34** (a) and **35** (b). Intermolecular hydrogen bond assigned by red dashed line (Carbon: grey; hydrogen: white; oxygen: red and nitrogen: blue).

Tetrazole ring can exist to be an equilibrium mixture of two tautomeric forms (1H and 2H-tetrazoles) "(Dabbagh & Lwowski, 2000)". 5-Aryloxy (1H) and/or (2H)-tetrazoles often show intermolecular hydrogen bond "(Noroozi Pesyan, 2011)". For instance, the crystal packing diagram of 5-(2,6-dimetylphenoxy)-(1H)-tetrazole (34) and 5-(2,6-diisopropylphenoxy)-(1H)-tetrazole (35) show intermolecular hydrogen bond (Fig. 22). In the compound 34, the crystal structure indicated that the tetrazole and phenyl rings are nearly perpendicular to each other, forming a dihedral angle of 95.5° (*versus* 92.06° from calcd. B3LYP/6-31G(d) and 6-31+G(d)). Because of the conjugation of O1 with tetrazole ring, the bond distance C1–O1 [1.322 Å] is slightly shorter than O1–C7 [1.399 Å]. These bond distances for C1–O1 and O1–C2 were obtained 1.333 and 1.419 Å with calculation by B3LYP/6-31G(d) method, respectively and also 1.332 and 1.420 Å derived with calculation by B3LYP/6-31+G(d) basis set, respectively. These data are in good agreement with experimental results (Table 8). In the compound 35, the crystal structure indicated that the tetrazole and phenyl rings are nearly perpendicular to each other, forming a dihedral angle of 85.91° (*versus* 107.2° from calcd. B3LYP/6-31G(d)). Because of the conjugation of O1 with tetrazole ring, the bond distance C2–O1 [1.3266(14) Å] is slightly shorter than O1–C7 [1.4257(13) Å]. These bond distances for C2–O1 and O1–C7 were obtained 1.332 and 1.423 Å with calculation by B3LYP/6-31+G(d) method, respectively and are in good agreement with experimental results. These bond distances were also obtained 1.322 and 1.422 Å with calculation by B3LYP/6-31G(d) method, respectively. The torsion angles between phenyl ring and each of methyl units on two isopropyl groups are -110.70°, 124.18° and 116.15° and 154.12°, respectively (Table 9).The selected parameters of bond length, angles and torsion angles of 34 and 35 derived by experimental and calculated results are shown in Tables 8 and 9.

The crystal packing of 34 exhibits an intermolecular N1–H1····N4 hydrogen bonds and comparized with the calculated at DFT (B3LYP) at 6-31–G(d) and 6-31+G(d) basis sets (Table 10). The crystal structure indicated that the bond distance value between donor – hydrogen (N1–H1) and hydrogen-acceptor (H1····N4) were found in results 0.861 and 1.959 Å, respectively. For instance, these bond distances were also found in results 1.033 for (N1–H1) and 1.814 for (H1····N4) by calculated at B3LYP/6-31G(d) and 1.031 for (N1–H1) and 1.809 for (H1····N4) B3LYP/6-31+G(d), respectively. The donor-acceptor distance value (N1····N4) was obtained 2.804 by experimental method. This parameter was found 2.842 and 2.838 Å by calculated methods B3LYP/6-31G(d) and 6-31+G(d), respectively. The angle of N1-H1····N4 was found 169.9, 172.9 and 172.1° by experimental, calculated B3LYP/6-31G(d) and B3LYP/6-31+G(d) basis sets, respectively. The results of calculated method (specially 6-31+G(d) basis set) are in good agreement with experimental results (Table 10).

The crystal packing of 35 also exhibits an intermolecular N3–H31····N6 hydrogen bonds and comparized with the calculated at DFT (B3LYP) at 6-31G(d) and 6-31+G(d) basis sets (Table 10). The crystal structure indicated that the bond distance value between donor – hydrogen (N3–H) and hydrogen-acceptor (H31····N6) were found in results 0.926 and 1.919 Å, respectively. For instance, these bond distances were also found in results 1.03 for (N3–H31) and 1.91 for (H31····N6) by calculated at B3LYP/6-31G(d) and 1.01 for (N3–H) and 1.93 for

Compd. 34			
Atom	Ex.	Calcd.[a]	Calcd.[b]
O1-C1	1.322	1.332	1.333
O1-C2	1.399	1.420	1.419
C1-N1	1.327	1.348	1.348
C1-N4	1.305	1.316	1.315
N1-N2	1.354	1.362	1.363
N1-H1	0.861	1.01	1.01
N2-N3	1.285	1.288	1.288
N3-N4	1.368	1.368	1.368
C2-C3	1.349	1.397	1.396
C2-C7	1.389	1.397	1.396
C3-C9	1.518	1.508	1.508
C7-C8	1.495	1.508	1.508
C1-O1-C2	117.3	117.6	117.6
O1-C1-N1	121.0	120.8	120.8
O1-C1-N4	129.3	130.05	130.05
C1-N1-H1	126.1	130.4	130.4
O1-C2-C3	117.8	117.8	117.8
O1-C2-C7	116.3	117.8	117.8
C2-C3-C9	120.4	121.2	121.2
C2-C7-C8	123.0	121.2	121.2
C2-O1-C1-N1	170.0	-180	-180
O1-C1-N1-H1	-0.8	0.0	0.0
O1-C2-C3-C9	4.4	4.9	4.9
O1-C2-C3-C4	-175.4	-175.6	-175.6
O1-C2-C7-C8	-5.7	-4.9	-4.9

[a] Calculated at B3LYP/6-31+G(d) basis set.
[b] Calculated at B3LYP/6-31G(d) basis set.

Table 8. The selected bond lengths (Å), angles (°) and torsion angles (ϕ) for **34**. Experimental and B3LYP/6-31+G(d) and B3LYP/6-31G(d).

(H····N6) B3LYP/6-31+G(d), respectively. The donor-acceptor distance value (N3····N6) was obtained 2.835 by experimental method. This parameter was found 2.941 and 2.912 Å by calculated methods B3LYP/6-31G(d) and 6-31+G(d), respectively. The angle of N3–H31···· N6 was found 169.1, 177.0 and 173.0° by experimental, calculated B3LYP/6-31G(d) and B3LYP/6-31+G(d) basis sets, respectively. The results of calculated method (specially 6-31+G(d) basis set) are in good agreement with experimental results (Table 10). Compounds **34** (entry no. CCDC-838541) and **35** (entry no. CCDC-819010) were deposited to the Cambridge Crystallographic Data Center and are available free of charge upon request to CCDC, 12 Union Road, Cambridge, UK (Fax: +44-1223-336033, *e-mail: deposit@ccdc.cam.ac.uk*).

Compd. 35			
Atom	Ex.	Calcd.[a]	Calcd.[b]
O1-C2	1.327	1.332	1.322
O1-C7	1.426	1.423	1.422
C2-N3	1.327	1.349	1.348
C2-N6	1.314	1.316	1.315
N3-N4	1.358	1.362	1.363
N3-H31	0.926	1.01	1.010
N4-N5	1.288	1.288	1.288
N5-N6	1.373	1.368	1.368
C7-C8	1.392	1.401	1.400
C7-C15	1.389	1.404	1.403
C8-C9	1.521	1.525	1.525
C15-C16	1.528	1.527	1.527
C2-O1-C7	114.49	118.64	118.39
O1-C2-N3	121.3	120.47	120.38
O1-C2-N6	128.5	130.51	130.57
C2-N3-H31	129	130.43	130.33
O1-C7-C8	116.8	118.67	118.72
O1-C7-C15	117.9	117.02	117.08
C7-C8-C9	120.8	123.14	123.08
C8-C9-H91	106.1	108.37	108.29
C10-C9-C11	111.7	111.27	111.41
C7-C15-C16	122.2	124.82	124.6
C15-C16-H161	106.3	104.94	105.01
C17-C16-C18	111.1	111.52	111.49
C7-O1-C2-N3	-174.8	178.5	178.7
O1-C2-N3-H31	-7.5	-0.2	-0.08
O1-C7-C8-C9	1.5	1.6	1.9
O1-C7-C8-C12	-176.5	-177.3	-177.3
O1-C7-C15-C16	-2.0	-1.1	-1.2
C7-C8-C9-C10	154.1	119.45	116.5
C7-C8-C9-C11	-80.2	-115.5	-118.5
C7-C15-C16-C17	-110.7	-63.8	-63.8
C7-C15-C16-C18	124.2	64.7	64.5

[a] Calculated at B3LYP/6-31+G(d) basis set.
[b] Calculated at B3LYP/6-31G(d) basis set.

Table 9. The selected bond lengths (Å), angles (°) and torsion angles (φ) for **35**. Experimental and B3LYP/6-31+G(d) and B3LYP/6-31G(d).

	D-H···A	D-H	H···A	D···A	D-H···A (degree, °)
Exp.[a] (34)	N1-H1···N4[b]	0.861	1.959	2.804	166.9
Calcd.[c] (34)		1.033	1.814	2.842	172.9
Calcd.[d] (34)		1.031	1.809	2.838	172.1
Exp.[a] (35)	N3-H31···N6[e]	0.926	1.919	2.835	169.1
Calcd.[c] (35)		1.03	1.91	2.941	177
Calcd.[d] (35)		1.01	1.93	2.912	174

[a] Experimental.
[b] Symmetry codes: (i) x, $-y+3/2$, $z+1/2$.
[c] Calculated at B3LYP/6-31G(d).
[d] Calculated at B3LYP/6-31+G(d).
[e] Symmetry codes: (i) x, $-y+3/2$, $z+1/2$.

Table 10. Experimental and calculated B3LYP/6-31+G(d) and B3LYP/6-31G(d) levels for hydrogen-bond geometry of **34** and **35** (Å, °)

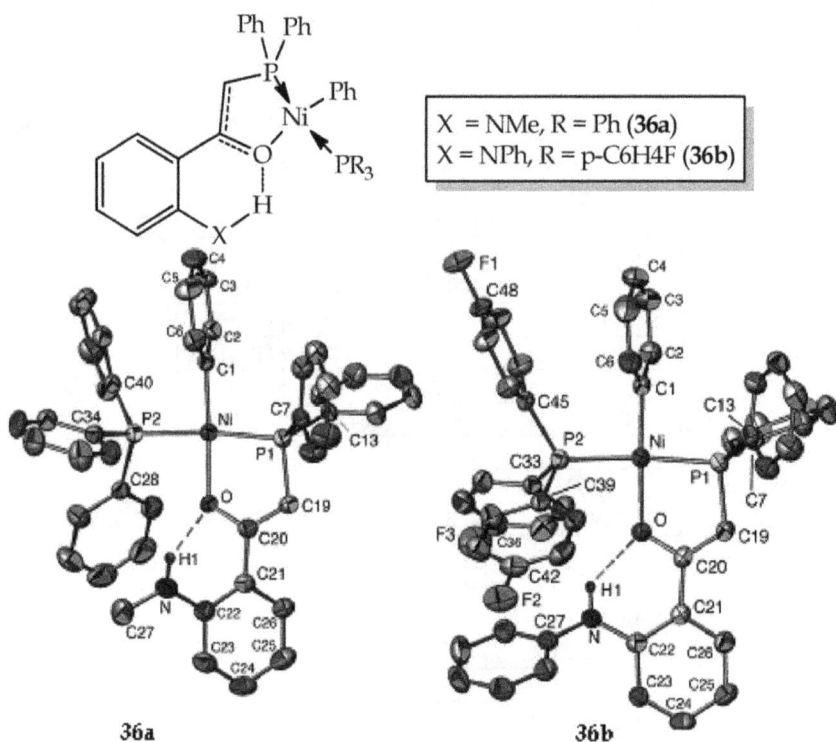

Figure 23. Formula and crystal structures of the compounds **36a** and **36b**.

Nickel(II) complexes containing specific phosphorus– oxygen chelating ligands are very efficient catalysts for the oligomerisation of ethylene to linear form "(Braunstein et al., 1994)". For instance, Nickel(II) diphenylphosphinoenolate complexes have been prepared from (ortho-

HX- substituted benzoylmethylene)triphenyl phosphoranes (X = NMe, NPh) and [Ni(1,5-cod)₂] in the presence of a tertiary phosphine (PPh₃ or P(p-C₆H₄F)₃) and their crystal structures have been studied by Braunstein et al. (structures of **36a** and **36b**). Formula and crystal structures of the compounds **36a** and **36b** are shown in Fig. 23. Crystallographic study of the complexes **36a** and **36b** establishes the presence of strong intramolecular hydrogen bonding between the enolate oxygen and the N–H functional group "(Braunstein et al., 2005)". The most notable feature in these structures is the strong intramolecular N–H····O hydrogen bonding: the calculated distance between the NH hydrogen atom and the oxygen atom of the enolate ligand is short: 2.18(5) Å in **36a** and 2.00(5) Å in **36b**, respectively "(Taylor & Kennard, 1982)".

Intramolecular hydrogen bond is also shown in alkoxyamines. These compounds and persistent nitroxide radicals are important regulators of nitroxide mediated radical polymerization (NMP). The formula and crystal structure of β-phosphorylated nitroxide radical (**37**) is shown in Fig. 24. Compound **37** show an eight-membered intramolecular hydrogen bond between P=O····H-O (versus N-O····H-O). The hydrogen bond distance for two enantiomers of **37** is different. The hydrogen bond distances of P=O····H-O in (*R*)- and (*S*)-**37** are 1.570 and 2.040 Å, respectively and favored. Instead, the hydrogen bond distance for N-O····H-O in (*R*)- and (*S*)-**37** are 3.070 and 3.000 Å, respectively and unfavored "(Acerbis et al., 2006)".

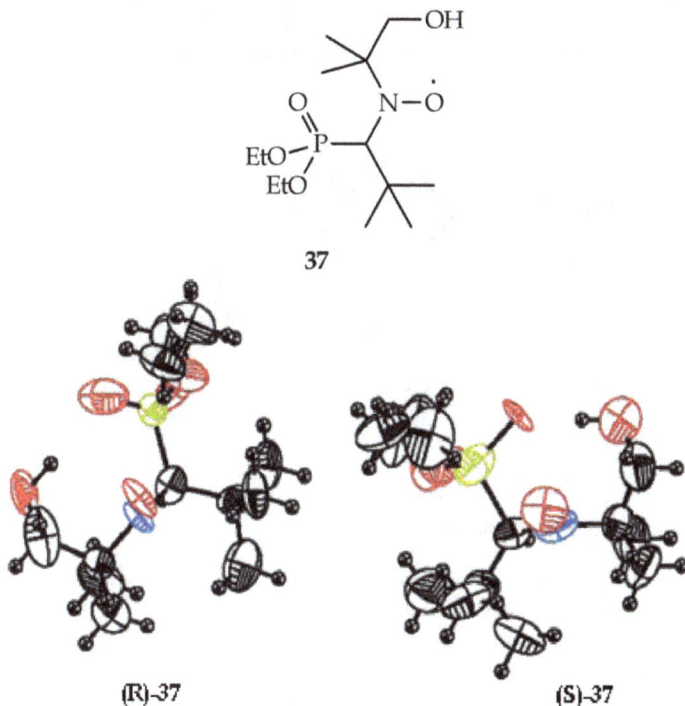

Figure 24. Formula and crystal structure of two enantiomers of compound **37** (O: red, N: blue and P: yellow).

1,8-diaminonaphthalene derivatives such as; *N*-(8-(dimethylamino)naphthalen-1-yl)-2-fluoro-*N*-methylbenzamide (**38**) is a proton sponge. An unusual strong intramolecular hydrogen bond was observed in the protonated **38**. In compound **38** in which a protonated amine group (**38-H⁺**) can act as a donor suitably positioned to engage in a strong intramolecular hydrogen bond with the amide nitrogen atom rather than with the carbonyl oxygen atom (Scheme 9). Crystal structure of the triflate salt of **38-H⁺** is shown in Fig. 25. The unit cell consists of two molecules of **38-H⁺**, two triflate counter ions, and one molecule of water. The dashed line indicates the proposed hydrogen bond between H1A and N2A. Selected bond lengths and angles are N2A–H1A = 2.17(4), N1A–N2A = 2.869(5), C14A–N2A = 1.369(5) (Å) and N2A–H1A–N1A = 136(3)° "(Cox et al., 1999)".

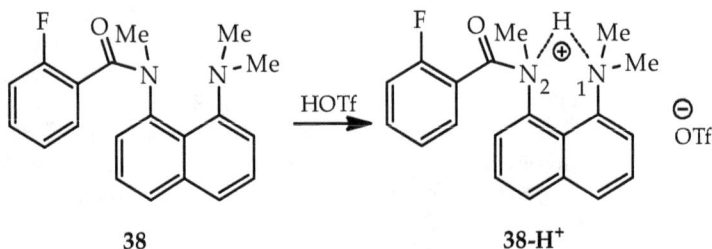

Scheme 9. Protonation of **38** in the presence of trifluoromethanesulfonic acid (TfOH).

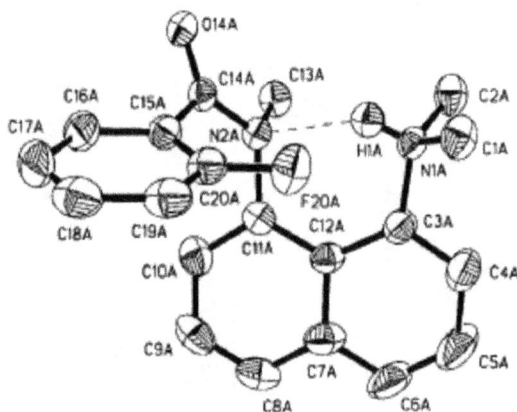

Figure 25. Crystal structure of the triflate salt of **38-H⁺** (50% ellipsoids and triflate counter ion is omitted).

2,4,6-Trisubstituted phenolic compounds such as 2,4,6-tri-*tert*-butyl phenol are as antioxidant "(Jeong et al., 2004)". Owing to the nature of the catalytic centres of galactose oxidase (GAO) and glyoxal oxidase (GLO), the *N,O*-bidentate pro-ligand, 2'-(4',6'-di-*tert*-butylhydroxyphenyl)-4,5-diphenyl imidazole (LH) (**39**) has been synthesized "(Benisvy et al., 2001)". The compound **39** possesses no readily oxidisable position (other than the phenol) and involves *o*- and *p*-substituents on the phenol ring that prevent radical coupling reactions. The compound **39** undergoes a reversible one-electron oxidation to generate the corresponding [LH]⁺ radical cation that possesses phenoxyl radical character. The unusual reversibility of the [LH]/[LH]⁺ redox couple is attributed to a stabilisation of [LH]⁺ by intramolecular O–H····N hydrogen bonding "(Benisvy et al., 2003)". The formula and crystal structure of **39** are shown in Fig. 26. Crystal structure of **39** shows an intra- and intermolecular hydrogen bonds in **39**. In respect of the chemical properties of **39**, there is a strong intramolecular hydrogen bond between the phenolic O–H group and N(5) of imidazole ring. The strength of this hydrogen bond, as measured by the O(1)····N(5) distance of 2.596(2) Å and the O(1)–H(1)····N(5) angle of 150.7°. Also, the N–H group of imidazole ring in **39** is involved in an intermolecular N–H····O hydrogen bond [N(2) ···· O(1S) 2.852(2) Å and N(2)–H(2A)····O(1S) 168.8°] to an adjacent trapped acetone molecule (**39·Me₂CO**) "(Benisvy et al., 2003)".

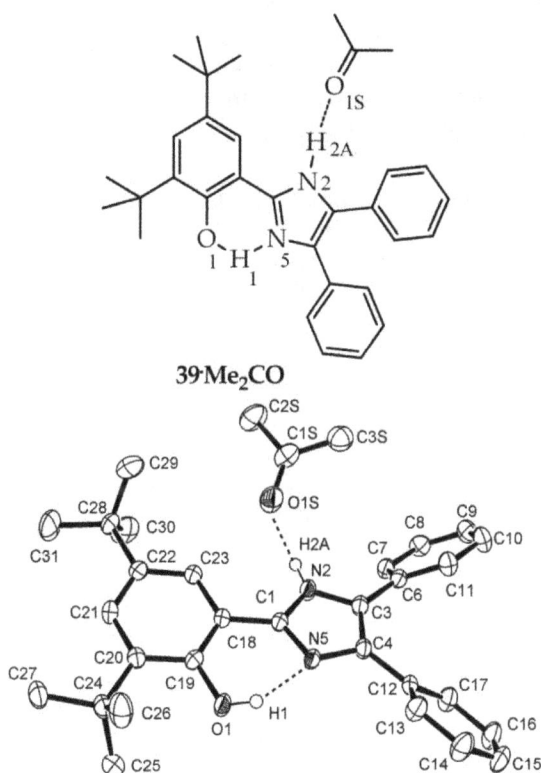

Figure 26. Formula and crystal structure of **39·Me₂CO**.

The azamacrocyclic ligand 1,4,7-triazacyclononane or TACN, **40**, has attracted considerable interest in recent years for its applications in oxidative catalysis. Another application of this compound was discussed by Pulacchini, et al. "(Pulacchini et al., 2003)". The incorporation of the 1,2-diaminocyclohexane moiety into a 1,4,7-triazacyclononane macrocyclic ligand was done by this research group, as it is an inexpensive starting material and both enantiomers are readily available. Moreover, this chiral framework has been included in a number of ligands that have been successfully applied in a range of asymmetric catalytic processes by Jacobsen et al. in metallosalen complexes "(Jacobsen & Wu, 1999)".

Crystal structure of **41** is shown in Fig. 27 and revealed the structure of the macrocyclic ligand in which the six-membered ring has chair conformation (Fig. 27). The asymmetric unit is completed by the two chloride ions and a water molecule in which all C–C, C–O and C–N bonds are unexceptional. Two short hydrogen bonding interactions of 2.724(4) Å between N(1)–H(01)····O(1) and 2.884(5) between N(2)–H(05)····O(1) within the macrocycle are then supplemented by an extensive hydrogen bonding network between the ammonium nitrogen atoms N(1) and N(2) the two chloride ions Cl(1) and Cl(2), as well as the water molecule of crystallisation, as shown in Fig. 28. The roles of the two chloride ions in the network are distinct with Cl(1) acting as a direct bridge between two macrocyclic moieties as well as linking to a third *via* a water molecule. In contrast, the second chloride ion, appears to essentially serve to template the macrocyclic ligand into the conformation observed *via* hydrogen bonding interactions with N(1)–H(01) and N(2)–H(05). The second chloride ion also links to other macrocyclic moieties *via* the water molecules.

The following hydrogen bond lengths (Å) were observed from the polymeric hydrogen bonding array in **41·2HCl·H₂O**.; N(1)–H(06)····Cl(1) 3.099(4), N(1)–H(01)····Cl(2) 3.185(4), N(1)–H(01)····N(2) 3.043(5), N(1)–H(01)····O(1) 2.724(4), N(2)–H(02)····Cl(1)#1 3.103(4), N(2)–H(05)····Cl(2) 3.108(3), N(2)–H(05)····O(1) 2.884(5), O(2)– H(04)····Cl(1) 3.271(4), O(2)–H(03)····Cl(2)#2 3.217(4) "(Pulacchini, (2003)".

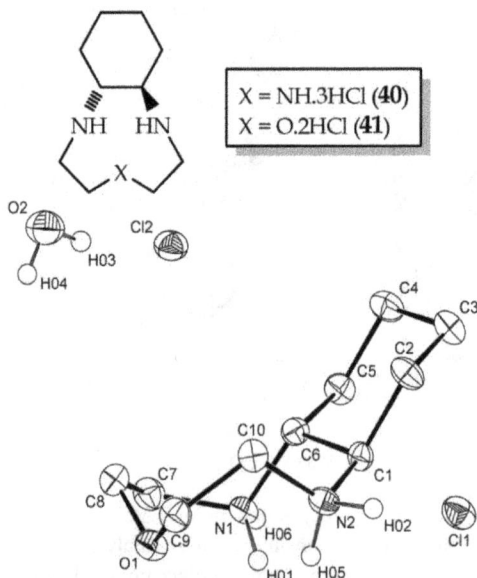

Figure 27. Formula structures of 40 and 41 and crystal structure of 41·2HCl·H₂O.

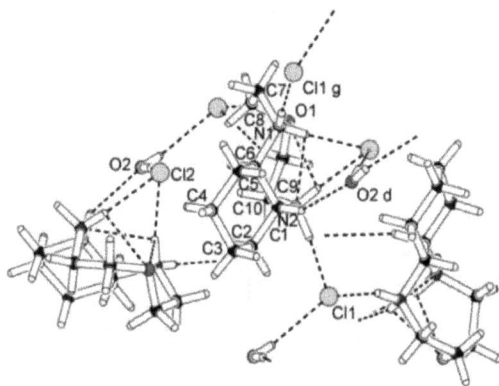

Figure 28. Polymeric hydrogen bonding network in **41·2HCl·H₂O** "(Pulacchini, (2003)".

In all thiohelicene crystals (see also Figs. 1 and 2) specific interactions were found involving sulfur "(Nakagawa et al., 1985; Yamada et al., 1981)" and hydrogen atoms at distances slightly shorter than the sum of van der Waals radii (1.80 Å for S and 1.20 Å for H). They are quite probably attractive, and, in all structures except **TH11** (hexathia-[11]-helicene **3**) they involve only atoms of terminal rings. In the case of the 5-ring system each molecule has two equivalent S···S interactions of 3.544 Å, while each **TH7** (tetrathia-[7]-helicene **1**) molecule is involved in four equivalent S···H contacts measuring 2.89 Å. All these interactions occur between enantiomeric pairs. Crystal structyre of pentathia-[9]-helicene (**TH9**, **42**) and crystal packing diagram of this compound including S···H contacts are shown in Figs. 29 and 30, respectively. For **42**, each molecule presents four equivalent S···H contacts at 2.87 Å, all with homochiral molecules giving rise to a quasi-hexagonal packing of tilted helices in planes parallel to the *ab* lattice plane. The crystal structure of **TH11** (**3**) is unusual because the asymmetric unit is formed by two complete molecules as opposed to half a molecule in all the lower racemic thiohelicenes. The packing environment of each of the two closely similar but crystallographically independent molecules, and of each of its halves, is unique: thus the C2 axes bisecting the central ring of each **TH11** (**3**) molecule are noncrystallographic. This situation is likely to arise in order to optimize the complex network of specific interactions involving S and H atoms. It leads to larger than expected asymmetric units and lower crystal symmetry, common occurrences in hydrogen bonded molecular systems. In the triclinic **TH11** (**3**) crystals four nonequivalent short S···S and an equal number of S···H interactions are found "(Caronna et al., 2001)". The essential geometric features of all these contacts in the racemic thiohelicene series and evidencing a remarkable consistency of the S···H interaction with expectations for weak hydrogen bonds have been reported "(Desiraju & Steiner, 2000)".

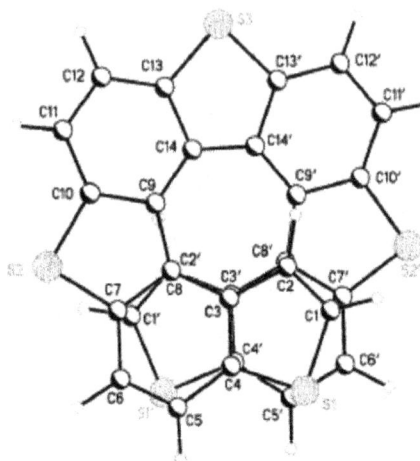

Figure 29. Crystal structure of **TH9** (**42**).

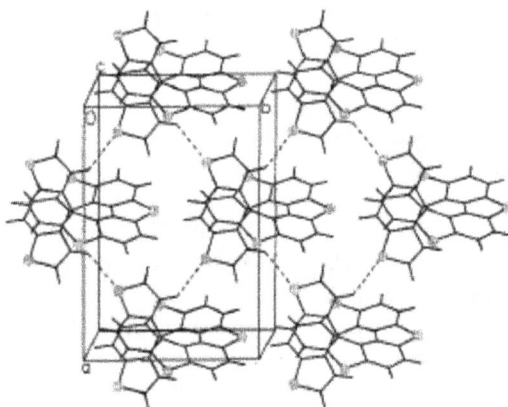

Figure 30. Crystal packing diagram of **42** in which each molecule consists of four equivalent S⋯H contacts.

The fused pyrimidines such as pyrimido[4,5-c]pyridazine-5,7(6H,8H)-diones, which are common sources for the development of new potential therapeutic agents, is well known "(Altomare et al., 1998; Brown, 1984; Hamilton, 1971)". Some of this class of compounds play new heterocyclizations based on $\left\{S_N^H\right\}$ methodology as N(2)-oxide and 3-alkylamino derivatives of 6,8-dimethylpyrimido[4,5-c]pyridazine-5,7(6H,8H)-dione "(Gulevskaya et al., 2003)".

Recently, the synthesis of 3-arylpyrimido[4,5-c]pyridazine-5,7(6H,8H)-diones (**43a–d**) and their sulfur analogs 3-aryl-7-thioxo-7,8-dihydropyrimido[4,5-c]pyridazin-5(6H)-ones **44a–d** have been reported "(Rimaz et al., 2010)". One of the most interesting intermolecular hydrogen bond in **43a–d** have been reported by our research group "(Rimaz et al., 2010)" (Figure 31). Owing to the less solubility of **43a–d** and **44a–d**, an attempt to achieve the single crystal of these compounds for investigation of the clustered water in their crystalline structure was failed. The ^{1}H NMR spectra of **43a–d** show two broad singlets in the range of $\delta = 7.00$–8.00 ppm that correspond to the protons of clustered water molecule in the **43a-d**. The chemical shift values of two variable protons of water in **43a–d** in ambient temperature are shown in Table 11. There are some reasons for demonstration and interpretation of this criterion. (i) One of the evidence is the mass spectra. The mass spectra of the compounds **43a–d** show not only the molecular ion fragment (M) but also the fragment of M+18. Therefore, the strength of hydrogen bond between the proton of H$_2$O (H$_a$) and oxygen atom of carbonyl group (C5=O$\cdots$ H$_a$–O) and also hydrogen bond between the N6–H of **43a–d** and oxygen atom of H$_2$O (N6–H$\cdots$ $\cdot$O–H$_a$) is considered more than that of the hydrogen bonding in the dimer form of **43a–d** (judging by the observation of the M+18 ion) (Fig. 32) "(Rimaz et al., 2010)". It seems that at least one molecule of water clustered and joined to **43** and **44** by two strong intermolecular hydrogen bonds and dissociated neither by DMSO molecules as a polar aprotic solvent nor in mass ionization chamber. Presumably, this intermolecular hydrogen bond is of quasi-covalent hydrogen bond type. There are some reports on literatures about quasi-covalent hydrogen bonds "(Dabbagh et al., 2007; Gilli et al., 1994, 2000, 2004; G. Gilli & P. Gilli, 2000; Golič et al., 1971; Madsen et al., 1999; Nelson, 2002; Steiner, 2002; Vishweshwar et al., 2004; Wilson, 2000)".

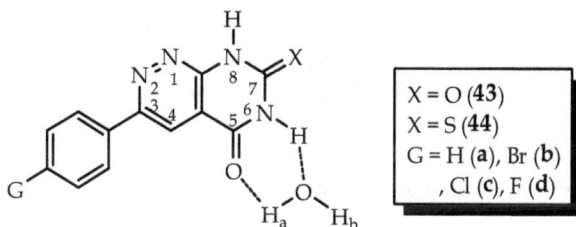

Figure 31. Formula structures of **43a–d·(H₂O)** and **44a–d·(H₂O)**.

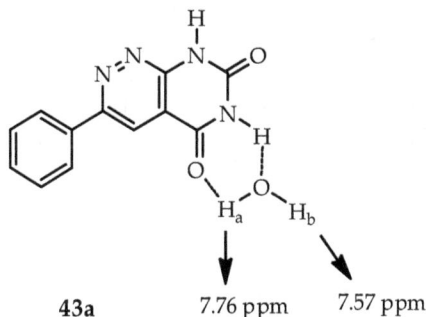

Figure 32. Representatively, strong intermolecular hydrogen bond and the chemical shifts of two hydrogen bonded protons of clustered water molecule with **43a** "(Rimaz et al., 2010)".

Compd.	δ (ppm)		
	Ha		Hb
43a	7.76		7.57
43b	7.75		7.61
43c	7.74		7.61
43d	7.75		7.61
44a		4.89	
44b		4.90	
44c		4.90	
44d		4.89	

[a] Two protons of water are equivalent in chemical shift appeared up-fielded as a broad singlet in **44a–d**.

Table 11. The chemical shift values of the two protons of a clustered water molecule in **43a–d** and **44a–d**[a] at ambient temperature "(Rimaz et al., 2010)".

The proton/deuterium exchange was examined on **43a–d** by adding one drop of D₂O. Interestingly, from hydrogen to fluorine substituent on phenyl ring in **43a–d** the exchange rate was decreased, and no deuterium exchanging of Ha and Hb was observed in **43d** while the amide protons were easily exchanged (Fig. 33). This phenomenon attributed to the fluorine atom that has made new intermolecular hydrogen bond with Ha and Hb of clustered water molecule in another molecule of **43d**. The intermolecular hydrogen bond of fluorine with the proton of clustered water (–F···Ha– and –F···Hb–) in **43d** inhibited the proton/deuterium exchanging of the clustered water protons. However, the electronegativity of fluorine atom caused deshielding of Ha and Hb on **43d** and blocked the proton/deuterium exchange (Fig. 33 and Scheme 10). Two conformational forms of **IA** and **IB** in **43d** are equivalent because of free rotation of phenyl ring about the C3–C9 and C12–F single bonds (Scheme 10) "(Rimaz et al., 2010)".

2.3. Crystal structure of some organic spiro compounds

Spiro compounds are very important and useful compounds and versatile applications. Many of heterocyclic spirobarbituric acids "(Kotha et al., 2005)", furo[2,3-d]pyrimidines "(Campaigne et al., 1969)" and fused uracils "(Katritzky & Rees, 1997; Naya et al., 2003)" are well known for their pharmaceutical and biological effects.

Recently, we have reported new spiro compound based on barbiturates; 5-alkyl and/or 5-aryl-1H, 1'H-spiro[furo[2,3-d]pyrimidine-6,5'-pyrimidine]2,2',4,4',6'(3H, 3'H, 5H)-pentaones which are dimeric forms of barbiturate (uracil and thiouracil derivatives) "(Jalilzadeh et al., 2011)". Reaction of 1,3-dimethyl barbituric acid (DMBA) with cyanogen bromide (BrCN) and acetaldehyde in the presence of triethylamine afforded 1,1',3,3',5-pentamethyl-1H,1'H-spiro[furo[2,3-d]pyrimidine-6,5'-pyrimidine]-2,2',4,4',6'(3H,3'H,5H)-pentaone (**46**) in excellent yield "(Jalilzadeh et al., 2011)". The formula structures of spiro compounds derived from barbituric acid (BA, **45**), DMBA **46** and 1,3-thiobarbituric acid (TBA, **47**) is shown in Fig. 34. Attempt for single crystallization of spiro compounds **45** and **47** were unsuccessful. The crystal structure and crystal packing diagram of **46** are shown in Figs. 35 and 36. This compound was crystalized in triclinic system. Selected crystallographic data for **46** is summarized at Table 12.

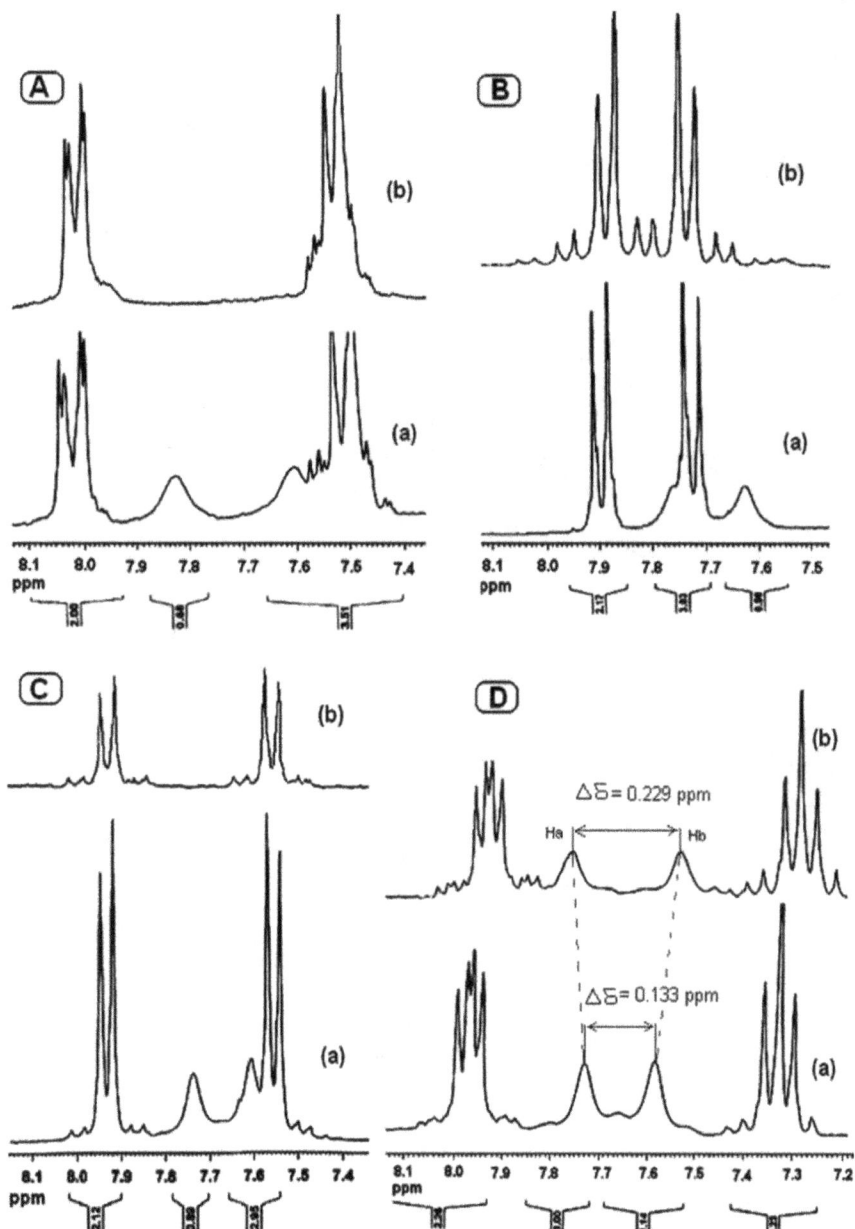

Figure 33. Proton/deuterium exchangeability of the H_a and H_b of clustered H_2O molecule in ^1H NMR spectra of **43a** (A), **43b** (B), **43c** (C) and **43d** (D). The assigned spectra are shown before (a) and after added D_2O (b). No exchange occurred in **43d** of clustered H_2O protons (D) "(Rimaz et al., 2010)".

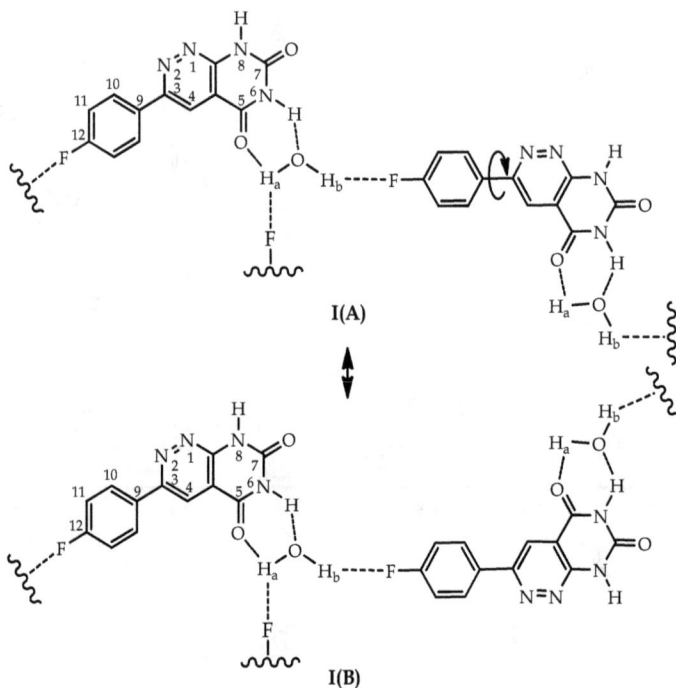

I(A)

I(B)

Scheme 10. Possible various types of intermolecular hydrogen bond of fluorine with a proton of a clustered were (-F····H_b- and -F····H_a-) in **43d**. This phenomenon presumably inhibited the proton/deuterium exchangeability of the clustered water protons.

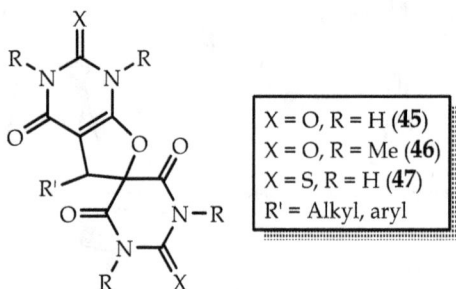

| X = O, R = H (**45**) |
| X = O, R = Me (**46**) |
| X = S, R = H (**47**) |
| R' = Alkyl, aryl |

Figure 34. Formula structures of **45-47**.

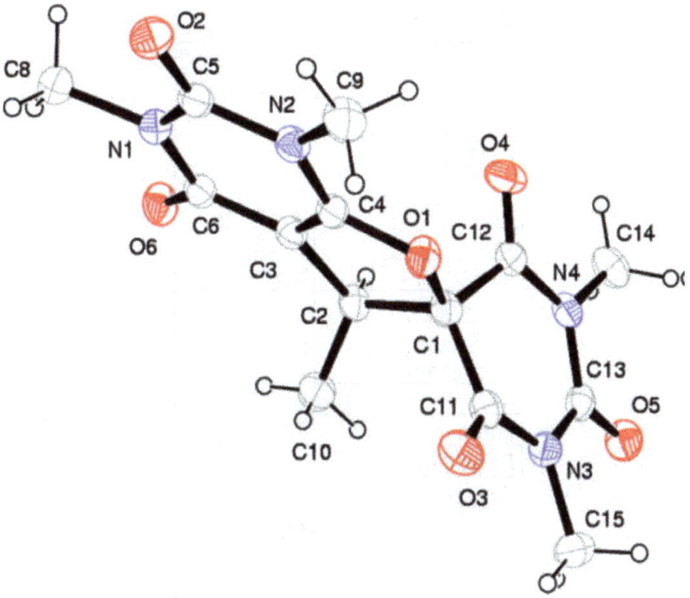

Figure 35. Crystal structure of **46.**

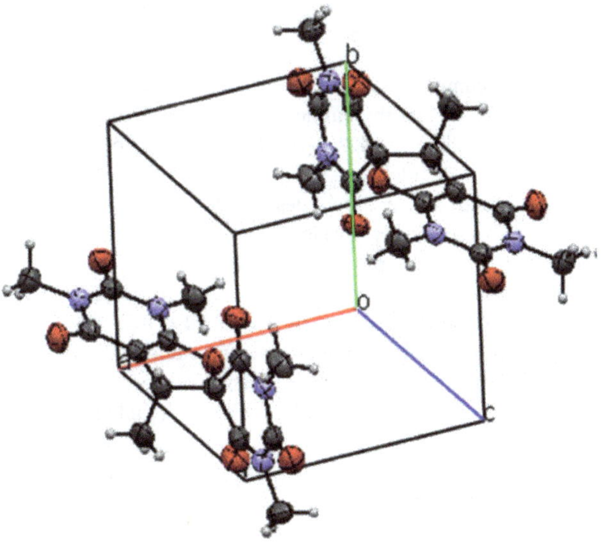

Figure 36. Crystal packing diagram of **46.**

Crystal data	
Emprical formula	$C_{14}H_{16}N_4O_6$
M	336.30
T	298 K
a (Å)	8.974 (5)
b (Å)	9.539 (5)
c (Å)	10.314 (5)
α (°)	64.782 (5)
β (°)	69.349 (5)
γ (°)	69.349 (5)
V (Å3)	725.8 (7)
Z	2
F(000)	352
Dx (mg m-3)	1.539
λ (Å)	0.71073
μ(mm-1)	0.12
Data collection	
Rint	0.062
θmax	29.0 °
θmin	2.3 °
Refinement	
R[F2 > 2σ(F2)]	0.067
wR(F2)	0.203
S	1.04

Table 12. Selected crystallographic data for **46**.

Another spiro barbiturate compound derived from the reaction of DMBA with BrCN and acetone in the presence of triethylamine is 1,1′,3,3′,5,5′-Hexamethylspiro[furo-[2,3-d]pyrimidine-6(5H),5′-pyrimidine]-2,2′,4,4′,6′(1H,3H,1′H,3′H,5H)-pentaone **(48)** "(Noroozi Pesyan et al., 2009)". Reaction of aldehydes with (thio)barbiturates is faster than ketones due to the reactivity and less hindrance in aldehydes. The formula and crystal structure of **48** is shown in Figs. 37 and 38, respectively. In Fig. 38, the fused 2,3-dihydrofurane ring has an envaloped conformation, and spiro pyrimidine ring has a half-chair conformation. Spiro pyrimidine ring is nearly perpendicular to 2,3-dihydro furan ring moiety as was observed earlier in the related compound. Torsion angles C2-C1-O4-C7 and C2-C1-C5-C6 are -99.39(3)° and 94.87(3) °, respectively. In the crystal, short intermolecular interaction O⋯C contacts between the carbonyl groups prove an existing of electrostatic interactions, which link the molecules into corrugated sheets parallel to *ab* plane (Table 13).

C8⋯O2i	2.835 (4)	C3⋯O5ii	2.868 (4)

Symmetry codes: (i) -x + 1, y -1/2, -z +3/2; (ii) x+1, y, z

Table 13. Selected interatomic distances (Å) in **48**.

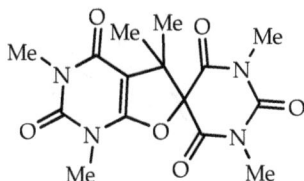

Figure 37. Formula structure of **48**.

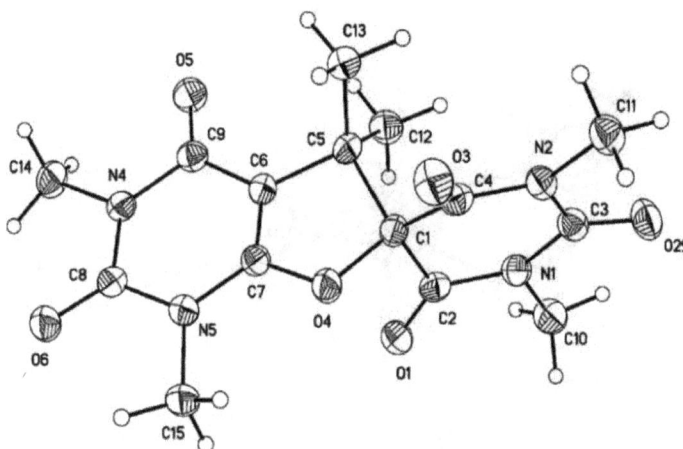

Figure 38. Crystal structure of **48**.

One of another interesting spiro barbiturate compound is the trimeric form of 1,3- DMBA; 5,6-dihydro - 1,3-dimethyl - 5,6 – bis - [1′,3′-dimethyl-2′,4′,6′-trioxo-pyrimid(5′,5′)yl]furo[2,3-d]uracil (**49**). This compound was first reported by electrochemical method "(Kato et al., 1974; Kato & Dryhurst, 1975; Poling & van der Helm, 1976)" and it has been reported the synthesis of **49** by chemical method for a first time two years ago "(Hosseini et al., 2011)". The formula and crystal structures of **49** are shown in Figs. 39 and 40, respectively. Crystals of **49** were obtained by slow evaporation of a solution of **49** in acetone at room temperature. The data were acquired using a STOE IPDS II diffractometer, data collection and cell refinement were processed using STOE X-AREA "(Stoe & Cie, 2002)" and data reduction was processed using STOE X-RED "(Stoe & Cie, 2002)" program. Program(s) used to refine structure was *SHELXL97* "(Sheldrick, 1997). Crystal data for **49**: Orthorhombic; $C_{18}H_{18}N_6O_9$; M = 462.38; Unit cell parameters at 293(2) K: a = 13.2422(4), b = 15.9176(6), c = 19.5817(6) Å; $\alpha = \beta = \gamma$ = 90°; V = 4127.5(2) Å^3; Z = 8; μ = 0.122 mm^{-1}; Total reflection number 4275; 304 parameters; λ = 0.71073 Å; 2916 reflections with $I > 2\sigma(I)$; R_{int} = 0.056; θ_{max} = 26.49°; $R[F2 > 2\ \sigma(F2)]$ = 0.048; $wR(F2)$ = 0.112; S = 1.02, F$_{000}$ = 1920 "(Hosseini et al., 2011)".

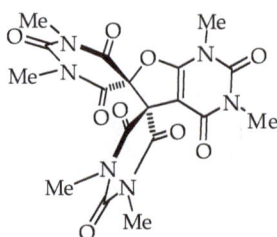

Figure 39. Formula structure of **49**.

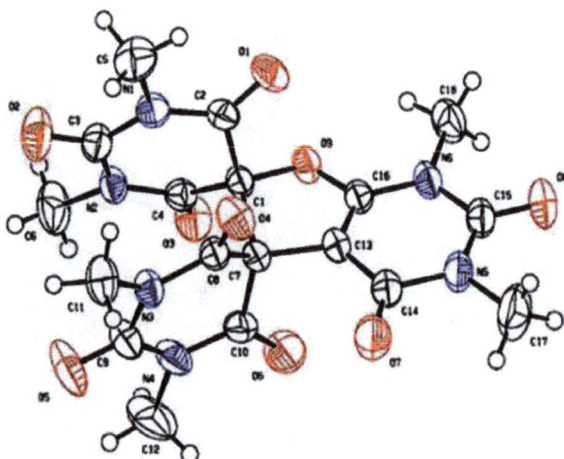

Figure 40. Crystal structure of **49**.

Amino acids derived from sugar are of extensive family of peptidomimetics "(Baron et al., 2004; Chakraborty et al., 2004)", an important sub-class of which incorporate an α-amino acid with a carbohydrate has anomeric effect. Such sugar amino acids may form spiro derivatives, some of which have been demonstrated to possess significant biological activity. For instance, the formula and crystal structure of (2'S,3aR,6S,6aR)-2,2,6-trimethyldihydro-3aH-spiro[furo[3,4-d][1,3]dioxole-4,2'-piperazine]-3',6'-dione (**50**) are shown in Figs. 41 and 42 "(Watkin et al., 2004)". This molecule show hydrogen bonds between N-H....O=C groups and are shown in crystal packing diagram, viewed along the c axis as dashed lines (Fig. 43).

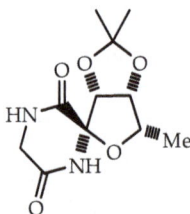

Figure 41. Formula structure of compound **50**.

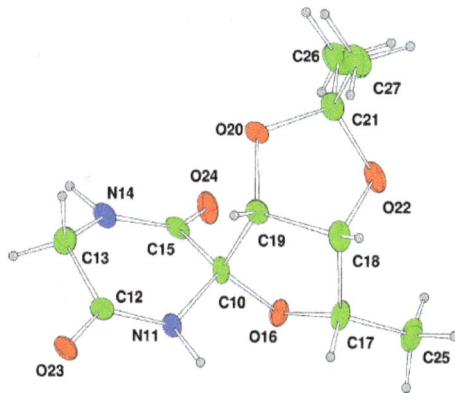

Figure 42. Crystal structure of compound **50** (Green: C, blue: N and red: O atom).

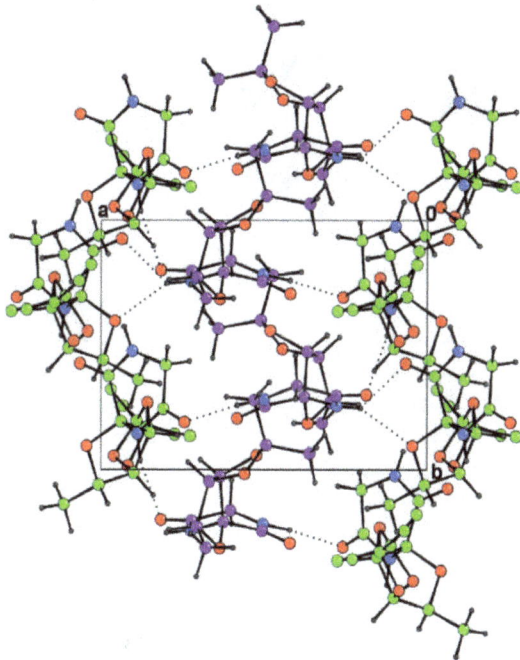

Figure 43. Crystal packing diagram of **50**.

Another interesting spiro linked barbituric acid to the cyclopentane ring moiety (spiro-nucleoside) possessing of hydroxyl and hydroxymethyl groups is (3S,2R)-3-hydroxy-2-hydroxymethyl-7,9-diazaspiro[4.5]decane-6,8,10-trione (**51**) (Figs. 44 and 45). Crystal structure of **51** shows *trans* stereochemical relationship of the two substituents hydroxyl and hydroxymethyl on cyclopentane ring moiety. The barbituric acid ring is almost planar, while the cyclopentane moiety adopts the C3'-*endo*-type conformation. Molecules of **51** interconnected by a two-dimensional network of hydrogen bonds build layers parallel to the *ab* plane. The hydrogen bond data for **51** is outlined at Table 14 "(Averbuch-Pouchot et al., 2002)".

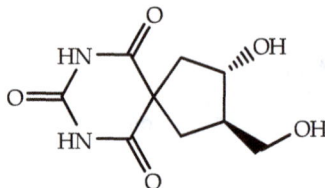

Figure 44. Formula structure of **51**.

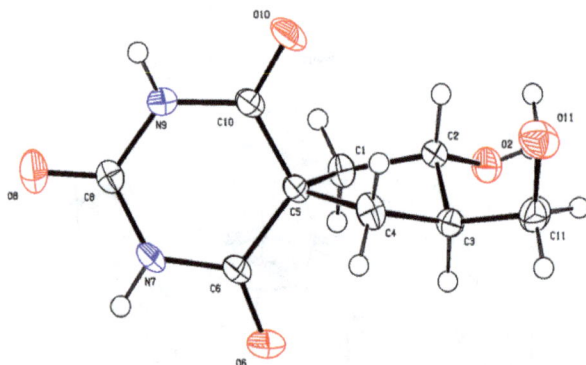

Figure 45. Crystal structure of **51**.

D−H····A	D−H	H····A	D····A	D−H····A
O11−H12····O6ᵛ	0.81	2.00	2.809 (2)	173
N7−H8····O10ⁱᵛ	0.86	1.99	2.840 (2)	170
N9−H9····O2ᵛⁱⁱ	0.85	2.04	2.8620 (10)	161

Symmetry codes: (iv) x, y–1, z; (v) x, y+1, z; (vii) x–1, y, z.

Table 14. Hydrogen-bond geometry in **51** (Å, º).

Hydantoins are very useful compounds due to their pharmaceutical behaviour such as; antitumor "(Kumar et al., 2009)", anticonvulsant "(Sadarangani et al., 2012)" and antidiabetic activity "(Hussain et al., 2009)". In the molecules of **52** and **53** (Figs. 46 and 47), the atoms in the hydantoin ring are coplanar. The crystal structures of **52** and **53** are stabilized by intermolecular N−H····O=C hydrogen bonds. The hydrogen bond lengthes and angles for **52** and **53** are summarized at Table 15. Crystal packing diagram of these molecules show the molecules are centrosymmetric dimer forms. The dihedral angle subtended by the 4-chloro- and 4-bromophenyl groups with the plane passing through the hydantoin unit are 82.98(4)° and 83.29(5)°, respectively. The cyclohexyl ring in both molecules adopts an ideal chair conformation and methyl group in an equatorial position "(Kashif et al., 2009)".

	D−H⋯A	D−H	H⋯A	D⋯A	D−H⋯A
52	N2−H2⋯O4ⁱ	0.84 (2)	2.04 (2)	2.8763 (15)	171.5 (19)
53	N2−H2⋯O4ⁱ	0.82 (3)	2.06 (3)	2.871 (2)	171 (3)

Symmetry code: (i) -x + 1;-y + 1;-z + 1.

Table 15. Hydrogen-bond geometry in **52** (Å, °).

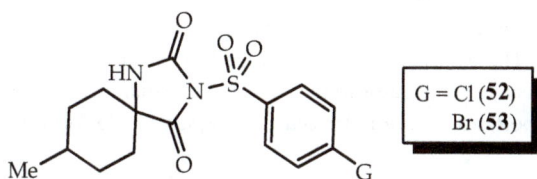

Figure 46. Formula structures of **52** and **53**.

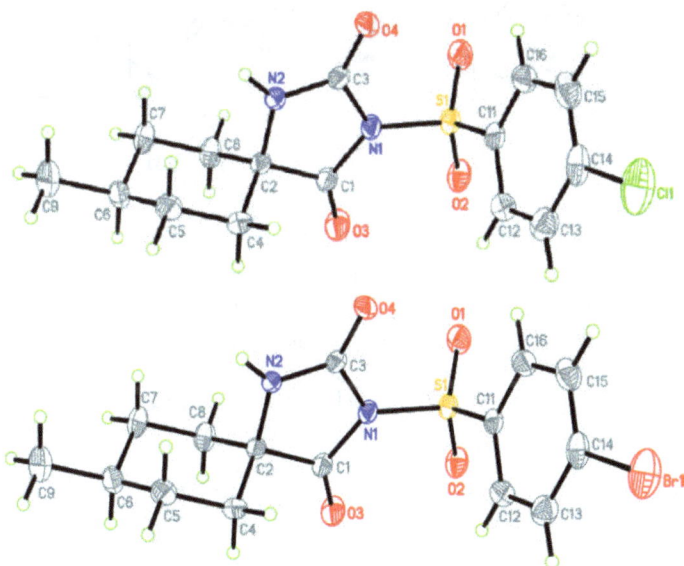

Figure 47. Crystal structures of **52** (top) and **53** (bottom).

Dihydropyridine are interesting and important systems because of their exceptional properties as calcium channel antagonists "(Si et al., 2006)" and as powerful arteriolar vasodilators "(Kiowski et al., 1990)". 4',4'-Dimethyl-2-methylsulfanyl-3,4,5,6,7,8-hexahydropyrido-[2,3-d]pyrimidine-6-spiro-1'-cyclohexane-2',4,6'-trione, **(54)**, has a markedly polarized molecular electronic structure, and the molecules are linked into a three-dimensional framework by a combination of N–H⋯O, C–H⋯O and C–H⋯A hydrogen bonds (Table 16). Two independent N–H⋯O hydrogen bonds generate a one-dimensional substructure in the form of a chain of rings; these chains are linked into sheets by the C–H····O hydrogen bonds, and the sheets are linked by C–H····𝛼 hydrogen bonds. Crystal packing diagram of **54** show four types of centrosymmetric ring. "(Low et al., 2004)" (Fig. 48). Compound **54** can exist in two zwitterionic forms of **54I** and **54II** (Scheme 11). For example, the bond lengths of N3–C4 and C4–O4 are both long for their types, the C4–C4A and C4A–C8A bonds are too similar in length to be characterized as single and double bonds, respectively. Also, the C8A–N8 bond, involving a three-coordinate N atom, is much shorter than the C8A–N1 bond, which involves a two-coordinate N atom. These observations, taken together, effectively preclude the polarized form (**54I**) as an effective contributor to the overall molecular electronic structure, instead pointing to the importance of the polarized vinylogous amide form (**54II**) "(Low et al., 2004)".

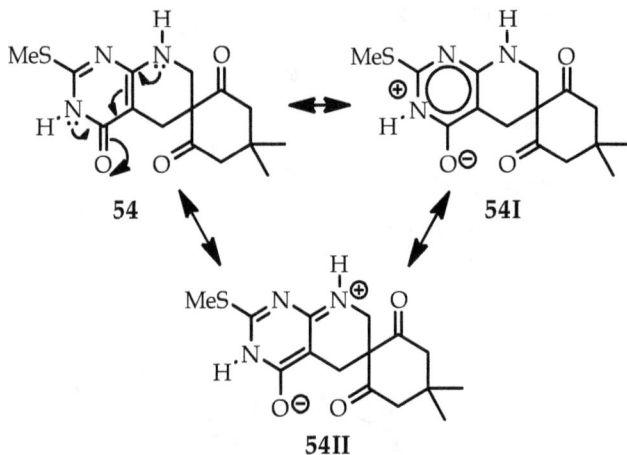

Scheme 11. Zwitterionic forms of **54**.

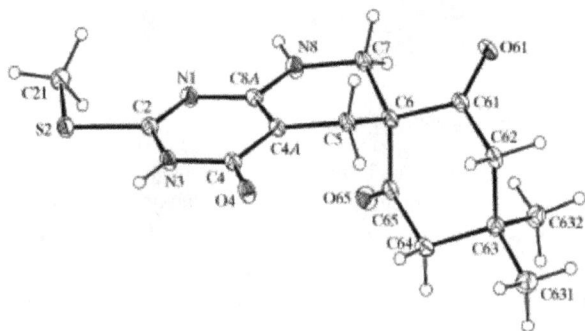

Figure 48. Crystal structure of **54**.

D–H····A	D–H	H····A	D····A	D–H····A
N3–H3····O4ⁱ	0.88	1.84	2.715 (2)	176
N8–H8····O65ⁱⁱ	0.88	2.10	2.965 (2)	166
C5–H5B....O61ⁱⁱⁱ	0.99	2.46	3.389 (2)	155
C64–H64A....Cg1ⁱᵛ	0.99	2.87	3.854 (2)	173

Symmetry codes: (i) 1 - x; 1 - y; 1 - z; (ii) -x; 1 - y; -z; (iii) -x; 2 - y;-z; (iv) 1 -x;-y;-z.
Cg1 is the centroid of the pyrimidinone ring.

Table 16. Hydrogen-bonding geometry (Å, °) for **54**.

3. Conclusion

In summary, X-ray single crystal diffraction analysis of the some helicenes and other helix molecules were discussed. In continuation, the crystal structure of some organic and organometallic compounds consists of intra- and/or intermolecular hydrogen bond were described. Finally, crystal structures of some new spiro compounds were analyzed.

Author details

Nader Noroozi Pesyan
Urmia University, Iran

Acknowledgement

The author gratefully acknowledge financial support by Research Council of Urmia University

4. References

Acerbis, S; Bertin, D; Boutevin, B; Gigmes, D; Lacroix-Desmazes, P; Le Mercier, C; Lutz, J-F; Marque, SRA; Siri, D & Tordo, P (2006). Intramolecular Hydrogen Bonding: The Case of β-Phosphorylated Nitroxide (=Aminoxyl) Radical, *Helve. Chim. Acta*, 89, 2119.

Altomare, C; Cellamare, S; Summo, L; Catto, M; Carotti, A; Thull, U; Carrupt, P-A; Testa, B & Stoeckli-Evans, H (1998). Inhibition of Monoamine Oxidase-B by Condensed Pyridazines and Pyrimidines: Effects of Lipophilicity and Structure-Activity Relationships, *J. Med. Chem.*, 41, 3812-3820.

Arai, S & Hida, M In *Adv. Heterocycl. Chem.*; Katritzky, AR, Ed.; Academic: San Diego, 1992; Vol. 55, p 261.

Averbuch-Pouchot, M-T; Durif, A; Renard, A; Kottera, M & Lhomme, J (2002). (3S,2R)-3-Hydroxy-2-hydroxymethyl-7,9-diazaspiro[4.5]decane-6,8,10-trione, *Acta Cryst.*, E58, o256-o258.

Azumaya, I; Uchida, D; Kato, T; Yokoyama, A; Tanatani, A; Takayanagi, H & Yokozawa, T (2004). Absolute Helical Arrangement of Stacked Benzene Rings: Heterogeneous

Double-Helical Interaction Comprising a Hydrogen-Bonding Belt and an Offset Parallel Aromatic–Aromatic-Interaction Array, *Angew. Chem. Int. Ed.* 43, 1360-1363.

Baron, R; Bakowies, D & van Gunsteren, WF (2004). Carbopeptoid Folding: Effects of Stereochemistry, Chain Length, and Solvent, *Angew. Chem. Int. Ed.*, 43, 4055-4059. doi:10.1002/anie.200454114

Benisvy, L; Blake, AJ; Collison, D; Davies, ES; Garner, CD; McInnes, EJL; McMaster, J; Whittaker, G & Wilson, C (2003). A Phenol–imidazole Pro-ligand That Can Exist as a Phenoxyl Radical, Alone and When Complexed to Copper(II) and Zinc(II), *Dalton Trans.*, 1975.

Benisvy, L; Blake, AJ; Davies, ES; Garner, CD; McMaster, J; Wilson, C; Collinson, D.; McInnes, EJL & Whittaker, G; (2001). A Phenoxyl Radical Complex of Copper(II), *Chem. Commun.*, 1824.

Bossi, A; Falciola, L; Graiff, C; Maiorana, S; Rigamonti, C; Tiripicchio, A; Licandro, E & Mussini, PR (2009). Electrochemical Activity of Thiahelicenes: Structure Effects and Electrooligomerization Ability, *Electrochimica Acta* 54 5083.

Bragg, WH (1925). The Investigation of Thin Films by Means of X-rays, *Nature*, 115, 266-269. doi:10.1038/115266a0

Braunstein, P; Chauvin, Y; Mercier, S; Saussine, L; DeCian, A & Fischer, J (1994). Intramolecular O–H···O–Ni and N–H···O–Ni Hydrogen Bonding in Nickel Diphenylphosphinoenolate Phenyl Complexes: Role in Catalytic Ethene Oligomerisation; Crystal Structure of [NiPH{Ph$_2$PCH$\equiv$C($\equiv$O)(o-C$_6$H$_4$NHPh)}(PPh$_3$)], *J. Chem. Soc., Chem. Commun.*, 2203. doi: 10.1039/C39940002203

Braunstein,P; Chauvin, Y; Mercier, S & Saussine, L (2005). Influence of Intramolecular N–H···O–Ni Hydrogen Bonding in Nickel(II) Diphenylphosphinoenolate Phenyl Complexes on the Catalytic Oligomerization of Ethylene, *C. R. Chimie*, 8, 31.

Brown, DJ (1984). In *Comprehensive Heterocyclic Chemistry*, vol. 3, (Eds: Katritzky, AR; Rees, CW), Pergamon Press: Oxford, pp 57.

Campaigne, E; Ellis, RL; Bradford, M & Ho, J (1969). Synthesis of Some Ureidodihydrofurans and Related Pyrimidones as Potential Antimalarials, *J. Med. Chem.* 12, 339.

Caronna, T; Catellani, M; Luzzati, S; Malpezzi, L; Meille, SV; Mele, A; Richter, C & Sinisi, R (2001). Molecular Crystal Architecture and Optical Properties of a Thiohelicenes Series Containing 5, 7, 9, and 11 Rings Prepared via Photochemical Synthesis, *Chem. Mater.* 13, 3906.

Casado, J; Patchkovskii, S; Zgierski, MZ; Hermosilla, L; Sieiro, C; Oliva, MM & Navarette, JTL (2008). Raman Detection of "Ambiguous" Conjugated Biradicals: Rapid Thermal Singlet-to-Triplet Intersystem Crossing in an Extended Viologen, *Angew. Chem., Int. Ed.*, 47, 1443.

Caspari, WA (1928). Crystallography of the Aliphatic Dicarboxylic Acids, *J. Chem. Soc. (London)*, 3235-3241.

Chakraborty, TK; Srinivasu, P; Tapadar, S & Mohan, BK (2004). Sugar Amino Acids and Related Molecules: Some Recent Developments, *Indian J. Chem. Sci.*, 116, 187-207.

Clays, K; Wostyn, K; Persoons, A; Maiorana, S; Papagni, A; Daul, CA & Weber, V (2003). Experimental Study of the Second-Order Non-linear Optical Properties of Tetrathia-[7]-helicene, *Chem. Phys. Lett.* 372, 438–442.

Collins, SK & Vachon, MP (2006). Unlocking the Potential of Thiaheterohelicenes: Chemical Synthesis as the Key, *Org. Biomol. Chem.* 4, 2518.

Corbridge, DEC (1990). *Phosphorous: An outline of it s chemistry, biochemistry and technology*. 4th ed., Elsevier, Amsterdam, Chapt. 14.

Cox, C; Wack, H & Lectka, T (1999). Strong Hydrogen Bonding to the Amide Nitrogen Atom in an "Amide Proton Sponge": Consequences for Structure and Reactivity, *Angew. Chem. Int. Ed.*, 38, 798. doi:10.1002/(SICI)1521-3773(19990315)38:6<798::AID-ANIE798>3.0.CO;2-W

Dabbagh, AH; Noroozi Pesyan, N; Najafi, CA; Brian, OP & Brian, RJ (2007). Diastereoselective Formation of 18-Membered Ring BINOL-hydrogenphosphonate Dimers-Quasi-covalent Hydrogen Bonds?, *Canadian J. Chem.*, 85, 466-474.

Dabbagh, HA & Lwowski, W (2000). Equilibria of the 5-Substituted-1,2-acylated Tetrazoles and Imidoyl Azides, *J. Org. Chem.*, 65, 7284-7290.

Dabbagh, HA; Noroozi-Pesyan, N; Patrick, BO & James, BR (2004). A One Pot Synthesis, and X-Ray Crystallographic and Computational Analyses of Methyl-2,4-dimethoxysalicylate — a Potential Anti-tumor Agent, *Can. J. Chem.*, 82, 1179.

Davies, C; Langler, RF; Krishnamohan, CV & Zaworotko, MJ (1997). A Supramolecular Carpet Formed via Self-assembly of Bis(4,4'-dihydroxyphenyl) sulfone, *Chem. Commun.*, 567.

de Broglie, M & Trillat, JJ (1925). Sur L'interprétation Physique Des Spectres X D'acides Gras, *Comptes Rendus Hebdomadaires Des Séances de L'Académie Des Sciences* 180, 1485.

Desiraju, GR & Steiner, T (2000). *The Weak Hydrogen Bond*; International Union of Crystallography, Oxford Science Publication: Oxford, U.K.

Dickinson, RG & Raymond, AL (1923). The Crystal Structure of Hexamethylene-Tetramine, *J. Am. Chem. Soc.*, 45, 22. doi:10.1021/ja01654a003

Fitjer, L; Gerke, R; Weiser, J; Bunkoczi, G & Debreczeni, JE (2003). Helical Primary Structures of Four-membered Rings: (M)-trispiro[3.0.0.3.2.2]tridecane, *Tetrahedron*, 59, 4443.

Fuchter, MJ; Weimar, M; Yang, X; Judge, DK & White, AJP (2012). An Unusual Oxidative Rearrangement of [7]-Helicene, *Tetrahedron Lett.* 53, 1108-1111. doi:10.1016/j.tetlet.2011.12.082

Garcia, MH; Florindo, FP; Piedade, MFM; Maiorana, S & Licandro, E (2009). New Organometallic Ru(II) and Fe(II) Complexes with Tetrathia-[7]-helicene Derivative Ligands, *Polyhedron*, 28, 621.

Gilli, G & Gilli, P (2000). Towards an Unified Hydrogen-Bond Theory, *J. Mol. Struc.*, 552, 1.

Gilli, P; Bertolasi, V; Ferreti, V & Gilli, G (1994). Covalent Nature of the Strong Homonuclear Hydrogen Bond. Study of the O–H····O System by Crystal Structure Correlation Methods, *J. Am. Chem. Soc.*, 116, 909.

Gilli, P; Bertolasi, V; Ferretti, V & Gilli, G (2000). Evidence for Intramolecular N-H····O Resonance-Assisted Hydrogen Bonding in β-Enaminones and Related Heterodienes. A

Combined Crystal-Structural, IR and NMR Spectroscopic, and Quantum-Mechanical Investigation, *J. Am. Chem. Soc.*, 122, 10405-10417.

Gilli, P; Bertolasi, V; Pretto, L; Ferretti, V & Gilli, G (2004). Covalent versus Electrostatic Nature of the Strong Hydrogen Bond: Discrimination among Single, Double, and Asymmetric Single-Well Hydrogen Bonds by Variable-Temperature X-ray Crystallographic Methods in β-Diketone Enol RAHB Systems, *J. Am. Chem. Soc.*, 126, 3845-3855.

Golič, L; Hadži, D & Lazarini, F (1971). Crystal and Molecular Structure of the Hydrogen-bonded Adduct of Pyridine N-oxide with Trichloroacetic Acid, *J. Chem. Soc. D: Chem. Commun.*, 860a. doi:10.1039/C2971000860a

Groen, MB; Schadenberg, H & Wynberg, H (1971). Synthesis and Resolution of Some Heterohelicenes, *J. Org. Chem.* 36, 2797, and references therein.

Gulevskaya, AV; Serduke, OV; Pozharskii, AF & Besedin, DV (2003). 6,8-Dimethylpyrimido[4,5-c]pyridazine-5,7(6H,8H)-dione: New Heterocyclizations Based on $\{S_N^H\}$ -Methodology. Unexpected Formation of the First iso-π-Electronic Analogue of the Still Unknown Dibenzo[a,o]pycene, *Tetrahedron*, 59, 7669-7679.

Guo, Z-Q; Chen, W-Q & Duan, X-M (2011). Seven-Membered Ring Excited-state Intramolecular Proton-Transfer in 2-Benzamido-3-(pyridin-2-yl)acrylic Acid, *Dyes and Pigments*, 92, 619-625.

Hamilton, GA (1971). In *Progress in Bioorganic Chemistry*, vol. 1, (Eds: Kaiser, ET; Kezdy, FJ), Wiley: New York, p 83.

Hosseini, Y; Rastgar, S; Heren, Z; Büyükgüngör, O & Noroozi Pesyan, N (2011). One-pot New Barbituric Acid Derivatives Derived from the Reaction of Barbituric Acids with BrCN and Ketones, *J. Chin. Chem. Soc.*, 58, 309.

Hussain, A; Hameed, S & Stoeckli-Evans, H (2009). 1-(4-Chlorophenylsulfonyl)-5-(4-fluorophenyl)-5-methylimidazolidine-2,4-dione, Acta Cryst., E65, o858–o859 and related references herein.

Hussain, A; Hameed, S & Stoeckli-Evans, H (2009). 1-(4-Methoxyphenylsulfonyl)-5-methyl-5-phenylimidazolidine-2,4-dione, Acta Cryst., E65, o1207–o1208 and related references herein.

Jacobsen, EN & Wu, MH (1999). *Comprehensive Asymmetric Catalysis Volume II*, ed. Jacobsen, EN; Pfaltz, A; Yamamoto, H, Springer-Verlag, Berlin-Heidleberg-New York, 1999, Chapt. 18.2, p. 649.

Jalilzadeh, M; Noroozi Pesyan, N; Rezaee, F; Rastgar, S; Hosseini, Y & Şahin, E (2011). New One-pot Synthesis of Spiro[furo[2,3-d]pyrimidine-6,5'-pyrimidine]pentaones and Their Sulfur Analogues, *Mol. Divers.*, 15, 721.

Jeffery, GA & Saenger, W (1991). *Hydrogen Bonding in Biological Structures*, Springer-Verlag, Berlin, Germany.

Jeffrey, GA & Saenger, W (1991). *Hydrogen Bonding in Biological Structures*, Springer-Verlag, Berlin, Chapt. 8.

Jeong, T-S; Kim, KS; Kim, J-R; Cho, K-H; Lee, S & Lee,WS (2004). Novel 3,5-Diaryl Pyrazolines and Pyrazole as Low-density Lipoprotein (LDL) Oxidation Inhibitor, *Bioorg. Med. Chem. Lett.*, 14, 2719-2723.

Johansson, MP & Patzschke, M (2009). Fixing the Chirality and Trapping the Transition State of Helicene with Atomic Metal Glue, *Chem. Eur. J.* 15, 13210.

Kashif, MK; Rauf, MK; Bolte, M & Hameed, S (2009). 3-(4-Chlorophenylsulfonyl)-8-methyl-1,3-diazaspiro[4.5]decane-2,4-dione, *Acta Cryst.*, E65, o1893.

Kashif, MK; Rauf, MK; Bolte, M & Hameed, S (2009).3-(4-Bromophenylsulfonyl)-8-methyl-1,3-diazaspiro[4.5]decane-2,4-dione, *Acta Cryst.*, E65, o1892.

Kato, S & Dryhurst, G (1975). Electrochemical Oxidation of Barbituric Acids at Low pH in the Presence of Chloride Ion, *J. Electroanal. Chem.* 62, 415-431.

Kato, S; Poling, M; van der Helm, D & Dryhurst, G (1974). Electrochemical Synthesis and Structure of a New Cyclic Barbiturate, *J. Am. Chem. Soc.* 96, 5255-5257.

Katritzky, AR & Rees, CW (eds.) (1997) Comprehensive Heterocyclic Chemistry, vol. 7. Pergamon Press, Oxford.

Katz, TJ & Pesti, J (1982). Synthesis of a Helical Ferrocene, *J. Am. Chem. Soc.* 104, 346.

Katz, TJ (2000). Syntheses of Functionalized and Aggregating Helical Conjugated Molecules, *Angew. Chem. Int. Ed.* 39, 1921.

Kiowski, W; Erne, P; Linder, L; Bühler, FR (1990). Arterial Vasodilator Effects of the Dihydropyridine Calcium Antagonist Amlodipine Alone and in Combination with Verapamil in Systemic Hypertension, *Am. J. Cardiol.*, 66,1469-1472.

Kotha, S; Deb, AC & Kumar, RV (2005). Spiro-annulation of Barbituric Acid Derivatives and Its Analogs By Ring-Closing Metathesis Reaction, *Bioorg. Med. Chem. Lett.*, 15, 1039.

Kumar, BCSA; Swamy, SN; Sugahara, K & Rangappa, KS (2009). Anti-tumor and Anti-angiogenic Activity of Novel Hydantoin Derivatives: Inhibition of VEGF Secretion in Liver Metastatic Osteosarcoma Cells, *Bioorg. Med. Chem.*, 17, 4928-4934.

Larsen, J; Dolbecq, A & Bechgaard, K (1996). Thiaheterohelicenes. 3. Donor Properties of a Series of Benzene-Capped Thiaheterohelicenes. Structure of a Tetrathianonahelicene and its TCNQ Salt, *Acta Chem. Scand.* 50, 83.

Liberko, CA; Miller, LL; Katz, TJ & Liu, L (1993). The Electronic Structure of Helicene-Bisquinone Anion Radicals, *J. Am. Chem. Soc.* 115, 2478.

Liu, L; Yang, B; Katz, TJ; Poindexter, MK (1991). Improved Methodology for Photocyclization Reactions, *J. Org. Chem.*, 56, 3769-3775.

Low, JN; Ferguson, G; Cobo, J; Nogueras, M; Cruz, S; Quirogae, J & Glidewell, C (2004). Three Hexahydropyridopyrimidine-spiro-cyclohexanetriones: Supra-molecular Structures Generated by O–H⋯O, N–H⋯O, C–H⋯O and C–H⋯π Hydrogen Bonds, and π – π Stacking Interactions, *Acta Cryst.*, C60, o438-o443.

Madsen, GKH; Wilson, C; Nymand, TM; McIntyre, GJ & Larsen, FK (1999). The Structure of Nitromalonamide: A Combined Neutron-Diffraction and Computational Study of a Very Short Hydrogen Bond, *J. Phys. Chem. A*, 103, 8684-8690.

Maiorana, S; Papagni, A; Licandro, E; Annunziata, R; Paravidino, P; Perdicchia, D; Giannini, C; Bencini, M; Clays, K & Persoons, A (2003). A Convenient Procedure for the Synthesis of Tetrathia-[7]-helicene and the Selective Functionalisation of Terminal Thiophene Ring, *Tetrahedron* 59, 6481–648.

Martín, N; Guldi, DM & Sánchez, L (2005). Hydrogen-Bonding Motifs in Fullerene Chemistry, *Angew. Chem. Int. Ed.*, 44, 5374.

Martin, RH (1974). Die Helicene, *Angew. Chem.* 86, 727.

McGillivray, LR & Atwood, JL (1997). A Chiral Spherical Molecular Assembly Held Together By 60 Hydrogen Bonds, *Nature*, 389, 469.

McKinlay, RM; Thallapally, PK; Cave, GWV & Atwood, JL (2005). Hydrogen-Bonded Supramolecular Assemblies as Robust Templates in the Synthesis of Large Metal-Coordinated Capsules, *Angew. Chem. Int. Ed.* 44, 5733-5736.

Miyasaka, M; Rajca, A; Pink, M & Rajca, S (2005). Cross-Conjugated Oligothiophenes Derived from the $(C_2S)_n$ Helix: Asymmetric Synthesis and Structure of Carbon-Sulfur [11]-Helicene, *J. Am. Chem. Soc.* 127, 13806.

Müller, A (1923). The X-ray Investigation of Fatty Acids, *J. Chem. Soc. (London)*, 123, 2043.

Müller, A (1928). X-ray Investigation of Long Chain Compounds (n-Hydrocarbons), *Proc. R. Soc. Lond.* A, 120, 437-459. doi:10.1098/rspa.1928.0158

Müller, A (1929). The Connection between the Zig-Zag Structure of the Hydrocarbon Chain and the Alternation in the Properties of Odd and Even Numbered Chain Compounds, *Proc. R. Soc. Lond.* A, 124, 317-321. Bibcode 1929RSPSA.124..317M. doi:10.1098/rspa.1929.0117.

Nakagawa, H; Obata, A; Yamada, K & Kawazura, H (1985). Crystal and Molecular Structures of Tetrathia[7]-heterohelicene: Racemate and Enantiomer, *J. Chem. Soc., Perkin Trans. 2*, 1899-1903. doi:10.1039/P29850001899

Nakano, K; Oyama, H; Nishimura, Y; Nakasako, S & Nozaki, K (2012). λ^5-Phospha[7]helicenes: Synthesis, Properties, and Columnar Aggregation with One-Way Chirality, *Angew. Chem.*, 124, 719.

Naya, SI; Miyama, H; Yasu, K; Takayasu, T & Nitta, M (2003). Novel Synthesis and Properties of 7,9-Dimethylcyclohepta[b]pyrimido[5,4-*d*]furan-8(7*H*),10(9*H*)-dionylium Tetrafluoroborate: Autorecycling Oxidation of Some Alcohols Under Photo-irradiation, *Tetrahedron*, 59, 1811.

Nelson, JH (2002). *Nuclear Magnetic Resonance Spectroscopy*, Prentice Hall: New Jersey, pp 160, Chapt. 6.

Newman, MS & Chen, CH (1972). The Resolution and Absolute Configuration of 7-Methylhexahelicene, *J. Org. Chem.* 37, 1312.

Newman, MS & Lednicer, D (1956). The Synthesis and Resolution of Hexahelkenel, *J. Am. Chem. Soc.* 78, 4765.

Newman, MS; Darlak, RS & Tsai, L (1967). Optical Properties of Hexahelicene, *J. Am. Chem. Soc.* 89, 6191.

Newman, MS; Lutz, WB & Lednicer, D (1955). A New Reagent for Resolution by Complex Formation: The Resolution of Phenanthro[3,4-*c*]phenanthrene, *J. Am. Chem. Soc.* 77, 3420.

Nishio, M (2005). CH/π Hydrogen Bonds in Organic Reactions, *Tetrahedron*, 61, 6923-6950.

Noroozi Pesyan, N (2006). A New and Simple Method for the Specification of Absolute Configuration of Allenes, Spiranes, Alkylidenecycloalkanes, Helicenes and Other Organic Complex Systems, *Chem. Educ. J. (CEJ)*, 9(1), (Serial No. 16), http://www.juen.ac.jp/scien/cssj/cejrnlE.html.

Noroozi Pesyan, N (2010). Crystal Structure of 1,9-Dimethyl-4,5-dihydro-6H-pyrido[3',2':4,5]thieno[2,3-f]pyrrolo[1,2-a][1,4]diazepin-6-one, *J. Struct. Chem.* 51, 991-995.

Noroozi Pesyan, N; Rastgar, S & Hosseini, Y (2009). 1,1',3,3',5,5'-Hexamethylspiro[furo-[2,3-d]pyrimidine-6(5H),5'-pyrimidine]-2,2',4,4',6'(1H,3H,1'H,3'H,5H)-pentaone, *Acta Cryst.* E65, o1444.

Noroozi-Pesyan, N (2011). Isotopic Effect on Tautomeric Behavior of 5-(2,6-Disubstituted-aryloxy)-tetrazoles, *Magn. Reson. Chem.*, 49, 592.

Nuckolls, C; Katz, TJ & Castellanos, L (1996). Aggregation of Conjugated Helical Molecules, *J. Am. Chem. Soc.* 118, 3767.

Nuckolls, C; Katz, TJ; Verbiest, T; Van Elshocht, S; Kuball, H-G; Kiesewalter, S; Lovinger, AJ & Persoons, A (1998). Circular Dichroism and UV-Visible Absorption Spectra of the Langmuir-Blodgett Films of an Aggregating Helicene, *J. Am. Chem. Soc.* 120, 8656.

Olovsson, I; Ptasiewicz-Bak, H; Gustafsson, T & Majerz, I (2001). Asymmetric Hydrogen Bonds in Centrosymmetric Environment: Neutron Study of Very Short Hydrogen Bonds in Potassium Hydrogen Dichloromaleate, *Acta Cryst.* B57, 311

Piper, SH (1929). Some Examples of Information Obtainable from the long Spacings of Fatty Acids, *Trans. Faraday Soc.* 25, 348-351. doi:10.1039/tf9292500348

Poling, M & van der Helm, D (1976). 5,6-Dihydro-1,3-dimethyl-5,6-bis[1',3'-dimethyl-2',4',6'-trioxopyrimid(5,5')yl]furo[2,3-d]uracil, *Acta Crystallogr., Sect. B: Struct. Sci.* 32, 3349.

Pulacchini, S; Sibbons, KF; Shastri, K; Motevalli, M; Watkinson, M; Wan, H; Whiting, A & Lightfoot, AP (2003). Synthesis of C2-Symmetric Aza- and Azaoxa-macrocyclic Ligands Derived From (1R,2R)-1,2-Diaminocyclohexane and Their Applications in Catalysis, *Dalton Trans.*, 2043.

Rajca, A & Miyasaka, M In: Miller, TJJ; Bunz, UHF (Eds.) (2007). *Functional Organic Materials. Syntheses, Strategies, and Applications*, Wiley-VCH,Weinheim, p. 547.

Rajca, A; Miyasaka, M; Pink, M; Wang, H & Rajca, S (2004). Helically Annelated and Cross-Conjugated Oligothiophenes: Asymmetric Synthesis, Resolution, and Characterization of a Carbon-Sulfur [7]-Helicene, *J. Am. Chem. Soc.* 126, 15211.

Rimaz, M; Khalafy, J; Noroozi Pesyan, N & Prager, RH (2010). A Simple One-Pot, Three Component Synthesis of 3-Arylpyrimido[4,5-c]pyridazine-5,7(6H,8H)-diones and their Sulfur Analogues as Potential Monoamine, Oxidase Inhibitors, *Aust. J. Chem.*, 63, 507.

Rimaz, M; Noroozi Pesyan, N & Khalafy, J (2010). Tautomerism and Isotopic Multiplets in the ^{13}C NMR Spectra of Partially Deuterated 3-Arylpyrimido[4,5-c]pyridazine-5,7(6H,8H)-diones and Their Sulfur Analogs – Evidence for Elucidation of the Structure Backbone and Tautomeric Forms, *Magn. Reson. Chem.*, 48, 276.

Robertson, JM (1936). An X-ray Study of the Phthalocyanines, Part II, *J. Chem. Soc.*, 1195.

Sadarangani, IR; Bhatia, S; Amarante, D; Lengyel, I & Stephani, RA (2012). Synthesis, Resolution and Anticonvulsant Activity of Chiral N-1'-ethyl, N-3'-(1-phenylethyl)-(R,S)-2'H,3H,5'H-spiro-(2-benzofuran-1,4'-imidazolidine)-2',3,5'-trione Diastereomers, *Bioorg. Med. Chem. Lett.*, 22, 2507-2509.

Saville, WB & Shearer, G (1925). An X-ray Investigation of Saturated Aliphatic Ketones, *J. Chem. Soc. (London)* 127, 591.

Schmuck, C (2003). Molecules with Helical Structure: How To Build a Molecular Spiral Staircase, *Angew. Chem. Int. Ed.* 42, 2448.

Severa, L; Adriaenssens, L; Vávra, J; Šaman, D; Císařová, I; Fiedler, P & Teplý, F (2010). Highly Modular Assembly of Cationic Helical Scaffolds: Rapid Synthesis of Diverse Helquats via Differential Quaternization, *Tetrahedron*, 66, 3537.

Sheldrick, GM (1997). *SHELXL97*, University of Göttingen, Göttingen, Germany.

Si, HZ; Wang, T; Zhang, KJ; Hu, ZD; Fan, BT (2006). QSAR Study of 1,4-Dihydropyridine Calcium Channel Antagonists-Based on Gene Expression Programming, *Bioorg. Med. Chem.*, 14, 4834-4841.

Smith, BM & March, J (2001). *Advanced Organic Chemistry, Reactions, Mechanism, and structure*, 5th ed., John Wiley & Sons, New York, p. 134.

Steiner, T (2002). The Hydrogen Bond in the Solid State, *Angew. Chem. Int. Ed.*, 41, 48.

Stoe & Cie, (2002). *X-AREA*. Stoe & Cie, Darmstadt, Germany.

Sudhakar, A & Katz, TJ (1986). Asymmetric Synthesis of Helical Metallocenes, *J. Am. Chem. Soc.* 108, 179.

Sudhakar, A & Katz, TJ (1986). Directive Effect of Bromine on Stilbene Photocyclizations. an Improved Synthesis of [7]-helicene, *Tetrahedron Lett.*, 27, 2231.

Taylor, R & Kennard, O (1982). Crystallographic Evidence for the Existence of C-H···O, C-H···N, and C-H···Cl Hydrogen Bonds, *J. Am. Chem. Soc.*, 104, 5063.

Trillat, JJ (1926). Rayons X et Composeés Organiques à Longe Chaine. Recherches Spectrographiques Sue Leurs Structures et Leurs Orientations, *Annales de physique* 6, 5.

Urbano, A (2003). Recent Developments in the Synthesis of Helicene-Like Molecules, *Angew. Chem. Int. Ed.* 42, 3986.

Vishweshwar, P; Babu, NJ; Nangia, A; Mason, SA; Puschmann, H; Mondal, R & Howard, JAK (2004). Variable Temperature Neutron Diffraction Analysis of a Very Short O-H···O Hydrogen Bond in 2,3,5,6-Pyrazinetetracarboxylic Acid Dihydrate: Synthon-Assisted Short O_{acid}-H···O_{water} Hydrogen Bonds in a Multicenter Array, *J. Phys. Chem. A*, 108, 9406-9416.

Wang, X-B; Woo, HK; Kiran, B & Wang, L-S (2005). Observation of Weak C–H···O Hydrogen Bonding to Unactivated Alkanes, *Angew. Chem. Int. Ed.* 44, 4968-4972.

Watkin, DJ; Müller, M; Blèriot, Y; Simonec, MI & Fleet, GWJ (2004). (2S,3R,4R,5S)-3,4-O-Isopropylidene-2-methyl-1-oxa-6,9-diazaspiro[4.5]decane-7,10-dione, *Acta Cryst.*, E60, o1820-o1822.

Wilson, CC (2000). *Single Crystal Neutron Diffraction From Molecular Materials* (1st edn), vol. 2, World Scientific Publishing Co. Pte. Ltd.: Singapore, Chapts. 4, 5 and 7.

Wynberg, H (1971). Some Observations on the Chemical, Photochemical, and Spectral Properties of Thiophenes, *Acc. Chem. Res.* 4, 65-71.

Yamada, K; Ogashiwa, S; Tanaka, H; Nakagawa, H & Kawazura, H (1981). [7],[9],[11],[13] and [15] Heterohelicenes Annelated With Alternant Thiophene and Benzene Rings. Syntheses and NMR Studies, *Chem. Lett.*, 343.

Yang, W-B; Lu, C-Z; Wu, C-D; Wu, D-M; Lu, S-F & Zhuang, H-H, (2003). Synthesis and Crystal Structure of Bis(barbiturato)triwater Complex of Copper(II), *Chinese J. Struct. Chem.* 22, 270.

Yoshida, N; Ito, N & Ichikawa, K (1997). Formation of a Three-Dimensional Hydrogen-Bonding Network by Self-assembly of the CuII Complex of a Semi-bidentate Schiff Base, *J. Chem. Soc., Perkin Trans. 2*, 2387.

Computational Modeling

Chapter 10

Unique and Effective Characterization of Space Groups in Computer Applications

Kazimierz Stróż

Additional information is available at the end of the chapter

1. Introduction

Any method for the solution of crystal structures or for Rietveld refinement of such structures needs at least the set of symmetry matrices of the given space group (SG). In data bases and concise standard descriptions of crystal structures, the complete list of atoms is reduced to atoms representatives and a space group symbol. Since symmetry operations are tensor objects and their matrix form depend on the origin and the orientation of the coordinate system, the symbol must identify not only the *space group type* but also the *space group description*. For this reason standard space group descriptions called *conventional* were tabulated since the year 1935. Contemporary descriptions are based on the 3rd series of the *International Tables for Crystallography* [1], frequently referred to as ITA83. In a common practice one takes a space-group symbol, usually an international Hermann-Mauguin (H-M) symbol, as the index and obtains the symmetry operators from the printed tables. This approach guarantees the conventional descriptions of space groups, 'standard' ordering of symmetry matrices, its geometric interpretation, etc. but has a disadvantage of being limited to the tabulated representations of space groups. Moreover, huge amount of data and a possibility of human errors in transferring the symmetry information to applications involving crystal symmetry [2,3] leads to the generally accepted conclusion, that the automated derivation of space-group information becomes essential when this information is routinely required, especially in the case of higher symmetry. In the years 1960-1980 well-documented algorithms [4,5] translating H-M symbols into a set of generators, which are then used to build a full set of symmetry matrices, were developed. Differences in the generated space group descriptions, caused by ambiguities in H-M symbols prompted in [6-8], brought to procedures based on explicit-origin generators, that is on symmetry operations with specified complete translation vectors not only on its characteristic components – glide or screw vectors. The explicit-origin generators were organized into lengthy [7,8] or concise space-group symbols and used for the SG generation. The latter symbol, known as the *Hall symbol*, was applied in [6,9]. It should be also noted, that according to the transformational concept [10], the one from equivalent sets of generators published in ITA83 can serve as the 'starting point' for any space group description in the ITA83-style.

The efficiency of space groups construction from the generators is also considered in the literature. Trivial approaches based on any symmetry operations, treated as the group generators, are redundant and completely ineffective. For example, the number of matrix multiplications and matrix comparisons needed for the generation of space group $Fm\bar{3}m$ has been estimated in [9] at 18528 and 1.7 million, respectively. Such huge number of mathematical operations can be drastically reduced at the expense of an algorithm complexity. In the approach described in reference [9], the improved algorithm was approximately 100 times faster than the simplistic method. The natural limit of such improvements corresponds to the non-redundant methods, that is procedures in which each matrix multiplication gives a new symmetry operation and thus matrix comparisons are not needed at all. Such method was based on representation of 32 crystallographic point groups in terms of cyclic groups and their products [7,8]. The 'composition series' method [11], successfully adapted to conventional space group descriptions in ITA83 [12], can also be modified to a non-redundant form [13]. This method is especially interesting, since it repeats the ITA83-type order of symmetry operations, assuming the published sets of generators are used. Basing on the composition series, all space group descriptions, that is conventional or not conventional, may be standardized [13-15].

For specific purposes justified in [8,9], the set of symmetry matrices should be complemented by the 'secondary symmetry information' which originates from the geometric characterizations of the symmetry operations given in ITA83 in the form of the operation symbols and the diagrams of symmetry elements. The symmetry operations are the foundation for the Hermann-Maugin symbols and to derive information important in understanding the Wyckoff positions, to find equivalent descriptions of crystal structures or to find transformations between different algebraic descriptions of the same space group.

The matrix-column pair of a space-group operator contains all the information that is needed for a geometrical characterization of the space-group operator and the corresponding symmetry element. Procedures for doing this are well documented in the literature [8,9,16-18]. Algorithms for the determination of (i) the type of rotation, (ii) the rotation angle ϕ given as $n = 360/ \phi$, (iii) the sense of rotation, (iv) the orthogonal decomposition of the translation vector of the space group operator into its intrinsic and location-dependent components, (v) the orientation and the location of a geometric element, (vi) the inversion point on a geometric element in case of a rotoinversion symmetry operation, are given in the above references.

Even though the recipes are clear, the geometric meaning of symmetry matrices needs some additional conventions [18] to obtain a unique form, not necessarily presented in algorithms. Difficulties are connected with the positive direction of an axis line or plane traces specifications, or the selection of a point which fixes a geometric element in the space.

Moreover, coding some glide vectors by the letters a, b, c, g causes that operation symbols are difficult to generate and are not generally understood, also in the community of crystallographers. Even though there are critical opinions like '*The symbols written under the heading Symmetry operations are actually descriptive symbols which repeat part of the information given by the diagram – actually they correspond to cosets representatives in the decomposition of the space group with respect to its translation subgroup....Solid state physicists would appreciate much more if these symbols are replaced by Seitz operators*' [19], the deriving of geometric descriptions, especially without any relations to the Crystallographic Tables, is crucial. It must be also remembered that symbols were used to give a scientific definition to a term '*symmetry element*' – the collective designation for a number of geometric concepts widely used by crystallographers, mineralogists and spectroscopists [20-22]. They were recommended by the *Ad-hoc committee on the nomenclature of symmetry* [21] for general use.

The analysis of symbol derivations as well as the contents of geometric features in the symbols, given in [15, 18, 23], suggests the possibility of defining new symbols, easier to calculate, universal and richer in geometric meaning.

Taking into account all above remarks, the most important features for the computer generation of space groups and their geometric characterization are: (i) a possibility of unique naming of any SG descriptions, (ii) an easy interpretation of SG symbols, that is a derivation of generators, (iii) effectiveness (non-redundancy) of group generation, (iv) reduced amount of predefined data, (v) full consistency with ITA83, (vi) an effective geometric characterization of obtained set of symmetry matrices. The generators selected by H. Wondratschek and used for the conventional space group descriptions in ITA83 meet some demands from the above list, but the geometric symbols need some modifications.

In this chapter a complete, effective and unique approach to the automated derivation of space-group information for a conventional as well as for a non-conventional space group description is proposed.

2. Transformational versus other space-group symbols

There are several systems of space groups naming. They can be divided into three categories in the context of group generations: (i) the *symbolic* one, containing symbols of 'space-group types', like the sequential or the Schönflies symbols, (ii) the *geometric* one, with the Hermann-Mauguin and the Shubnikov symbols which contain the glide or screw components of symmetry operations given in such a way that the nature of the symmetry elements, their orientation and relative location can be deduced from the symbol, (iii) the *algebraic* category with the Zacharasen, Schmuelli and Hall symbols of complete generators, most convenient for the derivation of 'specific groups'. The Hermann-Mauguin symbols are mainly used in connection with the International Tables for Crystallography for the designation of the conventional space group descriptions. The computer programs transforming these most informative symbols into specific space groups have limited possibilities, are rather complicated, sensitive to changes in space-group symbols and must contain additional conventions, since H-M symbols do not depend on the origin selection. On the other side, the algebraic SG symbols lose a clear geometric interpretation and should be selected from the 'economic' point of view – minimal stored data, a simple and non-redundant algorithm generation. The presented transformational approach to naming and deriving specific space groups, based on earlier works [10,15], meets the mentioned demands.

SG No	+1	+2	+3	+4	+5	+6	+7	+8	+9	+10
0	$P1$	$P\bar{1}$	$P2$	$P2_1$	$C2$	Pm	Pc	Cm	Cc	$P2/m$
10	$P2_1/m$	$C2/m$	$P2/c$	$P2_1/c$	$C2/c$	$P222$	$P222_1$	$P2_12_12$	$P2_12_12_1$	$C222_1$
20	$C222$	$F222$	$I222$	$I2_12_12_1$	$Pmm2$	$Pmc2_1$	$Pcc2$	$Pma2$	$Pca2_1$	$Pnc2$
30	$Pmn2_1$	$Pba2$	$Pna2_1$	$Pnn2$	$Cmm2$	$Cmc2_1$	$Ccc2$	$Amm2$	$Aem2$	$Ama2$
40	$Aea2$	$Fmm2$	$Fdd2$	$Imm2$	$Iba2$	$Ima2$	$Pmmm$	$Pnnn$	$Pccm$	$Pban$
50	$Pmma$	$Pnna$	$Pmna$	$Pcca$	$Pbam$	$Pccn$	$Pbcm$	$Pnnm$	$Pmmn$	$Pbcn$
60	$Pbca$	$Pnma$	$Cmcm$	$Cmca$	$Cmmm$	$Cccm$	$Cmma$	$Ccca$	$Fmmm$	$Fddd$
70	$Immm$	$Ibam$	$Ibca$	$Imma$	$P4$	$P4_1$	$P4_2$	$P4_3$	$I4$	$I4_1$
80	$P\bar{4}$	$I\bar{4}$	$P4/m$	$P4_2/m$	$P4/n$	$P4_2/n$	$I4/m$	$I4_1/a$	$P422$	$P42_12$
90	$P4_122$	$P4_12_12$	$P4_222$	$P4_22_12$	$P4_322$	$P4_32_12$	$I422$	$I4_122$	$P4mm$	$P4bm$
100	$P4_2cm$	$P4_2nm$	$P4cc$	$P4nc$	$P4_2mc$	$P4_2bc$	$I4mm$	$I4cm$	$I4_1md$	$I4_1cd$
110	$P\bar{4}2m$	$P\bar{4}2c$	$P\bar{4}2_1m$	$P\bar{4}2_1c$	$P\bar{4}m2$	$P\bar{4}c2$	$P\bar{4}b2$	$P\bar{4}n2$	$I\bar{4}m2$	$I\bar{4}c2$
120	$I\bar{4}2m$	$I\bar{4}2d$	$P4/mmm$	$P4/mcc$	$P4/nbm$	$P4/nnc$	$P4/mbm$	$P4/mnc$	$P4/nmm$	$P4/ncc$
130	$P4_2/mmc$	$P4_2/mcm$	$P4_2/nbc$	$P4_2/nnm$	$P4_2/mbc$	$P4_2/mnm$	$P4_2/nmc$	$P4_2/ncm$	$I4/mmm$	$I4/mcm$
140	$I4_1/amd$	$I4_1/acd$	$P3$	$P3_1$	$P3_2$	$R3$	$P\bar{3}$	$R\bar{3}$	$P312$	$P321$
150	$P3_112$	$P3_121$	$P3_212$	$P3_221$	$R32$	$P3m1$	$P31m$	$P3c1$	$P31c$	$R3m$
160	$R3c$	$P\bar{3}1m$	$P\bar{3}1c$	$P\bar{3}m1$	$P\bar{3}c1$	$R\bar{3}m$	$R\bar{3}c$	$P6$	$P6_1$	$P6_5$
170	$P6_2$	$P6_4$	$P6_3$	$P\bar{6}$	$P6/m$	$P6_3/m$	$P622$	$P6_122$	$P6_522$	$P6_222$
180	$P6_422$	$P6_322$	$P6mm$	$P6cc$	$P6_3cm$	$P6_3mc$	$P\bar{6}m2$	$P\bar{6}c2$	$P\bar{6}2m$	$P\bar{6}2c$
190	$P6/mmm$	$P6/mcc$	$P6_3/mcm$	$P6_3/mmc$	$P23$	$F23$	$I23$	$P2_13$	$I2_13$	$Pm\bar{3}$
200	$Pn\bar{3}$	$Fm\bar{3}$	$Fd\bar{3}$	$Im\bar{3}$	$Pa\bar{3}$	$Ia\bar{3}$	$P432$	$P4_232$	$F432$	$F4_132$
210	$I432$	$P4_332$	$P4_132$	$I4_132$	$P\bar{4}3m$	$F\bar{4}3m$	$I\bar{4}3m$	$P\bar{4}3n$	$F\bar{4}3c$	$I\bar{4}3d$
220	$Pm\bar{3}m$	$Pn\bar{3}n$	$Pm\bar{3}n$	$Pn\bar{3}m$	$Fm\bar{3}m$	$Fm\bar{3}c$	$Fd\bar{3}m$	$Fd\bar{3}c$	$Im\bar{3}m$	$Ia\bar{3}d$

Table 1. The Hermann-Mauguin symbols which can appear in the transformational space-group descriptions

The transformational approach was inspired by the fact, that in ITA83 all different descriptions of the same space group (settings, origin choices, cell choices) were generated from the same operations transformed to the new coordinate system. Thus, all multiple descriptions as well as any non-conventional description of a given space-group type may be constructed from one set of generators. In contrary to the generally accepted convention that all multiple descriptions are equivalent, one and only one description (and corresponding information: the conventional cell, origin, H-M symbol, etc.) serves as the reference for other SG descriptions. As a result, the number of H-M symbols is reduced to 230 and each such symbol likewise the sequential number or the Schönflies symbol denotes the space-group type and points at the 'starting' set of generators.

The selection of the reference descriptions is based on the following conventions. In monoclinic system the settings with *unique axis b, cell choice* 1 are chosen. Five groups with *R*

– centred cells are described in the *hexagonal axes*. For the space groups with two origin choices, the description with a centre as the origin is chosen.

Having settled the relation between a space-group type and its well established description, other descriptions of the same space-group type are easily identified by a *transformational space-group symbol* (TSG) obtained from the type symbol and explicitly given coordinate system transformation

$$\text{TSG} = \text{type symbol} \left(P_{11}, P_{21}, P_{31},\ P_{12}, P_{22}, P_{32},\ P_{13}, P_{23}, P_{33}\right) \left(p_1, p_2, p_3\right), \tag{1}$$

where parameters in the first parenthesis describe the unit cell transformation

$$(a', b', c') = (a, b, c) \begin{pmatrix} P_{11} & P_{12} & P_{13} \\ P_{21} & P_{22} & P_{23} \\ P_{31} & P_{32} & P_{33} \end{pmatrix} = (a, b, c)\, P, \tag{2}$$

and the remaining parameters define the origin shift

$$t = p_1 a + p_2 b + p_3 c. \tag{3}$$

In most cases some of the parameters are zero and an abbreviated 'short-hand' notation

$$\text{TSG} = \text{type symbol} \left(a', b', c'\right) \left(t\right) \tag{4}$$

is preferred. The shortening concept is similar to that used in storing the symmetry matrices as the coordinate triplets in ITA83. Moreover, the identity transformation (a, b, c) and the zero origin shift (0) are omitted in the symbol.

To be familiar with, some examples of the TSG symbols are given underneath:

Fdd2 (b/2+c/2, a/2+c/2, a/2+b/2) – a primitive description of the space group,
Pn$\bar{3}$ (-1/4,-1/4,-1/4) – *origin choice 1*,
5 (c,a,b) – *unique axis* c, *cell choice* 1, for the group with sequence number 5 or H-M symbol C2.

It must be remembered that the Bravais centring letter in an H-M symbol is valid only for the conventional space-group type and generally does not describe the centring vectors resulting from the axis transformation given in the transformational symbols. On the other hand, the use of a sequence number or a Schönflies symbol to point the space-group type is not very informative. The relation between H-M symbols in the ITA83 and the generators based on the composition series method is lost, but it can be restored by the following trick. Each space group type has one and only one set of generators identified by unique H-M symbol, Schönflies symbol or even by a sequential number. Thus, the group symbol identifies not only the group type, but also its selected matrix representation (including the axis labelling, the origin). Other group descriptions, conventional or not, need explicit information on the axis transformation and/or the origin shift enabled. In any case an original or transformed generators gives a complete set of symmetry matrices in an unambiguous and effective way.

3. Description of symmetry operations

Crystallographic groups are groups in the mathematical sense of the word *group*, i.e. they are the sets of elements which fulfil the group conditions. The elements of a space group are the symmetry operations of a crystal, which can be described in several ways. In ITA83 they are presented in four different forms:

i.   by a list of symmetry matrices of *general position* in the form of coordinate triplets,
ii. by a general-position diagram,
iii. by the diagram(s) of symmetry elements,
iv. by a list of geometric interpretations of symmetry matrices from (i).

The great importance of geometrical intuition and the geometric point of view on space groups, their symmetry elements, H-M symbols, Wyckoff positions, origin specifications is clear from the above list. The diagrams are rather complicated constructions and cannot be derived *on line* for any space group description, contrary to the geometric symbols for each symmetry matrix. In the computer applications an operation symbol may be considered as the 'gravity centre' of geometric characterization of a symmetry matrix.

For easer understanding of the further material, in the first part of this section the common facts about 'standard' algebraic and geometric descriptions of symmetry operations will be recalled. Next the new geometric symbolism will be proposed. A *dual symbol* of space-group operation is based on a dual symbol of point operation [23] and a point on *geometric elements* closest to the origin [18]. These modifications improve the informative properties and also reduce some conventions necessary for the standard symbols to be unique.

In the three-dimensional space the general linear mappings, called also the *affine mappings*, transform a point coordinates x, y, z into the coordinates $\tilde{x}, \tilde{y}, \tilde{z}$ of the image point by the system of equations

$$\tilde{x} = W_{11}x + W_{12}y + W_{13}z + w_1$$
$$\tilde{y} = W_{21}x + W_{22}y + W_{23}z + w_2 \tag{5}$$
$$\tilde{z} = W_{31}x + W_{32}y + W_{33}z + w_3$$

represented in the matrix form as

$$\tilde{x} = Wx + w \xrightarrow{\text{definition}} (W, w)x. \tag{6}$$

For computer applications a more convenient description uses so called augmented matrix $\mathcal{W}$ which combines the (3x3) matrix W with a (3x1) column matrix w and a row matrix (0 0 1). The augmenting of the coordinate column x, y, z by a fourth dummy coordinate, fixed at 1, leads to a more homogenous form

$$\tilde{x} = \mathcal{W}x. \tag{7}$$

Contrary to the affine mappings, which preserve the straight lines, planes and parallelism of such objects, a special case called *isometries* preserves also distances and in consequence the volumes and orthogonalities. In this situation the determinant $\det(W) = \pm1$. Moreover, for crystallographic symmetry operations related to the crystallographic coordinate system all elements of the matrix are integers reduced to the values {-1,0,1}, if such coordinate system is based on the shortest lattice vectors or on the Bravais cells. In the latter case, especially simple forms of **W** consisting of three or four non-zero elements are obtained. Therefore, in printed descriptions of space groups, like in ITA83, the space - consuming forms (**W**,**w**) or $\mathcal{W}$ are equivalently given as shorthand notation of the equation systems (5) and called the *coordinate triplets* $\tilde{x}, \tilde{y}, \tilde{z}$ or the Jones faithful notation.

There are also some advantages in notation (R|t) introduced by Seitz in [25] and adopted by solid-states physicists [26]. In this notation 'R' means a point operation **W** given symbolically and 't' is an explicitly given translation, equivalent to **w**. Thus, the diversity of R-symbols, corresponding to a set of different symmetry matrices occurring in the conventional space group descriptions, is reduced to 64 items (48 orthogonal matrices with three non-zero elements occuring in the cubic system and 16 additional matrices with four non-zero elements coming from the oriented hexagonal system). The Seitz symbols are very concise and with the help of multiplication tables they identify the product of symmetry operations

$$\left(R_1 | t_1\right)\left(R_2 | t_2\right) = \left(R_1 R_2 | R_1 t_2 + t_1\right) = \left(R | t\right), \tag{8}$$

but for explicit values of components **t**, the **W** matrices symbolized by R must be known. Two special actions R on **t** should be distinguished. Assuming the k order of R, that is the equality $\mathbf{W}^k = \mathbf{I}$ is held, these actions can be distinguished by equations

$$\left(R | t_{para}\right)^k = \left(I | k t_{para}\right) \text{and} \left(R | t_{ortho}\right)^k = \left(I | 0\right), \tag{9}$$

where vectors t_para and t_ortho are mutually perpendicular.

Recently, the set of geometric symbols for R used in ITA83 was compared with other sets of symbols of point-group operations [24]. Some symbols are purely symbolic, other contain geometric information and are more or less self-defined.

Relations between different spaces, their geometric invariants and group properties of such invariants were 'discovered' before the development of the crystallographic groups. Starting from the Erlangen program announced by Felix Klein in 1872, it is understood that different symmetries represented by the abstract structure-groups are consequences of different geometries and their invariants. Thus, there are two points of view on symmetry: one is algebraic and the second one is geometric. In the case of groups described in ITA83, the three items from the four-element list of equivalent symmetry descriptions possess the geometric nature. The geometric information has played important roles in many aspects of crystallography. The algorithms for the characterization of symmetry operations are the foundation for the Hermann-Maugin symbols. The 'secondary symmetry information' is important in understanding the Wyckoff positions, distinguishing partners in an enantiomorfic pair, finding equivalent descriptions of crystal structures or finding transformations between different algebraic descriptions of the same space group.

The description of a general procedure for deriving symbols for the symmetry operations was included in Chapter 11 of ITA83 [17]. A similar approach was used in [16], different modifications can be found in [7,8]. In all this approaches the basic concepts consist in extracting from **w** (**t** in Seitz notation) its characteristic part $\mathbf{w_g}$ interpreted as screw/glide vectors, finding a set of fixed points of pure rotations or reflections and thus orient and locate so called *geometric element* and find the sense of rotation angle symbolized by a number n = 2,3,4 or 6 (rotation angle = $360^\circ/n$). The scheme of calculation and some critical remarks is given below.

3.1. Type of symmetry operation

The type of symmetry operation is obtained by a modification of point operation symbols according to the characteristic part of translation vector **w**. It is obvious that the type of point operations is completely determined by a matrix part **W** of (**W,w**). Table 2 contains the classification of matrices **W** based on the analysis of their traces and determinants.

tr(**W**)	det(**W**)=1					det(**W**)=-1				
tr(**W**)	3	2	1	0	-1	-3	-2	-1	0	1
type	1	6	4	3	2	$\bar{1}$	$\bar{6}$	$\bar{4}$	$\bar{3}$	m
order k	1	6	4	3	2	2	6	4	6	2

Table 2. Classification of the point symmetry matrices

A vector **w**, if different from **0**, should be decomposed into orthogonal components: a *glide/screw part* $\mathbf{w_g}$ and a *location part* $\mathbf{w_l}$. While the first component changes operation types, namely: rotations into screw rotations and reflections into glide reflections, the second one is responsible for shifting the corresponding symmetry element from the origin. The part $\mathbf{w_g}$ may be derived by projecting **w** onto the space invariant under **W**, but this needs the metrical information about the coordinate system. The metric-free derivation uses the property (9)

$$\left(\mathbf{W,w}\right)^k = \left(\mathbf{W,w_g + w_1}\right)^k = \left(\mathbf{I}, k\mathbf{w_g}\right) \tag{10}$$

and the location part is the difference $\mathbf{w_l}$ = **w** - $\mathbf{w_g}$.

3.2. Sense of direction and sense of rotation

To complete the characterization of **W**, excluding the cases where **W** represents a two-fold rotation or a reflection, the sense of rotation must be determined. For this purpose and also for obtaining the compatibility between analytical descriptions of axes, a convention which fixes the positive direction must be adopted. Such convention was not explicitly described in [17] and in consequence other conventions are sometimes applied. For example, from the two equivalent descriptions of the same crystallographic direction [$1\bar{1}\bar{1}$] and [$\bar{1}11$] the latter is positively directed according to the rule applied in [13] and the opposite selection is compatible with geometric descriptions in ITA83. A systematic analysis of 'standard' symbols allowed to specify in [18] six conventions C1 – C6, which together with the algebraic procedures lead to unique symbols. Convention C2, in the detailed form is given in Table 3.

Stand up element	Two zeroes values	One zero	All non-zero values
u	[+ 0 0]	[0 + +] or [0 + -]	[+ - -]
v	[0 + 0]	[+ 0 +] or [- 0 +]	[- + -]
w	[0 0 +]	[+ + 0] or [+ - 0]	[- - +]
-			[+ + +]

Table 3. Selection of the positive direction from a pair [uvw] and [$\bar{u}\bar{v}\bar{w}$]. The symbols '-', '0', '+' are used for the positive, zero, negative values, respectively.

According to the convention in Table 3, positive directions characterize the positive product of the non-zero components. Such situation cannot occur if there are only two non-zero elements, which differ in the sign. In this case the rule can be stated as 'a negative component may precede a zero component, but never may follow it'.

Assuming a positive direction of the axis corresponding to a rotation or rotoinversion operation has been determined, one of two commonly used procedures [17, 27] should be applied to obtain the sense of rotation. The Boisen & Gibbs procedure given in [27] is more practical. Having the positive axis direction [uvw] derived from $\mathbf{W}$, the positive sense of rotation is obtained if one of the following conditions is fulfilled:

$$\text{if } v = w = 0 \text{ and } uW_{32} > 0 \text{ or } wW_{21} - vW_{31} > 0. \tag{11}$$

3.3. Orientation – location part

In the classical approaches to the considered topic, the orientation of a geometric element is determined together with its location in one step. Solving of three simultaneous equations of reduced operation ($\mathbf{W},\mathbf{w_l}$), given in a matrix form

$$\left(\mathbf{W} - \mathbf{I}\right)\mathbf{x_F} + \mathbf{w_1} = 0, \tag{12}$$

leads to the linear or planar sets of solutions (fixed points) for rotations or reflections, respectively. For a rotoinversion other than a pure inversion, the axis results from the equation

$$\left(\mathbf{W}^2 - \mathbf{I}\right)\mathbf{x_F} + \mathbf{W}\mathbf{w} + \mathbf{w} = 0. \tag{13}$$

In both cases the set of equations is indeterminate, since det($W - I$) in (12) or det($W^2 - I$) in (13) are equal zero. This leads to parametric solutions, which forms depend on the mode of calculation.

Although the procedure may also be applied to cases where space groups are given in non-conventional descriptions, there are situations where it is difficult to obtain a unique result even for coordinate triplets taken from ITA83 [18].

For example the matrix equation

$$
\begin{pmatrix} 0 & 0 & \bar{1} \\ \bar{1} & 0 & 0 \\ 0 & 1 & 0 \end{pmatrix}
\begin{pmatrix} x \\ y \\ z \end{pmatrix}
+
\begin{pmatrix} 1/6 \\ 1/3 \\ -1/6 \end{pmatrix}
=
\begin{pmatrix} x \\ y \\ z \end{pmatrix}
$$

gives three solutions presented here as the orientation-location parts:

1. $x, -x + 1/3, -x + 1/6$ if x is treated as the parameter,
2. $-y + 1/3, y, y - 1/6$ if y is selected as the parameter,
3. $-z + 1/6, z + 1/6, z$ if z represents the variable parameter.

Of course, these solutions describe the same symmetry axis, but differ in a selected parameter, in a positive direction and in a location point. Unique solutions require special conventions in programmable algorithms, based for example on the *row echelon forms* [9, 27], or in post-calculation standardization.

3.4. Pseudo-inverse and the point closest to origin

Another possibility originates from the linear algebra and from the concept of *pseudo-inverse* matrices. This mathematical formulation allows obtaining single points x_F from equations (12) and (13). Such unique points have simple interpretation; they represent points from the linear or planar sets closest to origin. The derivation of the pseudo-inverse matrix $(W-I)^+$ is rather cumbersome, but this idea inspires the new point of view on the location derivation and specification in the geometric descriptions.

4. Dual symbols for space-group operations

As mentioned in the previous section, the geometric interpretation of symmetry operations is vital for many crystallographic topics. But there are some disadvantages in the derivation and application of the pioneers symmetry operation symbols (Fischer, Koch in ITA83). In the published procedure, intended mainly for manual calculations and for a link between symmetry matrices and the diagrams of symmetry elements, there is a lack of conventions needed to cast this approach into a programmable algorithm and to obtain unique relations: given symmetry operation → the operation symbol. In the literature concerning this topic individual conventions are frequently added into specified algorithms [8,9], which causes that the derived symbols are unique, but not necessary agree with ITA83 symbols. Special difficulties arise in deriving the orientation-location part of a symmetry operation. A systematic analysis of ITA83 symbols resulted in extracting 6 conventions necessary for automatic generation of the orientation-location parts in an unambiguous way [14]. The novel approach presented in that publication contains the derivation of a *point closest to origin* x_c as an alternative to solving an indeterminate set of equations by the pseudo-inversion matrices. But a methodological advance connected with the unambiguity of x_c was next lost in the standardization step, in which the obtained result was transformed into a conventional point on a geometric element. The inconveniences of classical symbols can be summarized as follows:

i.   the procedure outlined in ITA83 for the classical names of the symmetry operations is not unique and must be supplemented by explicitly given conventions like in [18],
ii. the symbols lost some information in comparing with the source symmetry matrices, for example the point operation symbol 2 0,y,0 and space operations based on it may be obtained from two different matrices,
iii. symbols, especially involving the reflections, are limited to the conventional space groups descriptions,
iv. symbols are laborious for computer derivation and manipulation, *e.g.* they contain the symbols a, b, c, g for glide planes, the symbols x, y, z for free parameters.

Some of the listed problems are immanently involved with the specific form of the orientation-location part and they should be overcome or at least simplified by developing a new symbol of symmetry operation. The problems (ii) and partially (iii) and (iv) disappear, if the orientation of a geometric element is presented in the form of the orthogonal lattice splitting $(uvw)[hkl]$. Advantages of using splitting indices for the characterization of point operations were described in [23]. The benefits for space operations should be even greater. The conventions for a unique description of a geometric element location in (i) have no meaning for x_c. Moreover, by removing the a, b, c, g designations of glide planes, a new symbol will be more consistent.

The splitting indices describe the orientation of the symmetry axes or the symmetry planes in the same way as $[uvw](hkl)$. They, together with a locating point on the geometric element, fully correspond to the orientation-location parts. More precisely, the informative content of this construction exceeds the analogous information stored classically. Lattice rows $[uvw]$ and lattice planes (hkl) are orthogonal to each other and exchangeable in the reciprocal space. Specifications of the same lattice direction in direct and reciprocal spaces better characterize the axis system and simplify the crystallographic calculations, that is calculations in non-Cartesian systems. While two point operations given by coordinate triplets $\bar{x}, y, \bar{z}$ and $\bar{x}, \bar{x} + y, \bar{z}$ lead to a common geometric symbol 2 0,y,0, the splitting indices are $[010](010)$ in the former case and $[010](\bar{1}20)$ in the latter.

The lattice perpendicularities revealed by splitting indices are also important in the orthogonal splitting of the translational part $\mathbf{w}$ of a space-group operation. The $\mathbf{w}$ components are parallel/orthogonal to $[uvw]$ and also to (hkl), what was symbolically drawn in Figure 1.

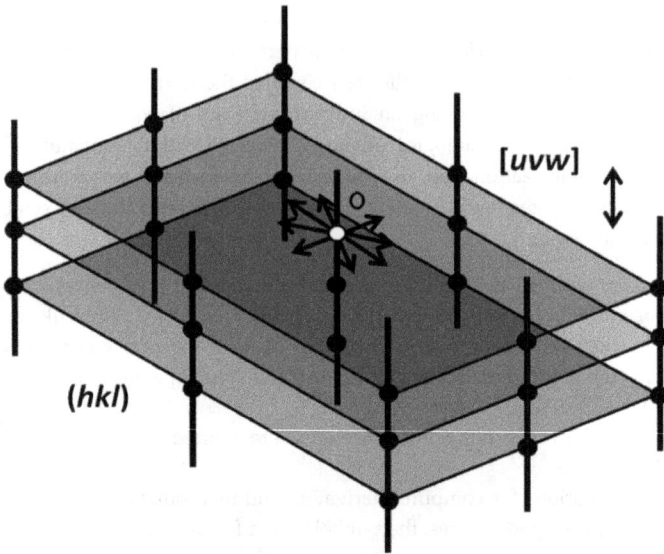

Figure 1. Origin - central lattice point of a dual symbol. Each crystallographic point operation (excluding identity and inversion) orthogonally splits the lattice into the lattice rows [uvw] and the lattice planes (hkl). The decomposition of a translation **w** in the space-group operations is also based on the lattice perpendicular property.

Let x_c denotes a special point, the point from set x_F closest to the coordinate system origin. Such points may be found on any geometric element (Fig.2). Thus, every 4x4 symmetry matrix $\mathcal{W}$, excluding only identity operation and pure translations, can be characterized by x_c. Vector x_c is always perpendicular to the geometric element and to the intrinsic part w_g, if not zero. Two vectors separated by the asterisk, that is $w_{g^*} x_c$ give complete geometric information about a space-group operation, assuming that the point –group operation was characterized by the dual symbol. For this reason, the new geometric symbol of a space-symmetry operation is also called the *dual symbol*, like in the case of point operations [18]. This should not lead to confusion, since the information concerning the point operation is separated from the characteristics involved with the space-symmetry operation. Such approach reflects the analogy to the algebraic difference between the point operation **W** and the space operation (**W**,**w**). Moreover, the perpendicularity between the lattice row and the lattice planes expressed by splitting indices is reflected also in the pair w_g, x_c. In consequence, the six components of two vectors $w_{g^*} x_c$ may be presented only by four simple ratios, since in every case w_g or x_c is parallel to the integer vector [uvw]. Thus, the dual symbol takes the form

$$\pm n^{\pm}\left[uvw\right](hkl)r_1 * r_2, r_3, r_4 \text{ for rotations and rotoinversions}$$
$$\text{and } m\left[uvw\right](hkl)r_1, r_2, r_3 * r_4 \text{ for reflections,} \tag{14}$$

where: the minus sign before n or a bar over n is the 'inversion sign' = det(**W**)

n – axis symbol, order of $\mathbf{W}_P$,

m – mirror plane (n =2, det(**W**) = -1),

$[uvw]$ – symmetry axis direction or a normal to the reflection plane,

$[hkl]$ – reflection plane or plane perpendicular to the direction of symmetry axis,

$r_1{}^* r_2, r_3, r_4$ – specification of vectors $\mathbf{w}_g = (r_1 u, r_1 v, r_1 w)$ and $\mathbf{x}_c = (r_2, r_3, r_4)$,

$r_1, r_2, r_3{}^* r_4$ – specification of vectors $\mathbf{w}_g = (r_1, r_2, r_3)$ and $\mathbf{x}_c = (r_4 u, r_4 v, r_4 w)$.

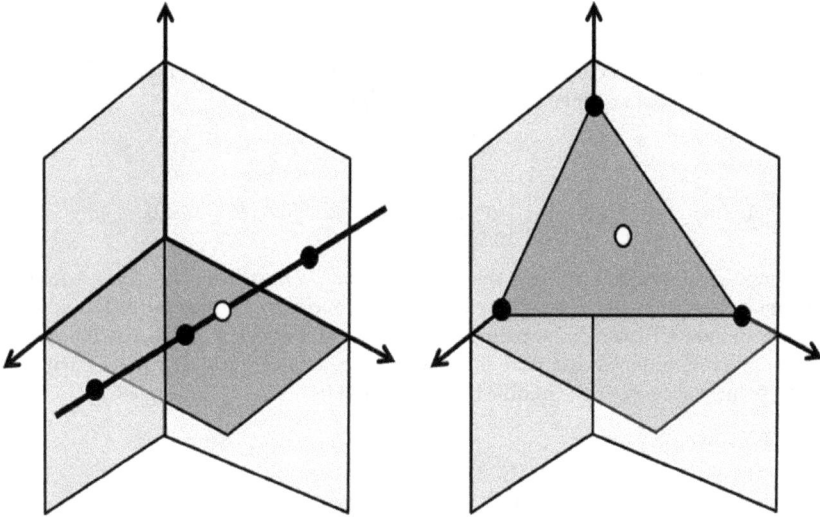

Figure 2. Characteristic points of geometric elements in general orientation. For a symmetry axis they defined the intersection of an axis with the basal planes. A symmetry plane is located by its intersection with the coordinate axes. Alternative fixing of geometric element may be based on the unique point $\mathbf{x}_c$ closest to the origin, schematically presented by open circles.

In most cases such complete form can be further reduced. Zero vectors are omitted. If the lattice row indices are equal to the lattice plane indices, a typical situation for a conventional description of space groups, the latter does not need to be specified. Generally, the indices (hkl) in dual symbols are related with the indices $[uvw]$ by the lattice metric. Since non-orthogonal symmetry matrices occur only in the oriented hexagonal system, the h symbol may designate a unique transformation matrix from $[uwv]$ to (hkl) according to the scheme:

$$[uvw]h \rightarrow \begin{pmatrix} 2 & \bar{1} & 0 \\ \bar{1} & 2 & 0 \\ 0 & 0 & 1 \end{pmatrix} \begin{pmatrix} u \\ v \\ w \end{pmatrix} = \begin{pmatrix} h \\ k \\ l \end{pmatrix}, \tag{15}$$

where the matrix elements are scaled to integers and the 3,3 element is arbitrarily set. The common divisor must be removed from the symbol (hkl), if present.

It may be surprising that the dual symbols do not contain the specification of an inversion point in the case of rotoinversion. This can be explained by extending the interpretation of

w_g also on such operations. Vector w_g 'measures' the intrinsic transformation of the point x_c into another point, both on a geometric element

$$w_g = (W, w)x_c - x_c. \tag{16}$$

Thus, w_g describes a classical screw/glide vector and a pure inversion of the point x_c. This means that

$$x_{inv} = x_c + w_g / 2 \tag{17}$$

so the calculation and specification of w_g is reasonable for all operation types. The relations and remarks may be better explained by a few examples.

Example 1

The symmetry matrix $\begin{bmatrix} 0 & 1 & 0 \\ -1 & 0 & 0 \\ 0 & 0 & -1 \end{bmatrix}, \begin{bmatrix} 0.5 \\ 0 \\ 0.5 \end{bmatrix}$ is described as $\bar{4}^+$ [001] ½ * ¼,-¼,0.

The obtained dual symbols contain the following data. The matrix describes a fourfold rotoinversion operation. Its geometric element is oriented along direction [001], which is orthogonal to the (001) plane. The point on the axis closest to origin x_c has coordinates: ¼,-¼,0. Mapping x_c is reduced to a pure inversion and generates the shift ½[0,0,1]. Thus, the inversion point is at ¼,-¼,¼, the middle between x_c and its image.

Example 2

The matrix $\begin{bmatrix} 0 & 1 & 0 \\ 1 & 0 & 0 \\ 0 & 0 & 1 \end{bmatrix}, \begin{bmatrix} 0.5 \\ 0.5 \\ 0 \end{bmatrix}$ is described as m[1-10] ½,½,0*.

In this case the matrix describes a glide symmetry operation. The symmetry plane has Miller indices (1-10) and goes through the origin. The glide vector is (½,½,0).

Example 3

The matrix $\begin{bmatrix} -1 & 1 & 0 \\ 1 & 1 & 0 \\ 0 & 0 & 1 \end{bmatrix}, \begin{bmatrix} 0 \\ 0 \\ 0.5 \end{bmatrix}$ is described as m[100](2-10) 0,0½* or shorter as m[100]h 0,0,½*.

In this case the matrix also describes a glide symmetry operation, but in a hexagonal system. The symmetry plane has Miller indices (2$\bar{1}$0) and goes through the origin. The glide vector is (0,0,½).

Example 4

The matrix $\begin{bmatrix} \bar{1} & \bar{1} & \bar{1} \\ 1 & 0 & 0 \\ 0 & 0 & 1 \end{bmatrix}, \begin{bmatrix} 0,5 \\ 0 \\ 0 \end{bmatrix}$ occurring in a primitive description of group No 227 is described as 3^+ [$\bar{1}\bar{1}3$](001) *1/6,1/6,0.

The matrix generates a three-fold rotation. The axis is oriented along [$\bar{1}\bar{1}3$]direction, which is perpendicular to (001) planes and is located by the point 1/6,1/6,0.

It may be noted from the above examples that – introducing the dual symbols – the inconveniences mentioned at the beginning of this section are avoided. The new notation displays the geometric features of a rotation, a rotoinversion, a screw rotation, a reflection, a glide reflection and even an inversion, in a homogeneous way. All such operations correspond to a geometric element (in the form of a single point, line or plane), for which a point closer to the origin always exists, a unique point x_c. The difference $(\mathbf{W},\mathbf{w})x_c - x_c$, defines an 'intrinsic transformation'. These two vectors extend the geometric description of a point operation to the description of a space group operation. The symbols are applicable to space groups presented conventionally as well as non-conventionally. The symbols are rather concise, especially in the context of geometric information, which they contain. The use of the splitting indices $[uvw](hkl)$ seems to be more applicable for space group then for point groups. Crystallographers interpret $[uvw]$ as a family of lattice rows and (hkl) as a family of planes. For practical purposes there is only one question: is the determination of x_c computationally simple?

5. Derivation of x_c points for space-group operations

Point x_c was geometrically defined as the point on the geometric element closest to the origin or equivalently, as the shortest position vector which tail is fixed at the origin and the head ends on the geometric element. Such definition is rather descriptive and not much useful for practical purposes. Generally, the derivation must be carried out in a non-Cartesian system for which the metric is not known. The splitting indices of the matrix part $\mathbf{W}$ make a geometric concept based on which the quantitative components of x_c may be calculated from the translation part $\mathbf{w}$ of a space-group operation. For any $\mathbf{W}$ (excluding identity or inversion operations) the indices of the orthogonal lattice splitting orient the reflection plane by (hkl) indices and also orient line $[uvw]$ in space perpendicular to this plane or orient the symmetry axis $[uvw]$ and specify some plane orthogonal to it. If $\mathbf{w} = \mathbf{0}$, the direction $[uvw]$ and the plane (hkl) intersect in the coordinate system origin O, like in Fig. 1. In this situation the head and the tail of x_c coincide with O, as for point operations. But if $\mathbf{w} \neq \mathbf{0}$, the head of x_c may move along line $[uvw]$ for reflections or may move on (hkl) plane for any type of rotations. Thus, the construction based on the lattice orthogonality given by the splitting indices enables finding x_c without complete metric information. The meaning of x_c for the inversion operation is obvious, since a geometric element in this situation is reduced to one point and for a pure translation $x_c = 0$.

The key to finding x_c on the (hkl) plane or on the $[uvw]$ line is the orbit, a set of points interrelated by the analysed operation. Contrary to general operation $(\mathbf{W},\mathbf{w})$, its reduced version $(\mathbf{W},\mathbf{w_l})$ leads to the cyclic groups and to a limited number of equivalent points in space. If an arbitrary point is located on direction $[uvw]$ for a symmetry plane, or on the (hkl) plane for a symmetry axis, then the centre of gravity of the orbit generated by this point and $(\mathbf{W},\mathbf{w_l})$ defines x_c. Thus, x_c describes the shift of the orbit generated by $(\mathbf{W},\mathbf{w_l})$ and points closest to the origin for all $(\mathbf{W},\mathbf{w_l})^n$ operations. Certainly, for $\mathbf{w_l} = 0$ the gravity centre coincides with the origin. Systematic derivation of x_c for all types of space-group operations is clearer with the help of sketches presented in Figure 3. The intrinsic translation parts $\mathbf{w_l}$

and the projection planes (*hkl*) are treated as already known. The considered orbits are one- or two-dimensional, assuming the origin is located in the fixed point of an rotoinversion operation. Not all points on the orbit are necessary to derive x_c. For an orbit in the form of square its centre is determined by points in two opposite vertices. The centre of the hexagon may be reached by rotating one side by 60°. In final formulae the rotation sense is guaranteed by the presence of **W**, but in geometric constructions the rotation sense must be taken into account. A systematic derivation of the formulae for x_c is followed below.

Figure 3a

Identity operation (**I**,0) can be interpreted as the reduced operation of a pure translation (**I**,**w**). Its geometric element contains all points. Point (0,0,0) is the closest to the origin. Thus, x_c is defined as the zero vector, not dependent on **w**. Another, somewhat artificial but more consistent with the rest of figures approach, is based on the selection of specific point **-w**/2 for the representation of translation

1. Vector to the special point, $\overrightarrow{01}$ = **x**= - **w**/2
2. Vector to its image, $\overrightarrow{02}$ = **Wx**+**w** = - **w**/2

Point x_c = $(\overrightarrow{01} + \overrightarrow{02})$/2 = **0**

Figure 3b

The origin is located at the inversion point. The combination of the inversion with the translation is equivalent to shifting the inversion point by a half of the translation vector, hence

1. Vector to any point, $\overrightarrow{01}$ = **x**
2. Vector to its image, $\overrightarrow{02}$ = **Wx**+**w** = -**x**+**w**

The new position of inversion point x_c = $(\overrightarrow{01} + \overrightarrow{02})$/2 = **w**/2

Figures 3c and 3d

The orbit generated by a reduced two-fold operation or reduced reflection contains two points.

1. Vector to any point, $\overrightarrow{01}$ = **x**
2. Vector to its image, $\overrightarrow{02}$ = **Wx**+**w**₁ = -**x**+**w**₁

The shortest vector x_c = $(\overrightarrow{01} + \overrightarrow{02})$/2 = **w**₁/2 lies on (*hkl*) for a two-fold rotation or on [*uvw*] for a reflection

Figures 3e and 3f

In the case of four–fold rotation or four-fold rotoinversion the calculations are as follows:

1. Vector to any point, $\overrightarrow{01}$ = **x**
2. Vector to its image, $\overrightarrow{02}$ = **Wx**+**w**₁
3. Vector to the image of image, $\overrightarrow{03}$ = **W**(**Wx**+**w**₁)+ **w**₁ = -**x** +**Ww**₁+**w**₁, since in this case **W**²**x** = -**x**

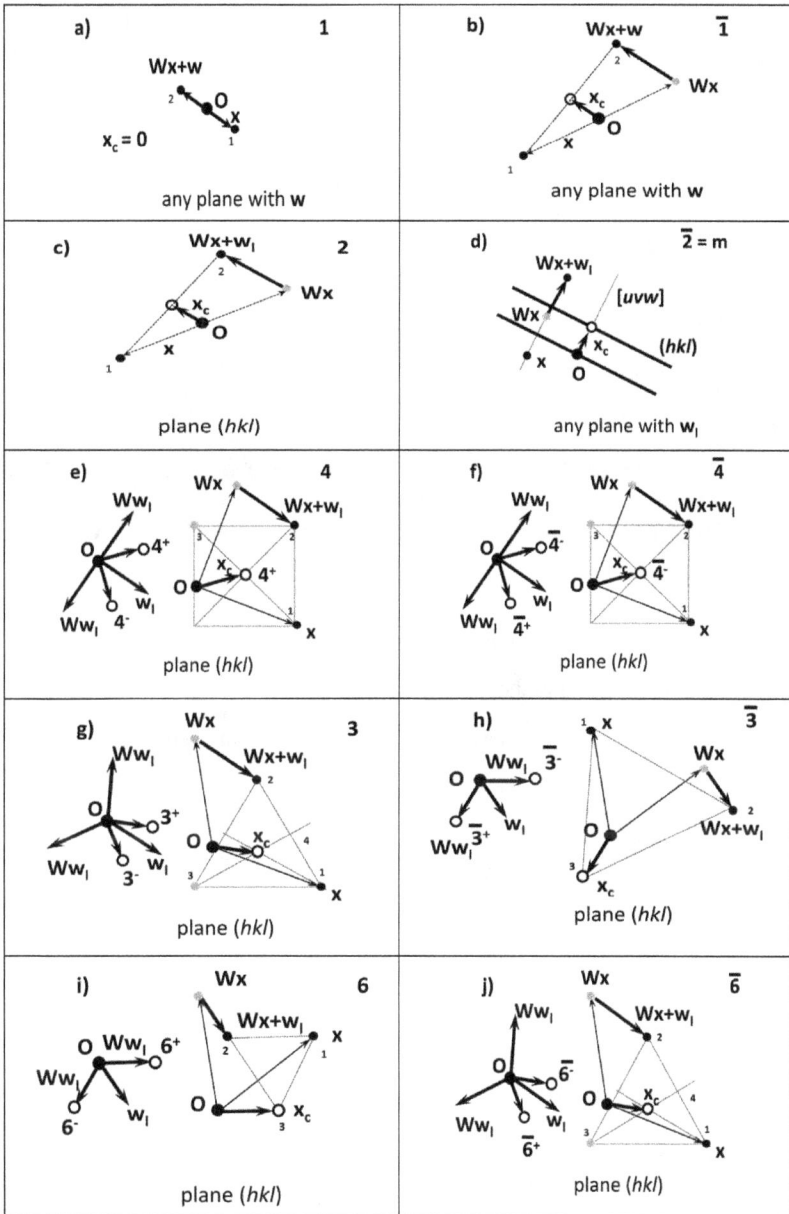

Figure 3. Sketches for derivation points x_c based on any point x and its image(s). On all drawings the projection plane is specified, the origin and x_c are marked as large filled and empty circles, respectively. Other explanations are given in the text.

The centre of gravity for all symmetry equivalent points is the same as for two points of opposite vertices. Thus, the new position of the rotation point is $x_c = (\overrightarrow{01} + \overrightarrow{03})/2 = (\mathbf{W}w_I + w_I)/2$

Figures 3g and 3j

In the case of three–fold rotation or six-fold rotoinversion, the location vectors from the origin to the vertices of equilateral triangle are given by any point x and its two images:

1. Vector to any point, $\overrightarrow{01} = x$
2. Vector to its image, $\overrightarrow{02} = \mathbf{W}x + w_I$
3. Vector to the image of image, $\overrightarrow{03} = \mathbf{W}(\mathbf{W}x + w_I) + w_I = \mathbf{W}^2x + \mathbf{W}w_I + w_I$

The centre of gravity determines the new position of the rotation point

$$x_c = (\overrightarrow{01} + \overrightarrow{02} + \overrightarrow{03})/3 = (\mathbf{W}w_I + 2w_I)/3,\ \text{since } \mathbf{W}^2x + \mathbf{W}x + x = 0.$$

Figures 3h and 3i

In the case of six–fold rotation or three-fold rotoinversion, the location vectors from the origin to the vertices of regular hexagon are given by any point x and its consecutive images. But a simpler way may be based on a rotation of a hexagon side by 60°

1. Vector to any point, $\overrightarrow{01} = x$
2. Displacement vector between x and its image, $\overrightarrow{12} = \mathbf{W}x + w_I - x$, a hexagon side
3. Rotation of $\overrightarrow{12}$ by 60° $= \mathbf{W}(\mathbf{W}x + w_I - x) = \mathbf{W}^2x + \mathbf{W}w_I - \mathbf{W}x$

The centre of gravity determines the new position of the rotation point

$$x_c = (\overrightarrow{01} + \mathbf{W}\overrightarrow{12}) = \mathbf{W}w_I,\ \text{since } \mathbf{W}^2x - \mathbf{W}x + x = 0,\ \text{if } \mathbf{W} \text{ describe rotation by } 60°.$$

The obtained results are summarized in Table 4.

Axis symbol	x_c	Axis symbol	x_c
1	0	$\bar{1}$	$w/2$
$2, \bar{2}(m)$	$w_I/2$	$4, \bar{4}$	$(\mathbf{W}w_I + w_I)/2$
$3, \bar{6}$	$(\mathbf{W}w_I + 2w_I)/3$	$\bar{3}, 6^*$	$\mathbf{W}w_I$

* erroneously stated in [18].

Table 4. Formulae for calculation points x_c.

It may be seen from Table 4 that having classified the matrix part $\mathbf{W}$ and having decomposed translation part w, the derivation of x_c for any space-group operation is unique and extremely simple. Calculations are based on the standard matrix arithmetic: multiplication of a matrix by a vector, adding two vectors or multiplying a vector by a constant. The geometric interpretation of specified rules as well as their explanation is also simple. A comparison of derived formulae with those described in article [18] reveals an incorrect equation for the axis symbol 6 in the published data. The error was overlooked, since in all conventional descriptions of the space groups, the origin is located on the hexagonal axis.

6. Predefined data for space groups generation

For practical and technical convenience the amount of predefined data needed for all 230 space group generation in computer applications should be minimal and the generation process should be most effective. In the case of space group generation in ITA83 style the latter demand is obtained by the composition series method, which on the other hand leads to some redundancy in the sets of non-translational 'generators selected'. This extension may be avoided by reselecting the generators in such a way, that their number is minimal but they can easily restore the original composition series. The concept leads to extending slightly the set of predefined point symmetry operations and to significant reduction of the generators specification for each group, remembering that SG belonging to the same crystal class are based on the same point generators. Since the arithmetic class may be extracted from a space-group symbol or a space-group number, the space generators may be composed of the point generators and individually given translations. Next, the obtained list should be slightly modified to produce a full list of generators needed in a very effective composition method.

This paragraph specifies complete data needed for effective and unique generation of all space groups:

a.   18 matrices (point operations)
b.   37 ranges of space group numbers which correspond to conventional point group descriptions specified by its generators – numbers of matrices from list (a)
c.   230 specification of Bravais letter and one or two translation parts which together with point operations from (b) define space group generators

Matrices needed for restoring the composition series are compiled in Table 5. According to [18] and possible future extensions of the algorithms, they are coded by single numbers. Only proper rotations should be explicitly defined.

No	Code	Symbol	No	Code	Symbol	No	Code	Symbol	No	Code	Symbol
0	16484	1	5	7410	2 [1-10]	10	16482	m x,y,0	15	5223	-6+ 0,0,z; 0,0,0
1	3200	2 [001]	6	14459	6+ 0,0,z	11	16322	m x,0,z	16	11865	-3+ 0,0,z;0,0,0
2	3360	2 [010]	7	7817	3+ 0,0,z	12	11784	-4+ 0,0,z; 0,0,0	17	8866	-3+ x,x,x;0,0,0
3	7898	4+ [001]	8	10816	3+ x,x,x	13	7412	m x,-x,z			
4	12270	2 [110]	9	3198	-1	14	12272	m x,x,z			

Table 5. Symmetry matrices used as point groups generators

Matrices of improper rotation haves numbers increased by 9 and codes complemented to 19682: No = No(proper)+9, code=19682-code (proper). The obvious relation **PntGen** = -1*PntGen (proper) may be also applied. A code is converted into the 3x3 matrix **PntGen** by the following procedure:

```
For k = 3 To 1 Step -1
For j = 3 To 1 Step -1
PntGen(k, j) = (code Mod 3) - 1
code = code \ 3 'integral division
Next j
Next k
```

The relations between the space groups from Table 1 and corresponding point groups are compiled in Table 6. The sequence of point-group generators taken from Table 5 is associated with a space group by its number, according to the order of space-group types unchanged since the first edition of Tables in 1935. A range of space-group numbers compatible with a given list (point group) is defined by its maximal value (SG column). Some subranges or individual SG numbers are needed for the alternative description of point groups, what is seen in the last 5 rows.

SG max	PG symbol	Generators	SG max	PG symbol	Genera-tors	SG max	PG symbol	Genera-tors	SG max	PG symbol	Genera-tors
1	1	-	82	$\bar{4}$	12	161	3m1	7,10	199	23	1,8
2	$\bar{1}$	9	88	4/m	3,9	167	$\bar{3}$m1	16,4	206	m$\bar{3}$	1,17
5	2	2	98	422	3,2	173	6	6	214	432	1,8,4
9	m	11	110	4mm	3,11	174	$\bar{6}$	15	220	$\bar{4}$3m	1,8,13
15	2/m	2,9	122	$\bar{4}$2m	12,2	176	6/m	6,9	230	m$\bar{3}$m	1,17,4
24	222	1,2	142	4/mmm	3,2,9	182	622	6,4	115-120	$\bar{4}$m2	12,11
46	mm2	1,11	146	3	7	186	6mm	6,10	149,151, 153	312	7,5
74	mmm	1,2,9	148	$\bar{3}$	16	190	$\bar{6}$m2	15,10	157,159	31m	7,13
80	4	3	155	321	7,4	194	6/mmm	6,4,9	162-163	$\bar{3}$1m	16,5
									189-190	$\bar{6}$2m	15,4

Table 6. 32 point groups (+ 5 additional orientations), its generators and the maximal numbers of space groups to which the group corresponds

There are 73 *symmorphic* space groups, which in the conventional description need only additional information on the centring type. In other groups the space generators must contain partial translations. A proper selection of the origin makes that only one or two non-zero vectors must be specified. Based on the 'conventional origins' from ITA83, the information necessary to convert point generators into the space generators was compiled in Table 7.

The lack of Bravais letter in the items of Table 7 means the primitive lattice or the symbol P. Only translation parts different from the zero vector are specified; their ordering positions are fixed by the asterisks. In order to overcome ratios, translation vectors are multiplied by 4, or by 6 in the trigonal and hexagonal crystal systems. Moreover, in the latter systems only the z-component differs from zero and thus is specified. The items can be treated as a concise information on the system of primitive as well as non-primitive translations including: lattice centring, origin specification, relative position of symmetry elements differing space groups in the arithmetic class.

Based on the unique identifier of a space-group type, which can be related to the space-group sequence number, the list of space-group generators organized in the composition series can be obtained from condensed data included in Tables 5-7.

SG No	+1	+2	+3	+4	+5	+6	+7	+8	+9	+10
0				020	C		002	C	C002	
10	020	C	002	022	C002		002002	*220	202022	C002002
20	C	F	I	I202022		002002	*002	*200	002200	*022
30	202202	*220	002220	*222	C	C002002	C*002	A	A*020	A*200
40	A*220	F	F*111	I	I*220	I*200		220202	*002	220200
50	200	200222	202202	200002	*220	220022	002022	*222	220020	222002
60	202022	202020	C002002	C022022	C	C*002	C020020	C200002	F	F330303
70	I	I*220	I202022	I020020		001	002	003	I	I021
80		I		002	200	022	I	I311		220220
90	001	221221	002	222222	003	223223	I	I021203		*220
100	002002	222222	*002	*222	002	002220	I	I*002	I021	I021002
110		*002	*220	*222		*002	*220	*222	I	I*002
120	I	I*203		*002	200200	200202	*220	*222	200020	200022
130	002	002002	202200	202202	002220	222222	202020	202022	I	I*002
140	I131202	I131200		2	4	R		R		
150	24	2	42	4	R			*3	*3	R
160	R*3		*3		*3	R	R*3		1	5
170	2	4	3			3		12	54	24
180	42	3		*3	33	3		33		3
190		*3	33	3		F	I	202	I202	
200	220	F	F330	I	202	I202		**222	F	F022*313
210	I	202*133	202*311	I202*311		F	I	**222	F*222	I202*111
220		220*002	**222	220*220	F	F**222	F312*312	F132*310	I	I202*311

Table 7. Bravais symbols and the translation parts of generators

7. Composition series and space groups generation

It is well known and seen in Table 6 that maximally three generators are necessary for obtaining all symmetry operations in the point groups, even in group $m\bar{3}m$. But for effective generation of groups based on a series of normal subgroups, the listed set of generators should be extended to the complete composition series. The following action leads to the non-translational sets of generators used in ITA83:

i.   generators 4 and $\bar{4}$ (matrices 3 and 12) are preceded by their squares.
ii. generator $\bar{3}$ is replaced by the 3 and the generator $\bar{1}$ is added as the last generator.

iii.   generators 6 and $\bar{6}$ (matrices 6 and 15) are replaced by their third and second powers.
iv.   in the case of a cubic system the two-fold rotation in z direction (matrix 1) is followed
      by the two-fold rotation in y direction (matrix 2). The situation is a little more
      cumbersome for the space groups: the translation part of generator 2[010] occurring in
      the composition series is not explicitly specified and must be obtained by a cyclic
      permutation of the translation components $(t_x, t_y, t_z) \rightarrow (t_z, t_x, t_y)$ given for 2[001]
      generator.

The construction of composition series of generators is shown on the example of the
tetragonal group $P42/nnm$ in its initial description. According to Table 1 the input, from
which the non-modified generators shown in Table 8 have been obtained, is reduced to
number '134'. Comparing this space-group number with the data contained in Table 3,
132<= 134 < 136, gives the point group $4/mmm$ and point generators 3,2,9, which matrix
forms are derived from the integers 7898, 3360 and 3198. The 'space' information is
presented in Table 7. The given group number corresponds to string '202202', and defines
translation parts for the first and the second generator. According to rule (i) in the above list,
the first generator of composition series is obtained by squaring the generator with their
matrix part no 3. As a result, four generators described in Table 8 are obtained.

$$\begin{pmatrix} \bar{1} & 0 & 0 \\ 0 & \bar{1} & 0 \\ 0 & 0 & 1 \end{pmatrix}\begin{pmatrix} 1/2 \\ 1/2 \\ 0 \end{pmatrix} \quad \begin{pmatrix} 0 & \bar{1} & 0 \\ 1 & 0 & 0 \\ 0 & 0 & 1 \end{pmatrix}\begin{pmatrix} 1/2 \\ 0 \\ 1/2 \end{pmatrix} \quad \begin{pmatrix} \bar{1} & 0 & 0 \\ 0 & 1 & 0 \\ 0 & 0 & \bar{1} \end{pmatrix}\begin{pmatrix} 1/2 \\ 0 \\ 1/2 \end{pmatrix} \quad \begin{pmatrix} \bar{1} & 0 & 0 \\ 0 & \bar{1} & 0 \\ 0 & 0 & \bar{1} \end{pmatrix}\begin{pmatrix} 0 \\ 0 \\ 0 \end{pmatrix}$$

Table 8. Composition series of generators for the space group type No 134, consistent with ITA83.

Having completed a composition series set of generators, the derivation of all symmetry
matrices or more precisely obtaining the *coset representatives* is a trivial task. Each non-
translational generator doubles, or triples in the case of 3⁺ operation, the current set of
symmetry operations and thus the process is non-redundant and needs only one matrix
multiplication per one new symmetry operation item.

The generating process needs only one list of 4x4 matrices for storing as well the original or
transformed generators as new generated operations. It is assumed that the GensNo
generators are located in the upper part of the **SymMtx** list starting from position 49, the
current number of symmetry operators **SymsNo** is set to 1 and the unit matrix is introduced
as the first element of **SymMtx**. Let the procedure **NewSymMtx** with three input indices
gen, oldsym, newsym multiplies the matrix 48+gen by the matrix **oldsym**, reduces
translation components modulo 1 and locates the new symmetry operation in the **newsym**
position. The complete space group generation procedure can be as follows:

For i = 1 To GensNo   'main loop

If SymMtx(48+i,1,1) + SymMtx(48+i,2,2) + SymMtx(48+i,3,3) = 0 Then   'test for 3+
triple = True
Else
triple = False

Endif

For j = 1 To SymsNo 'generation of the successive normal subgroups

Call NewSymMtx(i, j, SymsNo + j)

If triple = True Then Call NewSymMtx(i, SymsNo + j, 2 * SymsNo + j)

Next j

If triple = True Then SymsNo = SymsNo * 3 Else SymsNo = SymsNo * 2 'current number of symmetry operation

Next i 'end of main loop

The advantages of the composition series method are clear from the above scheme – every new symmetry matrix is obtained by a single multiplication without checking for redundancy. Since the complete list of generators corresponds to the explicitly given ones in ITA83, the obtained result repeats the corresponding conventional space group description in ITA83, including also the order of the symmetry matrices. This is also the base for the other unique space group descriptions, irrespective of whether they are conventional or not, since before the process starts, the generators can be recalculated according to any given coordinate transformation T by standard equation $G' = T^{-1}GT$. Such approach is especially simple in connection with the *translational space group symbols*, where transformations are described explicitly.

8. Deriving dual symbols for symmetry operations

The derivation of a dual symbol from a symmetry operation (W,w) involves the following:

a. Characterizing the point operation W by: (i) rotation type, that is 1, 2, 3, 4, 6, $\bar{1}$, m, $\bar{3}$, $\bar{4}$, $\bar{6}$, (ii) sense of rotation, (iii) orthogonal lattice splitting indices $[uvw](hkl)$

b. Characterizing the space operation (W,w) by: (i) orthogonal splitting of the translation part $w = w_g + w_l$, (ii) point x_c on geometric element closest to the origin.

The type of point operation is easily recognized from the determinant and trace of W using a look-up Table 2. Splitting indices are the same for both matrices W and $-W$. Let W_p denotes the proper rotation matrix $det(W)W$. A lattice vector $[uvw]$ parallel to the rotation axis as well its reciprocal (hkl) are defined by any non-zero column and any non-zero row of the matrices constructed in dependence on the order k of W_p:

$$k = 2 : I + W_p, k = 3 : I + W_p + W_p{}^2, k = 4 : I + W_p{}^2 \text{ and } k = 6 : I + W_p{}^3$$

where I denotes a unit matrix.

Next, a positive scalar product is required, if necessary by multiplying (hkl) by -1. If vectors $[uvw]$ and (hkl) have common divisors, indices should be divided. At last all indices can change the signs in order to obtain the positive axis direction according to Table 3. Having determined the direction $[uvw]$ of an axis, the sense of rotation is simply determined from the inequalities (11). These are all steps needed for a complete geometric description of a point symmetry operation. Practically, the conventions are reduced to the one rule about the positive direction of $[uvw]$.

The orthogonal splitting of a translation part $\mathbf{w}$ depends on the axis direction $[uvw]$. The same results are obtained for $(\mathbf{W},\mathbf{w})$ and $(-\mathbf{W},\mathbf{w})$, thus calculations can be reduced to the pair $(\mathbf{W_p},\mathbf{w})$. Since $(\mathbf{W_p},\mathbf{w})^k = (\mathbf{I},k\mathbf{w_g})$, the k-th part of the resulting translation defines the intrinsic component of the translation, and in consequence also the location component $\mathbf{w_l} = \mathbf{w} - \mathbf{w_g}$. In the presented procedure such decomposition is also valuable for a rotoinversion operation. As usual, for a reflection the numerical values of $\mathbf{w_g}$ and $\mathbf{w_l}$ must be interchanged. The central problem of deriving the point $\mathbf{x_c}$ (an invariant relative to a reduced operation and closest to the origin) is to find such point on the $[uvw]$ lattice vector or on the (hkl) plane for a rotation and rotoinversion in the first case and for a reflection in the latter. Geometrical considerations showed that $\mathbf{x_c}$ are rather simply calculated from $\mathbf{w_l}$ according to relations given in Table 4.

The derivation of $\mathbf{x_c}$ ends the calculation needed to describe the space symmetry operation. But vectors $\mathbf{x_c}$ in the case of reflections and $\mathbf{w_g}$ in the case of rotations or rotoinversions are parallel to $[uvw]$ and may be represented by a single ratio r. Thus, an intrinsic component $\mathbf{w_g}$ and a shift vector (orthogonal to the geometric element) are presented by four simple ratios. An asterisk mark serves as a separation mark between vectors and thus a dual symbol of space-group operation takes the form of $\pm n \pm [uvw](hkl)\ \mathbf{w_g}^*\mathbf{x_c}$.

9. Origin problem

In the non-symmorphic space groups the symmetry elements are not constrained to pass through the origin. The selection of a reasonable origin for a coordinate system relative to non-intersecting symmetry elements is not unique. The symbol of space-group type, like Hermann-Mauguin symbol, fixes in space only relative positions of symmetry elements. Absolute positions need complete translational parts in the space-group generators. The presented technique of space group derivation based on predefined generators favours one and the only one origin for each space group, even if this group is tabulated in ITA83 relative to two origins.

Finding a transformation between two descriptions of the same space group differing only by the origin shift is arithmetically at least cumbersome. In this case, similarly like in other space-group considerations, the geometric information is very practical. A typical way of resolving the mentioned problem consists of a geometrical interpretation of the symmetry matrices in both descriptions and of deduction of the transformation from the diagram of symmetry elements in ITA83. Such analysis is impossible for a non-conventional space group description, but in every case may be carried out on dual symbols.

Since the dual symbols described in the preceding section are easy to obtain, they should be routinely derived together with the space group generation. Their role in the origin control is rather evident, but we illustrate this feature by means of an example. Let column 2 of Table 9 lists the dual symbols of the $P4_2/nnm$ operations obtained from the generators presented in Table 8. The items 1, 7, 9, 15 have reduced $\mathbf{x_c}$ parts. It is visible that the origin is located at the inversion point, the intersection of 2[110] and m[110]. A full symmetry of the origin is 2/m (this is the second origin choice tabulated in ITA83).

A higher symmetry of the origin can be obtain by placing it on $\bar{4}$. The origin shifted from that in the first column by (1/4,-1/4,0) leads to a group description symbolized by TSG = P42/nnm (1/4,-1/4,0) (this is the first origin choice in ITA83). The result is listed in column 3 of Table 9. It can be seen that rotoinversion operations 11,12 contain the intrinsic part ½ and the inversion point according to (17) in (0,0,1/4).

Another possibility to select the origin in a high symmetry point is to put it on $\bar{4}$, but exactly at the inversion point. In this situation the origin is shifted by (1/4,-1/4,1/4) in comparison with column 2. The symmetry matrices of the group description TSG = P42/nnm (1/4,-1/4,1/4) presented by dual symbols are given in the last column of Table 9.

No	TSG		
	P42/nnm	P42/nnm (¼,-¼,0)	P42/nnm (¼,-¼,¼)
1	1	1	1
2	2 [001] * 1/4,1/4,0	2 [001]	2 [001]
3	4+[001] 1/2 * 1/4,1/4,0	4+[001] 1/2 * 0,1/2,0	4+[001] 1/2 * 0,1/2,0
4	4-[001] 1/2 * 1/4,1/4,0	4-[001] 1/2 * 1/2,0,0	4-[001] 1/2 * 1/2,0,0
5	2 [010] * 1/4,0,1/4	2 [010] * 0,0,1/4	2 [010]
6	2 [100] * 0,1/4,1/4	2 [100] * 0,0,1/4	2 [100]
7	2 [110]	2 [110] 1/2 *	2 [110] 1/2 * 0,0,1/4
8	2 [1-10] * 1/4,1/4,0	2 [1-10] * 1/4,1/4,0	2 [1-10] * 1/4,1/4,1/4
9	-1	-1 * 1/4,1/4,0	-1 * 1/4,1/4,1/4
10	m [001] 1/2,1/2,0 *	m [001] 1/2,1/2,0 *	m [001] 1/2,1/2,0 * 1/4
11	-4+[001] 1/2 * 1/4,-1/4,0	-4+[001] 1/2 *	-4+[001]
12	-4-[001] 1/2 * -1/4,1/4,0	-4-[001] 1/2 *	-4-[001]
13	m [010] 1/2,0,1/2 *	m [010] 1/2,0,1/2 * 1/4	m [010] 1/2,0,1/2 * 1/4
14	m [100] 0,1/2,1/2 *	m [100] 0,1/2,1/2 * 1/4	m [100] 0,1/2,1/2 * 1/4
15	m [110]	m [110]	m [110]
16	m [1-10] 1/2,1/2,0 *	m [1-10]	m [1-10]

Table 9. Different descriptions of the space group type P42/nnm

In order to show different group-subgroup relations, other descriptions of space groups may be desirable. Contrary to rather difficult manipulation based on coordinate triplets [28], the determination of shift vectors with the help of geometric information is simple. For this purpose the classical symbols of symmetry operations as well dual symbols are similarly useful, but the latter may be also applicable in a non-conventional space-group description. Since the multiplication of symmetry matrices is based on modulo 1 arithmetic, the origin control should involve only the generators of space groups.

10. Axes problem

A rigorous classification of space groups, that is their specific descriptions, into space-group type can be given in an algebraic or a geometric way. Typically, the matrix algebra and the group-theoretical approach is preferred. For this classification, each space group description is referred to a primitive base and an origin. Two space groups **G** and **G'** belong to the same space-group type if a transformation pair **P**, **p** exists, for which the 3x3 matrix has integral elements with det(**P**) = 1 and the **p** vector consists of three real numbers, such that **G** is transformed into **G'** (see, Wondratschek in ITA83). This definition is very simple, but finding the transformation between two sets of matrices may be a real challenge.

Let's modify the above equivalence definition for the practical purposes. Now **G'** means a conventionally described *space-group type*, represented by a unique set of generators or symmetry matrices. **G** is still referred to a primitive base. **G** belongs to the space-group type **G'** if a transformation pair **P**, **p** exists, for which the 3x3 matrix has integral elements with det(**P**) = 1,2,3 or 4 and the **p** vector consists of three real numbers, such that **G** is transformed into **G'**. The first step in determining the type of group **G** is to refer it to a centred Bravais base by a proper selection of coordinate axes. It is simple with the help of dual symbols what will be illustrated by the group description TSG = $I4_122$ (1,0,0; 0,1,0; ½,1/2,1/2) (1/4,1/4,0) given in Table 10.

1	1 0 0 0	2	-1 0 -1 0,5	3	0 -1 -1 0,25	4	0 1 0 0,75
	0 1 0 0		0 -1 -1 0,5		1 0 0 0,25		-1 0 -1 0,75
	0 0 1 0		0 0 1 0		0 0 1 0,5		0 0 1 0,5
	1		2 [-1-12](0 0 1) * 1/4,1/4,0		4+[-1-12](0 0 1) 1/4 * 0,1/2,0		4-[-1-12](0 0 1) 1/4 * 1,0,0
5	-1 0 0 0,25	6	1 0 1 0,75	7	0 1 1 0	8	0 -1 0 0,5
	0 1 1 0,25		0 -1 0 0,75		1 0 1 0		-1 0 0 0,5
	0 0 -1 0,5		0 0 -1 0,5		0 0 -1 0		0 0 -1 0
	2 [010](0 2 1) 1/2 * 1/8,-1/8,1/4		2 [100](2 0 1) 1 * - 1/8,3/8,1/4		2 [110](1 1 1)		2 [1-10] * 1/4,1/4,0

Table 10. Symmetry matrices and dual symbols of the group description TSG = $I4_122$ (1,0,0; 0,1,0; ½,1/2,1/2) (1/4,1/4,0)

Items 1,2,3 and 4 define symmetry axis 4_1 parallel to [$\bar{1}\bar{1}2$]. Matrices (5,6) and (7,8) describe two pairs of orthogonally oriented twofold axes, according to property $u_1h_2 + v_1k_2 + w_1l_2 = u_2h_1 + v_2k_1 + w_2l_1 = 0$ between the corresponding splitting indices. A similar test shows the orthogonality between 4_1 and all twofold axes. Thus, two orthogonal bases and two transformation matrices may be constructed from [uvw] indices. The first transformation matrix **P** with columns [100], [010], [$\bar{1}\bar{1}2$] has det(**P**) = 2 and leads to I-centred basis. The equivalent F-centred Bravais tetragonal cell is involved with P in the form [$1\bar{1}0$], [110] and [$\bar{1}\bar{1}2$]. The application of the first transformation matrix is equivalent to the selection of the conventional axes for the tetragonal space group.

Moreover, the arithmetic type $422I$ of the analysed group is determined. From the predefined data one can find that only two space-group types, namely the groups with sequence numbers 97 and 98, belong to this arithmetic type. Group 97 is symmorphic. Thus,

4_1 axis points directly onto type 98 space group. The origin shift **p** may be determined by comparing points x_c in dual symbols of the space-group type generators and symmetry operations of the analysed group with identical matrix parts. The generators of group type $I4_122$ are characterized by dual symbols as: 4+[001] 1/4 * -1/4,1/4,0 and 2 [010] * 1/4,0,3/8. Items 3 and 5 from Table 10 transformed according to **P** matrix give 4+[001] 1/4 * 0,1/2,0 and 2 [010] 1/2 * 1/4,0,1/8. The origin shift **p** = -1/4,-1/4,0 predicted from the 4_1^+ operation equals both descriptions.

The analysed example does not attempt to be a kind of algorithmic approach, but was included to show the advantages of representing symmetry operations in both forms, as a symmetry matrix and as a dual symbol.

11. Conclusions

It was assumed that, the choice of the known or modified algorithms should be motivated by obtaining functional relations f_1 (*SG symbol, ordering number*) = *symmetry matrix* and f_2 (*symmetry matrix*) = *geometric description* on the assumption that the needed conventions are reduced to a minimum. This unique arithmetic and geometric description of space groups cannot be obtained at the cost of excessive amount of the predefined data, ineffective or sophisticated algorithms. Moreover, in the case of conventional axes and origins the results cannot differ from that contained in ITA83. It appears that such practical purposes have been achieved.

The commonly accepted Hermann-Mauguin symbol is very informative and useful for controlling group orientation, but is not dependent on the origin choice and is not applicable for a non-conventional space group description. The absolute position of symmetry elements needs an explicit origin specification, by giving a complete information about the translation part of generators. The coding of such generators leads to the elaborated symbols (Zachariassen/Shmueli) or to the concise symbols (Hall), but involved with many conventions and rather sophisticated interpretation of symbols. Moreover, in universal approaches based on arbitrary generators most of the computing time is spent on tests for closure or for redundancy of generated operations.

The '*transformational concept*' presented in this chapter introduces an 'absolute' description of each space group type contrary to the multiple standards for some space group types in ITA83. Thus, all groups belonging to a given SG type may be derived from the same set of optimally selected generators, assuming that the transformation from the SG type to a needed description is known. The TSG symbol was aimed at pointing in a computer program at the set of generators, prepared as 'composition series', and at giving an explicit definition of necessary transformation of these generators. There is no need to interpret the TSG symbol or tests the generators. If generators published in ITA83 serve as predefined type generators, nearly 100% of computing time is spent on non-redundant SG derivation. Programs based on a transformational concept reproduce all conventional descriptions given in ITA83; in the non-conventional cases they lead to descriptions in the ITA83 style.

From this point of view, such algorithms standardize any SG descriptions and their geometric interpretations and may be treated as an electronic extension of the printed ITA83 tables. In this context the classification of space groups into conventionally versus non-conventionally described, and the evolutionary changes between different editions of Tables are unsubstantial.

The generation schema based on composition series leads also to a considerable reduction of predefined data, what is an important feature in the implementing and testing of SG derivation programs. Since all the space groups which are based on a given point group are generated in the same way, only the translational parts of generators should be stored. Using small improvements discussed in the text, the non-redundant number of composition series generators is 3, and thus the non-redundant number of translations for each space-group type is 0,1 or maximum 2.

The introduced dual symbol of symmetry operation sheds some light on the lattice properties and thus changes a rather informative character of the geometric symbols into a valuable tool for finding the transformation between different descriptions of the same groups. Since the symbol is universal, convention free, easy to derive and to manipulate on a computer, it is advised to generate space group descriptions simultaneously in two forms, as coordinate triplets and as dual symbols. Any new information can be obtained from the additional Seitz symbol.

The simplicity of the considered approach is evident and seems to be valuable also for teaching purposes. Algorithms were implemented in the Visual Basic, in the Excel environment. All procedures need less than 600 programming lines including 200-line subroutine for transforming a symmetry matrix into a dual symbol. This procedure may be also used as stand-alone item to characterize any symmetry matrix. The source code included in the Excel worksheet may be obtained from the author on request.

Author details

Kazimierz Stróż
Faculty of Informatics and Materials Science, University of Silesia, Katowice, Poland

12. References

[1] Han Th ed. (1983) *International Tables for Crystallography* Vol. A. *Space-group Symmetry,* Dordrecht: *Cluwer Academic Publishers*

[2] Abad-Zapatero C, O'Donell T J (1987) *TABLES, a program to display space-group symmetry information in three dimensions,* J. Appl. Cryst. 20:532-535

[3] Sakurai T, Kobayashi K, Horiki T, Furukawa M, Naitou K (1989) *CRYST – a system to display 3D images of crystal structure, symmetry operations and crystal forms,* J. Appl. Cryst. 22:633-639

[4] Larson A C (1969) Acta Cryst. A25, S1

[5]   Burzlaff H, Hountas A. (1982) J. Appl. Cryst. 15: 464-46
[6]   Hall S R (1981) Acta Cryst. A37: 517-525
[7]   Zachariasen W H (1967) *Theory of X-ray Diffraction in Crystals*, ch. II. New York:Dover
[8]   Shmueli U (1984) *Space-group algorithms. I. The space group and its symmetry elements*, Acta Cryst. A40: 559-567
[9]   Grosse – Kunstlieve R W (1999) *Algorithms for deriving crystallographic space-group information*, Acta Cryst. A55: 383-395.
[10] Stróż K (2006) *Transformational space-group symbols*, Acta Cryst. A63: 77-79.
[11] Ledermann W (1957) *Introduction to the theory of finite groups*, 3rd ed. Edinburgh: Oliver and Boyd.
[12] Fokkema D S (1983) In *International Tables for Crystallography*, Vol.A, pp xiv-xv, Dordrecht:Reidel
[13] Stróż K (1997) *SPACER: a program to display space-group information for a conventional and nonconventional coordinate system*, J. Appl. Cryst. 30:178:181
[14] Stróż K (1997) *On unique description of space groups*, in. Proc. of the XVII Conf. on Appl. Cryst., Singapore: World Scientific
[15] Stróż K (2007) *Structure of crystallographic space groups and algorithms of their generation*, Katowice: Editorial Board of Silesian University (in Polish)
[16] Wondratschek H, Neubüser J (1967), Acta Cryst. 23: 349-352.
[17] Fischer W, Koch E (1983) *Symmetry operations*, in *International Tables for Crystallography*, Vol.A, chapter 11, Dordrecht:Reidel
[18] Stróż K (2007) *A systematic approach to the derivation of standard orientation-location parts of symmetry-operation symbols*, Acta Cryst. A63: 447-454)
[19] Kopský V (2001) *Is a revision of Vol. A of the International Tables desirable or necessary?* in *Advances in Structure Analysis*, edited by Kužel R, Hašek J, Praha: CSCA.
[20] De Wolf P M, et all (1989) *Definition of symmetry elements in space groups and point groups. Report of the International Union of Crystallography Ad-Hoc Committee on the Nomenclature of Symmetry*, Acta Cryst. A45: 494-499
[21] De Wolf P M, et all (1992) *Symbols for symmetry elements and symmetry operations. Final report of the International Union of Crystallography Ad-Hoc Committee on the Nomenclature of Symmetry*, Acta Cryst. A45: 727-732
[22] Flack H D, Wondratschek H, Hahn Th, Abrahams S C (2000) *Symmetry elements in space groups and point groups. Addenda to two IUCr reports on the nomenclature of symmetry*, Acta Cryst. A56: 96-98
[23] Stróż K (2011) *Space of symmetry matrices with elements 0,±1 and complete geometric description; its properties and application*, Acta Cryst. A67: 421-429.
[24] Litvin D B, Kopský V(2011) *Seitz notation for symmetry operations of space groups*, Acta Cryst. A67: 415-418.
[25] Seitz, F. (1934). *Z. Kristallogr.* 88, 433-459.
[26] Burns G, Glazer A M (2007) *Space groups for solid State Physicists*, 2nd ed. New York: Academic Press.
[27] Boisen M B Jr, Gibbs G V (1990) *Mathematical crystallography, reviews in mineralogy.* Vol.15, Washington: Mineralogical Society of America.

[28] Sands D E (1982) *Vectors and tensors in crystallography*, Massachusetts: Addison-Wesley Publishing Company

Computational Potential Energy Minimization Studies on the Prion AGAAAAGA Amyloid Fibril Molecular Structures

Jiapu Zhang

Additional information is available at the end of the chapter

1. Introduction

X-ray crystallography, NMR (Nuclear Magnetic Resonance) spectroscopy, and dual polarization interferometry, etc are indeed very powerful tools to determine the 3D structure of a protein (including the membrane protein), though they are time-consuming and costly. However, for some proteins, due to their unstable, noncrystalline and insoluble nature, these tools cannot work. Under this condition, mathematical and physical theoretical methods and computational approaches allow us to obtain a description of the protein 3D structure at a submicroscopic level. This Chapter presents some practical and useful mathematical optimization computational approaches to produce 3D structures of the Prion AGAAAAGA Amyloid Fibrils, from an energy minimization point of view.

X-ray crystallography finds the X-ray final structure of a protein, which usually need refinements in order to produce a better structure. The computational methods presented in this Chapter can be also acted as a tool for the refinements.

All neurodegenerative diseases including Parkinson's, Alzheimer's, Huntington's, and Prion's have a similarity, which is they all featured amyloid fibrils (en.wikipedia.org/wiki/Amyloid and references (Nelson et al., 2005; Sawaya et al., 2007; Sunde et al., 1997; Wormell, 1954; Gilead and Gazit, 2004; Morley et al. 2006; Gazit, 2002; Pawar et al., 2005; and references therein). A prion is a misshapen protein that acts like an infectious agent (hence the name, which comes from the words protein and infection). Prions cause a number of fatal diseases such as 'mad cow' disease in cattle, scrapie in sheep and kuru and Creutzfeldt-Jakob disease (CJD) in humans. Prion diseases (being rich in β-sheets (about 43% β-sheet) (Griffith, 1967; Cappaia and Collins, 2004; Daude, 2004; Ogayar and Snchez-Prez, 1998; Pan et al., 1993; Reilly, 2000) belong to neurodegenerative diseases. Many experimental studies such as (Brown, 2000; Brown, 2001; Brown, 1994; Cappai and Collins, 2004; Harrison et al., 2010; Holscher, 1998; Jobling et al., 2001; Jobling et al., 1999; Kuwata et al., 2003; Norstrom and Mastrianni, 2005; Wegner et al., 2002; Laganowsky et al., 2012; Jones et al., 2012; Sasaki et al., 2008; Haigh et al., 2005; Kourie et al., 2003; Zanuy et al., 2003; Kourie, 2001; Chabry et al., 1998; Gasset et al., 1992) have shown that the normal hydrophobic region (113-120) AGAAAAGA of prion proteins is an inhibitor/blocker of prion diseases. PrP lacking this palindrome could not convert to prion diseases. The presence of residues 119 and 120 (the two last residues within the motif AGAAAAGA) seems to be crucial for this inhibitory effect. The replacement of Glycine at residues 114 and 119 by Alanine led to the inability of the peptide to build fibrils but it nevertheless increased. The A117V variant is linked to the GSS disease. The physiological conditions such as pH (Cappai and Collins, 2004) and temperature (Wagoner et al., 2011) will affect the propensity to form fibrils in this region. The 3D atomic resolution structure of PrP (106-126), i.e. TNVKHVAGAAAAGAVVGGLGG, can be looked as the structure of a control peptide (Cheng et al., 2011; Lee et al., 2008). Ma and Nussinov (2002) established homology structure of AGAAAAGA and did its molecular dynamics simulation studies. Recently, Wagoner et al. computer simulation studied the structure of GAVAAAAVAG of mouse prion protein (Wagoner, 2010; Wagoner et al., 2011). Furthermore, the author computationally clarified that prion AGAAAAGA segment indeed has an amyloid fibril forming property (Fig. 1).

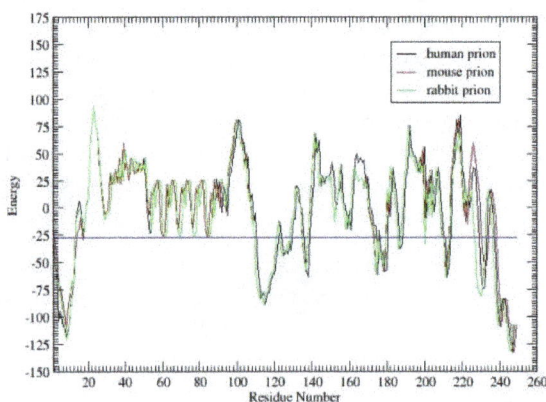

Figure 1. Prion AGAAAAGA (113-120) is surely and clearly identified as the amyloid fibril formation region, because its energy is less than the amyloid fibril formation threshold energy of -26 kcal/mol (Zhang et al., 2007).

However, to the best of the author's knowledge, there is little X-ray or NMR structural data available to date on AGAAAAGA (which falls just within the N-terminal unstructured region (1.–123) of prion proteins) due to its unstable, noncrystalline and insoluble nature. This Chapter will computationally study the molecular modeling (MM) structures of this region of prions.

2. Molecular structures of prion AGAAAAGA amyloid fibrils

"Amyloid is characterized by a cross-β sheet quaternary structure" and "recent X-ray diffraction studies of microcrystals revealed atomistic details of core region of amyloid" (en.wikipedia.org/wiki/Amyloid and references (Nelson et al., 2005; Sawaya et al., 2007; Sunde et al., 1997; Wormell, 1954; Gilead and Gazit, 2004; Morley et al., 2006; Gazit, 2002; Pawar et al., 2005; and references therein). All the quaternary structures of amyloid cross-β spines can be reduced to the one of 8 classes of steric zippers of (Sawaya et al., 2007), with strong van der Waals (vdw) interactions between β-sheets and hydrogen bonds (HBs) to maintain the β-strands.

A new era in the structural analysis of amyloids started from the 'steric zipper'- β-sheets (Nelson et al., 2005). As the two sheets zip up, HPs (Hydrophobic Packings) (& vdws) have been formed. The extension of the 'steric zipper' above and below (i.e. the β-strands) is maintained by HBs (but there is no HB between the two β-sheets). This is the common structure associated with some 20 neurodegenerative amyloid diseases, ranging from Alzheimer's and type-II diabetes to prion diseases. For prion AGAAAAGA amyloid fibril structure, basing on the common property of potential energy minimization of HPs, vdws, and HBs, we will present computational molecular structures of prion AGAAAAGA amyloid fibrils.

2.1. Review on materials and methods, and results of MM models

2.1.1. Hybrid method of steepest descent – conjugate gradient with simulated annealing

X-ray crystallography finds the X-ray final structure of a protein, which usually need refinements using a simulated annealing protocol in order to produce a better structure. Thus, it is very amenable to use simulated annealing (SA) to format the models constructed. Zhang (2011a, 2011d) presents a hybrid method of global search SA with local steepest descent (SD), conjugate gradient (CG) searches. The hybrid method is executed with the following three procedures. (1) Firstly the SD method and then the CG method are executed. These two local search methods are traditional optimization methods. The former has nice convergence but is slow when close to minimums. The latter is efficient but its gradient RMS and GMAX gradient (Case et al., 2010) do not have a good convergence. (2) When models cannot be optimized further, we employ standard SA global search procedure. (3) Lastly, the SD and CG methods are used to refine the models. The PDB (Berman et al., 2000) templates used in (Zhang, 2011a, 2011d) are 2OKZ.pdb, 2ONW.pdb, 2OLX.pdb, 2OMQ.pdb, 2ON9.pdb, 2ONV.pdb, 2ONA.pdb, 1XYO.pdb, 2OL9.pdb, 2OMN.pdb, 2ONX.pdb, 2OMP.pdb, 1YJP.pdb of (Sawaya et al., 2007), but only the 2OMP and 1YJP template-based three MM-Models (Fig. 6a~6c in (Zhang, 2011a)) are successfully passed through the SDCG-SA-SDCG computational procedures.

2.1.2. Hybrid method of discrete gradient with simulated annealing

Zhang et al. (2011a, 2011d) used 3FVA.pdb as the pdb template to build two MM-Models (Figs. 11~12 in (Zhang et al., 2011a)). The Models were built using a hybrid SA Discrete

Gradient (DG (Bagirov et al., 2008)) method. Then the Models were optimized using SDCG-SA-SDCG methods as in (Zhang, 2011a).

2.1.3. Computational method of canonical dual global optimization theory

Zhang et al. (2011, 2011d) used 3NHC.pdb, 3NVF/G/H/E.pdb templates to build several MM-Models (Figs. 9~11 in (Zhang et al., 2011b), and Figs. 5~8 in (Zhang, 2011b)). These Models were built in the use of canonical dual global optimization theory (Gao et al., 2012; Gao and Wu, 2012; Gao, 2000) and then refined by SDCG-SA-SDCG methods as in (Zhang, 2011a).

2.2. New material and method, and new MM-models

2.2.1 New material

This Chapter uses a suitable pdb file template 3NHD.pdb (the GYVLGS segment 127-132 from human prion with V129 (Apostol et al., 2010) from the Protein Data Bank to build MM-models of AGAAAAGA amyloid fibrils for prions.

2.2.2. New computational method - computational method of simulated annealing evolutionary computations

The computational methods used to build the new MM-Models will be simulated annealing evolutionary computations (SAECs), where SAECs were got from the hybrid algorithms of (Abbass et al., 2003) by simply replacing the DG method by the SA algorithm of (Bagirov and Zhang, 2003) and numerical computational results show that SAECs can successfully pass the test of more than 40 well-known benchmark global optimization problems (Zhang, 2011c).

2.2.3. New MM-models

The atomic-resolution X-ray structure of 3NHD.pdb is a steric zipper, with strong vdw interactions between β-sheets and HBs to maintain the β-strands (Fig. 2).

By observations of the 3rd column of coordinates of 3NHD.pdb and Fig. 2, G(H) chains (i.e. β-sheet 2) of 3NHD.pdb can be calculated from A(B) chains (i.e. β-sheet 1) by Eq. 1 (where T is Transpose of column vector) and other chains can be calculated by Eqs. 2~3:

$$G(H) = ((-1, 0, 0)^T, (0, 1, 0)^T, (0, 0, -1)^T) A(B) + (-20.5865, 9.48, 0.0)^T, \qquad (1)$$

$$K(L) = G(H) + (0.0, 0.0, 9.59)^T, I(J) = G(H) + (0.0, 0.0, -9.59)^T, \qquad (2)$$

$$C(D) = A(B) + (0.0, 0., 9.59)^T, E(F) = A(B) + (0.0, 0.0, -9.59)^T. \qquad (3)$$

Basing on the template 3NHD.pdb from the Protein Data Bank, three prion AGAAAAGA palindrome amyloid fibril models - an AGAAAA model (Model 1), a GAAAAG model

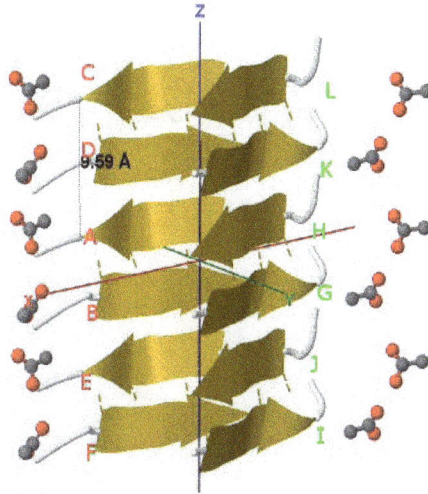

Figure 2. Protein fibril structure of human V129 prion GYVLGS (127–132) (PDB ID: 3NHD). The dashed lines denote the hydrogen bonds. A, B ... K, L denote the chains of the fibril.

(Model 2), and an AAAAGA model (Model 3) - will be successfully constructed in this Chapter. Because the template is a segment of 6 residues, the three shorter prion fragments are selected. This Chapter does not perform calculations on the full AGAAAAGA. Chains AB of Models 1~3 were respectively got from AB chains of 3NHD.pdb using the mutate module of the free package Swiss-PdbViewer (SPDBV Version 4.01) (http://spdbv.vital-it.ch). It is pleasant to see that almost all the hydrogen bonds are still kept after the mutations; thus we just need to consider the vdw contacts only. Making mutations for GH chains of 3NHD.pdb, we can get the GH chains of Models 1~3. However, the vdw contacts between Chain A and Chain G, between B chain and H chain are too far at this moment (Figs. 3~5).

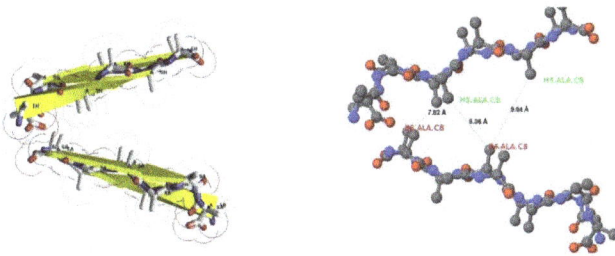

Figure 3. At initial state, the vdw contacts between AB chains (β-sheet 1) and GH chains (β-sheet 2) of Model 1 are very far.

Seeing Figs. 3~5, we may know that for Model 1 at least 3 vdw interactions B6.ALA.CB-H3.ALA.CB-B4.ALA.CB-H5.ALA.CB should be maintained (their distances in Fig. 3 are 7.82, 8.36, 9.04 angstroms respectively), for Model 2 at least 3 vdw interactions G4.ALA.CB-

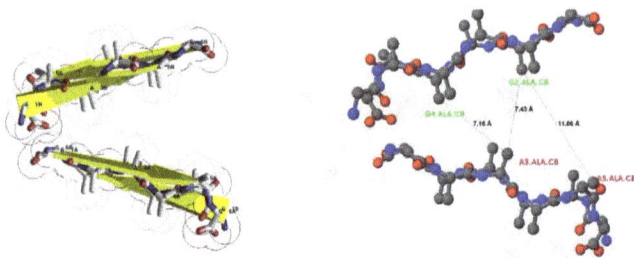

Figure 4. At initial state, the vdw contacts between AB chains (β-sheet 1) and GH chains (β-sheet 2) of Model 2 are very far.

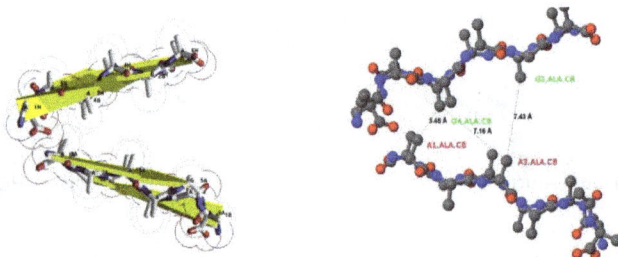

Figure 5. At initial state, the vdw contacts between AB chains (β-sheet 1) and GH chains (β-sheet 2) of Model 3 are very far.

A3.ALA.CB-G2.ALA.CB-A5.ALA.CB should be maintained (their distances in Fig. 4 are 7.16, 7.43, 9.31 angstroms respectively), and for Model 3 at least 3 vdw interactions A1.ALA.CB-G4.ALA.CB-A3.ALA.CB-G2.ALA.CB should be maintained (their distances in Fig. 5 are 3.45, 7.16, 7.43 angstroms respectively). For Model 1, fixing the coordinates of B6.ALA.CB and B4.ALA.CB, letting the coordinates of H3.ALA.CB and H5.ALA.CB be variables, we may get a simple Lennard-Jones (LJ) potential energy minimization problem just with 6 variables (see Eq. 9). Similarly, for Model 2 fixing the coordinates of A3.ALA.CB and A5.ALA.CB, letting the coordinates of G4.ALA.CB and G2.ALA.CB be variables, we may get a simple LJ potential energy minimization problem just with 6 variables (see Eq. 10); for Model 3, fixing the coordinates of A1.ALA.CB and A3.ALA.CB, letting the coordinates of G4.ALA.CB and G2.ALA.CB be variables, we may get a simple LJ potential energy minimization problem with 6 variables (see Eq. 11).

The vdw contacts of atoms are described by the LJ potential energy:

$$V_{LJ}(r) = 4\varepsilon \left[(\sigma/r)^{12} - (\sigma/r)^6 \right], \tag{4}$$

where ε is the depth of the potential well and σ is the atom diameter; these parameters can be fitted to reproduce experimental data or deduced from results of accurate quantum chemistry calculations. The $(\sigma/r)^{12}$ term describes repulsion and the $-(\sigma/r)^6$ term describes attraction. If we introduce the coordinates of the atoms whose number is denoted by N and let $\varepsilon=\sigma=1$ be the reduced units, the Eq. 4 becomes into

$$f(x) = 4 \sum_{i=1}^{N} \sum_{j=1,j<i}^{N} (1/t_{ij}^6 - 1/t_{ij}^3), \tag{5}$$

where $t_{ij} = (x_{3i-2} - x_{3j-2})^2 + (x_{3i-1} - x_{3j-1})^2 + (x_{3i} - x_{3j})^2$, $(x_{3i-2}, x_{3i-1}, x_{3i})$ is the coordinates of atom i, $N \geq 2$. The minimization of LJ potential $f(x)$ on R^n (where $n = 3N$) is an optimization problem:

$$\text{min } f(x) \text{ subject to } x \in R^{3N}. \tag{6}$$

Similarly as Eq. 4, i.e. the potential energy for the vdw interactions between β-sheets:

$$V_{LJ}(r) = A/r^{12} - B/r^6, \tag{7}$$

the potential energy for the HBs between the β-strands has the formula

$$V_{HB}(r) = C/r^{12} - D/r^{10}, \tag{8}$$

where A, B, C, D are given constants. Thus, the amyloid fibril molecular modeling problem can be deduced into the problem to solve the mathematical optimization problem Eq. 6. Seeing Fig. 6, we may know that the optimization problem Eq. 6 reaches its optimal value at the bottom of the LJ potential well, where the distance between two atoms equals to the sum of vdw radii of the atoms. In this Chapter, the sum of the

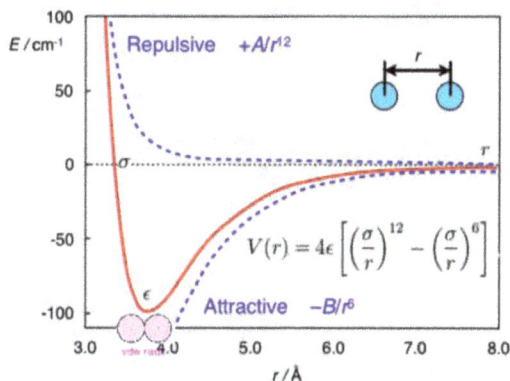

Figure 6. The Lennard-Jone Potential (Eqs.4 and 7) (This Fig. can be found in website homepage.mac.com/swain/CMC/DDResources/mol_interactions/molecular_interactions.html).

vdw radii is the twice of the vdw radius of Carbon atom, i.e. 3.4 angstroms. The optimization problem Eq. 6 is a nonconvex complex optimization problem. By the observation from Fig. 6, we may solve its simple but equal convex-and-smooth least square optimization problem (or the so-called distance geometry problem or sensor network problem) with a slight perturbation if data for three atoms violate the triangle inequality. The following three optimization problems for Models 1~3 respectively are:

$$\text{min } f(x) = \tfrac{1}{2}\{ (x_{11}+16.359)^2 + (x_{12}-9.934)^2 + (x_{13}+3.526)^2 -3.4^2\}^2 + \tfrac{1}{2}\{ (x_{21}+9.726)^2 + (x_{22}-8.530)^2 + (x_{23}+3.613)^2 -3.4^2\}^2 + \tfrac{1}{2}\{ (x_{11}+9.726)^2 + (x_{12}-8.530)^2 + (x_{13}+3.613)^2 -3.4^2\}^2 \text{ with} \tag{9}$$
$$\text{initial solution } (-12.928, 12.454, 3.034; -6.635, 14.301, 2.628),$$

min f(x)= ½{ $(x_{11}+8.655)^2 + (x_{12}-8.153)^2 + (x_{13}-1.770)^2 -3.4^2$}^{2} + ½{ $(x_{21}+8.655)^2 + (x_{22}-8.153)^2 + (x_{23}-1.770)^2 -3.4^2$}^{2} + ½{ $(x_{21}+2.257)^2 + (x_{22}-6.095)^2 + (x_{23}-3.078)^2 -3.4^2$}^{2} with initial solution (-13.909, 12.227, -0.889; -7.439, 14.419, -2.033), (10)

min f(x)= ½{ $(x_{11}+15.632)^2 + (x_{12}-9.694)^2 + (x_{13}-0.687)^2 -3.4^2$}^{2} + ½{ $(x_{11}+8.655)^2 + (x_{12}-8.153)^2 + (x_{13}-1.770)^2 -3.4^2$}^{2} + ½{ $(x_{21}+8.655)^2 + (x_{22}-8.153)^2 + (x_{23}-1.770)^2 -3.4^2$}^{2} with initial solution (-13.909, 12.227, -0.889; -7.439, 14.419, -2.033). (11)

We may use any optimization algorithms or packages to easily solve problems Eqs. 9~11 and get their respective global optimal solutions (-13.062, 9.126, -3.336; -12.344, 6.695, -2.457), (-11.275, 6.606, 3.288; -5.461, 7.124, 2.424), (-12.149, 8.924, 1.229; -9.256, 11.007, 3.517), which were got by the SAEC algorithms in this Chapter. Input these global optimal solutions into Eq. 1, take average and tests then we get Eq. 12:

$$G(H) = ((-1, 0, 0)^T, (0, 1, 0)^T, (0, 0, -1)^T) A(B) +(-20.2788, -0.0821, 0.5609)^T. \qquad (12)$$

By Eq. 12, we can get close vdw contacts in Figs. 7~9.

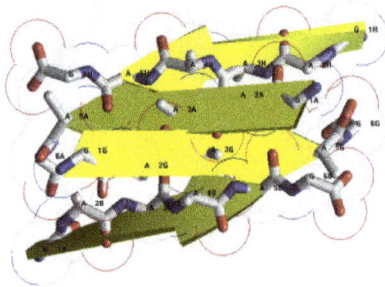

Figure 7. After LJ potential energy minimization, the vdw contacts of Model 1 become very closer (the distances are illuminated by the overlap of border of CB atoms' surface).

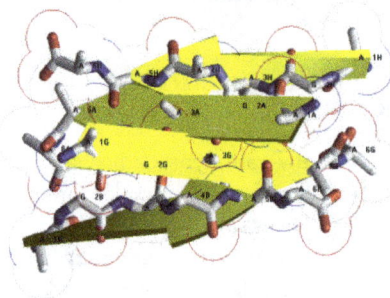

Figure 8. Fig. 8: After LJ potential energy minimization, the vdw contacts of Model 2 become very closer (the distances are illuminated by the overlap of border of CB atoms' surface).

From Figs. 3~5 to Figs. 7~9, we may see that the Optimization algorithm works and the computational experiences show us we had better at least define two sensors and two anchors in order to form a zipper between the two β-sheets. Next, in order to remove very close bad contacts, we relax Figs. 7~9 by a slight SDCG-Optimization in the use of Amber 11 (Case et al., 2010) and we get the optimized MM-Models 1~3. The other CDEF and LKJI chains can be got by parallelizing ABGH chains in the use of mathematical Eqs. 2~3. The new amyloid fibril models are useful for the drive to find treatments for prion diseases in the field of medicinal chemistry. The computational algorithms presented in this Chapter and their references therein are useful in materials science, drug design, etc.

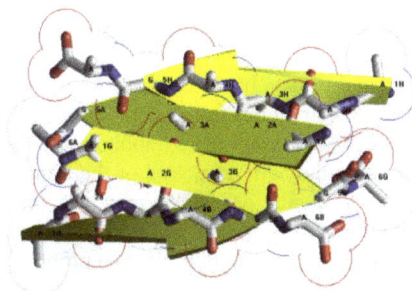

Figure 9. After LJ potential energy minimization, the vdw contacts of Model 3 become very closer (the distances are illuminated by the overlap of border of CB atoms' surface).

Because Eqs. 9~11 are optimization problems with 6 variables only and these optimization problems are to minimize fourth-order polynomials, the proposed SAEC method and other computational methods can easily get the same optimal solutions to optimize the above three models.

2.2.4. The practical LBFGS quasi-Newtonian method

Energy minimization (EM), with the images at the endpoints fixed in space, of the total system energy provides a minimum energy path. EM can be done using SD, CG, and LBFGS (Limited-memory Broyden-Fletcher-Goldfarb-Shanno). SD is robust and easy to implement but it is not most efficient especially when closer to minimum. CG is slower than SD in the early stages but more efficient when closer to minimum. The hybrid of SD-CG will make SD more efficient than SD or CG alone. However, CG cannot be used to find the minimization energy path, for example, when "forces are truncated according to the tangent direction, making it impossible to define a Lagrangian" (Chu et al., 2003). In this case, the powerful and faster quasi-Newtonian method (e.g. the LBFGS quasi-Newtonian minimizer) can be used (Chu et al., 2003; Liu and Nocedal, 1989; Nocedal and Morales, 2000; Byrd et al., 1995; Zhu et al., 1997). We briefly introduce the LBFGS quasi-Newtonian method as follows.

Newton's method in optimization explicitly calculates the Hessian matrix of the second-order derivatives of the objective function and the reverse of the Hessian matrix (Dennis et al., 1996). The convergence of this method is quadratic, so it is faster than SD or CG. In high dimensions, finding the inverse of the Hessian is very expensive. In some cases, the Hessian is a non-invertible matrix, and furthermore in some cases, the Hessian is symmetric indefinite. Qusi-Newton methods thus appear to overcome all these shortcomings.

Quasi-Newton methods (a special case of variable metric methods) are to approximate the Hessian. Currently, the most common quasi-Newton algorithms are the SR1 formula, the BHHH method, the widespread BFGS method and its limited /low-memory extension LBFGS, and Broyden's methods (http://en.wikipedia.org/wiki/Quasi-Newton_method). In Amber (Case et al., 2010) and Gromacs (van der Spoel et al., 2010), LBFGS is used, and the hybrid of LBFGS with CG - a Truncated Newton linear CG method with optional LBFGS Preconditioning (Nocedal and Morales, 2000) - is used in Amber (Case et al., 2010).

2.3. New thinking about the construction of 3D-structure of a protein

If a NMR or X-ray structure of a protein has not been determined and stored in PDB bank yet, we still can easily get the 3D-structural frame of the protein. For example, before 2005 when we did not know the NMR structure of rabbit prion protein, we could get its homology model structure using the NMR structure of the human prion protein (PDB id: 1QLX) as the template (Zhang et al., 2006). We may use the homology structure to determine the 3D-structural frame of a protein when its NMR or X-ray structure has not been determined yet. The determination is an optimization problem described as follows.

"Very often in a structural analysis, we want to approximate a secondary structural element with a single straight line" (Burkowski, 2009: page 212). For example, Fig. 10 uses two straight lines that act as the longitudinal axis of β-Strand A (i.e. A chain), β-Strand B (i.e. B chain) respectively. Each straight line should be positioned among the Cα atoms so that it is closest to all these Cα atoms in a least-squares sense, which is to minimize the sum of the squares of the perpendicular

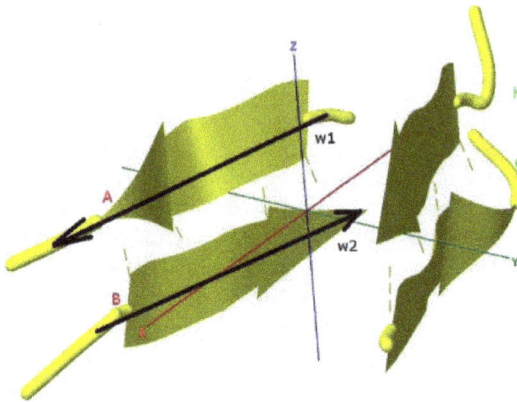

Figure 10. The 3D-structural frame of AB chains of Model 1 in Fig. 3 with two β-Strands.

distances (d_i) from the Cα atoms to the strand/helix axis:

$$S^* = \min S = \sum_{i=1}^{N} ||d_i||^2. \tag{13}$$

Define the vector $w=(w_x, w_y, w_z)^T$ for the axis. Then d_i represents the perpendicular vector going from Cα atom a to the axis:

$$\| d_i \|^2 = \| a^{(i)} \|^2 \sin^2\theta_i = \| a^{(i)} \|^2 (1-\cos^2\theta_i) = \| a^{(i)} \|^2 \{ 1- (a^{(i)\mathrm{T}} w)^2 / (\| a^{(i)} \|^2 \| w \|^2) \} = \{ a^{(i)}_x{}^2 + a^{(i)}_y{}^2 + a^{(i)}_z{}^2 \} \{ 1- (a^{(i)}_x w_x + a^{(i)}_y w_y + a^{(i)}_z w_z)^2 / [(a^{(i)}_x{}^2 + a^{(i)}_y{}^2 + a^{(i)}_z{}^2)(w_x{}^2 + w_y{}^2 + w_z{}^2)] \}. \quad (14)$$

According to Eqs. 13~14, for the β-Strand A – β-Strand B of AB chains, we get the following two optimization problems for Model 1 respectively:

$$\min S_A = ((-16.196)^2 + 8.315^2 + 1.061^2) \{ 1- (-16.196 w_x + 8.315 w_y + 1.061 w_z)^2 /[((-16.196)^2 + 8.315^2 + 1.061^2) (w_x{}^2 + w_y{}^2 + w_z{}^2)] \} +$$

$$((-12.977)^2 + 6.460^2 + 1.908^2) \{ 1- (-12.977 w_x + 6.460 w_y + 1.908 w_z)^2 /[((-12.977)^2 + 6.460^2 + 1.908^2) (w_x{}^2 + w_y{}^2 + w_z{}^2)] \} +$$

$$((-9.178)^2 + 6.745^2 + 1.448^2) \{ 1- (-9.178 w_x + 6.745 w_y + 1.448 w_z)^2 /[((-9.178)^2 + 6.745^2 + 1.448^2) (w_x{}^2 + w_y{}^2 + w_z{}^2)] \} +$$

$$((-6.455)^2 + 4.112^2 + 1.558^2) \{ 1- (-6.455 w_x + 4.112 w_y + 1.558 w_z)^2 /[((-6.455)^2 + 4.112^2 + 1.558^2) (w_x{}^2 + w_y{}^2 + w_z{}^2)] \} +$$

$$((-3.006)^2 + 5.750^2 + 1.782^2) \{ 1- (-3.006 w_x + 5.750 w_y + 1.782 w_z)^2 /[((-3.006)^2 + 5.750^2 + 1.782^2) (w_x{}^2 + w_y{}^2 + w_z{}^2)] \} +$$

$$((-1.226)^2 + 2.750^2 + 0.233^2) \{ 1- (-1.226 w_x + 2.750 w_y + 0.233 w_z)^2 /[((-1.226)^2 + 2.750^2 + 0.233^2) (w_x{}^2 + w_y{}^2 + w_z{}^2)] \}, \quad (15)$$

$$\min S_B = ((-0.959)^2 + 2.950^2 +(- 4.817)^2) \{ 1- (-0.959 w_x + 2.950 w_y -4.817 w_z)^2 /[((-0.959)^2 + 2.950^2 +(- 4.817)^2) (w_x{}^2 + w_y{}^2 + w_z{}^2)] \} +$$

$$((-3.465)^2 + 4.999^2 +(-2.846)^2) \{ 1- (-3.465 w_x + 4.999 w_y -2.846 w_z)^2 /[((-3.465)^2 + 4.999^2 + (-2.846)^2) (w_x{}^2 + w_y{}^2 + w_z{}^2)] \} +$$

$$((-7.213)^2 + 4.412^2 +(-3.340)^2) \{ 1- (-7.213 w_x + 4.412 w_y -3.340 w_z)^2 /[((-7.213)^2 + 4.412^2 + (-3.340)^2) (w_x{}^2 + w_y{}^2 + w_z{}^2)] \} +$$

$$((-9.954)^2 + 7.078^2 +(-3.168)^2) \{ 1- (-9.954 w_x + 7.078 w_y -3.168 w_z)^2 /[((-9.954)^2 + 7.078^2 + (-3.168)^2) (w_x{}^2 + w_y{}^2 + w_z{}^2)] \} +$$

$$((-13.660)^2 + 6.241^2 +(-3.137)^2) \{ 1- (-13.660 w_x + 6.241 w_y -3.137 w_z)^2 /[((-13.660)^2 + 6.241^2 + (-3.137)^2) (w_x{}^2 + w_y{}^2 + w_z{}^2)] \} +$$

$$((-16.702)^2 + 8.507^2 +(-3.074)^2) \{ 1- (-16.702 w_x + 8.507 w_y -3.074 w_z)^2 /[((-16.702)^2 + 8.507^2 + (-3.074)^2) (w_x{}^2 + w_y{}^2 + w_z{}^2)] \}. \quad (16)$$

We solve Eqs. 15~16 (taking the average of the coordinates of Cα atoms as initial solutions), getting their optimal solutions $\mathbf{w1}$ = (-10.751, 6.428, 1.411)$^\mathrm{T}$, $\mathbf{w2}$ = (-7.960, 4.579, -2.256)$^\mathrm{T}$ respectively (Fig. 10). We may use the vectors $\mathbf{w1}$, $\mathbf{w2}$ and Eq. 12 to construct Chains GH and then build an optimal Model 1 (Aqvist, 1986; Abagyan and Maiorov, 1988; Orengo et al., 1992; Young et al., 1999; Foote and Raman, 2000). In (Burkowski, 2009: pages 213-216), $w_x{}^2 + w_y{}^2 + w_z{}^2 =1$ (i.e. w is a unit vector) is restrained and Eq. 13 becomes into a problem to seek the smallest eigenvalue (S^*) and its corresponding eigenvector $\mathbf{w}$ of the following matrix:

$$((\sum_{i=1}^{N}(a_y{}^{(i)})^2 + (a_z{}^{(i)})^2, -\sum_{i=1}^{N}a_x{}^{(i)}a_y{}^{(i)} , -\sum_{i=1}^{N}a_z{}^{(i)}a_x{}^{(i)})^\mathrm{T}, (-\sum_{i=1}^{N}a_x{}^{(i)}a_y{}^{(i)} , \sum_{i=1}^{N}(a_z{}^{(i)})^2 + (a_x{}^{(i)})^2, - \sum_{i=1}^{N}a_y{}^{(i)}a_z{}^{(i)})^\mathrm{T}, (-\sum_{i=1}^{N}a_z{}^{(i)}a_x{}^{(i)}, -\sum_{i=1}^{N}a_y{}^{(i)}a_z{}^{(i)}, \sum_{i=1}^{N}(a_x{}^{(i)})^2 + (a_y{}^{(i)})^2)^\mathrm{T}).$$

This matrix is symmetric and positive definite, and its eigenvectors form an orthogonal basis for the set of atoms under consideration. In physics, it is called the inertial tensor involving studies of rotational inertia and its eigenvectors are called the principle axes of inertia. Furthermore, we may also notice that Eq. 13 can be rewritten as

$$\min \left(\sum_{i=1}^{N} ||d_i||^2 \right)^2 \quad \text{subject to} \quad w^T w = 1, \tag{17}$$

where $||d_i||^2 = (a^{(i)}_x{}^2 + a^{(i)}_y{}^2 + a^{(i)}_z{}^2)(w_x{}^2 + w_y{}^2 + w_z{}^2) - (a^{(i)}_x w_x + a^{(i)}_y w_y + a^{(i)}_z w_z)^2$. Thus, Eq. 17 can be easily solved by the canonical dual global optimization theory (Gao et al., 2012; Gao and Wu, 2012; Gao, 2000), by the ways of solving the canonical dual of Eq. 17 or solving the quadratic differential equations of the prime-dual Gao-Strang complementary function (Gao et al., 2012; Gao and Wu, 2012; Gao, 2000) through some ordinary or partial differential equation computational strategies.

3. Conclusions

To date the hydrophobic region AGAAAAGA palindrome (113-120) of the unstructured N-terminal region (1-123) of prions has little existing experimental structural data available. This Chapter successfully constructs three molecular structure models for AGAAAAGA palindrome (113-120) by using some suitable template 3NHD.pdb from Protein Data Bank and refinement of the Models with several optimization techniques within AMBER 11. These models should be very helpful for the experimental studies of the hydrophobic region AGAAAAGA palindrome of prion proteins (113-120) when the NMR or X-ray molecular structure of prion AGAAAAGA peptide has not been easily determined yet. These constructed Models for amyloid fibrils may be useful for the goals of medicinal chemistry.

This Chapter also introduces numerous practical computational approaches to construct the molecular models when it is difficult to obtain atomic-resolution structures of proteins with traditional experimental methods of X-ray and NMR etc, due to the unstable, noncrystalline and insoluble nature of these proteins. Known structures can be perfectly reproduced by these computational methods, which can be compared with contemporary methods. As we all know, X-ray crystallography finds the X-ray final structure of a protein, which usually need refinements using a SA protocol in order to produce a better structure. SA is a global search procedure and usually it is better to hybrid with local search procedures. Thus, the computational methods introduced in this Chapter should be better than SA along to refine X-ray final structures.

Author details

Jiapu Zhang
The University of Ballarat, Australia

Acknowledgement

This research was supported by a Victorian Life Sciences Computation Initiative (VLSCI) grant numbers VR0063 & 488 on its Peak Computing Facility at the University of Melbourne, an initiative of the Victorian Government. The author appreciates the editors for their comments and helps to improve this paper.

4. References

Abagyan R.A., Maiorov V.N. (1988). "A simple quantitative representation of polypeptide chain folds: Comparison of protein tertiary structures". *J. Biomol. Struct. Dyn.* Vol 5, pp 1267-1279.

Abbass H.A., Bagirov A.M., Zhang J.P. (2003). "The discrete gradient evolutionary strategy method for global optimization". *The IEEE Congress on Evolutionary Computation (CEC2003, Australia)*, IEEE-Press, Vol 1, pp 435-442.

Apostol M.I., Sawaya M.R., Cascio D., Eisenberg D. (2010). "Crystallographic studies of prion protein (PrP) segments suggest how structural changes encoded by polymorphism at residue 129 modulate susceptibility to human prion disease". *J. Biol. Chem.* Vol 285, pp 29671-29675.

Aqvist J. (1986). "A simple way to calculate the axes of an α-helix". *Computers & Chemistry* Vol 10, pp 97-99.

Bagirov A.M., Karasozen B., Sezer M. (2008). "Discrete gradient method: a derivative free method for nonsmooth optimization". *J. Opt. Theor. Appl.* Vol 137, pp 317-334.

Bagirov A.M., Zhang J.P. (2003). "Comparative analysis of the cutting angle and simulated annealing methods in global optimization". *Optimization* Vol 52, No 4-5, pp 363-378.

Berman H.M., Westbrook J., Feng Z., Gilliland G., Bhat T. N., Weissig H., Shindyalov I. N., Bourne P. E. (2000). "The protein data bank". *Nucleic Acids Res.* Vol 28, pp 253-242.

Brown D.R. (2000). "Prion protein peptides: optimal toxicity and peptide blockade of toxicity". *Mol. Cell. Neurosci.* Vol 15, pp 66-78.

Brown D.R. (2001). 'Microglia and prion disease". *Microsc. Res. Tech.* Vol 54, pp 71-80.

Brown D.R., Herms J., Kretzschmar H.A. (1994). "Mouse cortical cells lacking cellular PrP survive in culture with a neurotoxic PrP fragment". *Neuroreport* Vol 5, pp 2057-2060.

Byrd R.H., Lu P., Nocedal J. (1995). "A limited memory algorithm for bound constrained optimization". *SIAM J. Scientif. Statistic. Comput.* Vol 16, pp 1190-1208.

Burkowski F.J. (2009). *Structural Bioinformatics: An Algorithmic Approach*, CRC Press, ISBN 9781584886839 (Hardcover).

Cappai R., Collins S.J. (2004). "Structural biology of prions", in *Prions – A Challenge for Science, Medicine and the Public Health System* (Rabenau H.F., Cinatl J., Doerr H.W. (eds)), Basel, Karger, Vol 11, pp 14-32.

Case, D.A., Darden, T.A., Cheatham, T.E., Simmerling, III C.L., Wang, J., Duke, R.E., Luo, R., Walker, R.C., Zhang, W., Merz, K.M., Roberts, B.P., Wang, B., Hayik, S., Roitberg, A., Seabra, G., Kolossvary, I., Wong, K.F., Paesani, F., Vanicek, J., Liu, J., Wu, X., Brozell, S.R., Steinbrecher, T., Gohlke, H., Cai, Q., Ye, X., Wang, J., Hsieh, M.-J., Cui, G., Roe, D.R., Mathews, D.H., Seetin, M.G., Sagui, C., Babin, V., Luchko, T., Gusarov, S., Kovalenko, A., Kollman, P.A. (2010). *AMBER 11*, University of California, San Francisco.

Chabry J., Caughey B., Chesebro B. (1998). "Specific inhibition of in vitro formation of protease-resistant prion protein by synthetic peptides". *J Biol. Chem.* Vol 273 No 21, pp 13203-13207.

Chan J.C.C.(2011). "Steric zipper formed by hydrophobic peptide fragment of Syrian hamster prion protein". *Biochem.* Vol 50, No 32, pp:6815-6823.

Cheng H.M., Tsai T.W.T., Huang W.Y.C., Lee H.K., Lian H.Y., Chou F.C., Mou Y., Chu J., Trout B.L., Brooks B.R. (2003). "A super-linear minimization scheme for the nudged elastic band method". *J. Chem. Phys.* Vol 119, pp 12708-12717.

Daude N. (2004). "Prion diseases and the spleen". *Viral Immunol.* Vol 17, pp 334-349.

Dennis J.E., Robert J.R. Schnabel B. (1996). *Numerical Methods for Unconstrained Optimization and Nonlinear Equations*, SIAM.

Foote J., Raman A. (2000). "A relation between the principal axes of inertia and ligand binding". *Proc. Natl. Acad. Sci. USA* Vol 97, pp 978-983.

Gao D.Y. (2000). Duality *Principles in Nonconvex Systems: Theory, Methods and Applications*, Kluwer Academic Publishers, Dordrecht, Boston, London, ISBN 9780792361459.

Gao D.Y., Ruan N., Pardalos P.M. (2012). "Canonical dual solutions to sum of fourth-order polynomials minimization problems with applications to sensor network localization", in *Sensors: Theory, Algorithms, and Applications* (editors: Vladimir L. Boginski, Clayton W. Commander, Panos M. Pardalos and Yinyu Ye), Springer Optimization and Its Applications Vol. 61, ISBN 9780387886183, pp 37-54.

Gao D.Y., Wu C.Z. (2012). "Triality theory for general unconstrained global optimization problems". *J. Glob. Optim.* arXiv:1104.2970v2.

Gasset M., Baldwin M.A., Lloyd D.H., Gabriel J.M., Holtzman D.M., Cohen F., Fletterick R., Prusiner S.B. (1992). "Predicted alpha-helical regions of the prion protein when synthesized as peptides form amyloid". *Proc. Natl. Acad. Sci. USA* Vol 89 No 22, pp 10940-10944.

Gazit E. (2002). "A possible role for pi-stacking in the self-assembly of amyloid fibrils". *FASEB J.* Vol 16 No 1, pp 77-83.

Gilead S., Gazit E. (2004). "Inhibition of amyloid fibril formation by peptide analogues modified with alpha-aminoisobutyric acid". *Angew. Chem. Int. Ed. Engl.* Vol 43 No 31, pp 4041-4044.

Griffith J.S. (1967). "Self-replication and scrapie". *Nature* Vol 215, pp 1043-1044.

Haigh C.L., Edwards K., Brown D.R. (2005). "Copper binding is the governing determinant of prion protein turnover". *Mol. Cell. Neurosci.* Vol 30 No 2, pp 186-196.

Harrison C.F., Lawson V.A., Coleman B.M., Kim Y.S., Masters C.L., Cappai R., Barnham K.J., Hill A.F. (2010). "Conservation of a glycine-rich region in the prion protein is required for uptake of prion infectivity". *J. Biol. Chem.* Vol 285 No 26, pp 20213-20223.

Holscher C., Delius H., Burkle A. (1998). "Overexpression of nonconvertible PrP^C delta114-121 in scrapie-infected mouse neuroblastoma cells leads to trans-dominant inhibition of wild-type PrP^{Sc} accumulation". *J. Virol.* Vol 72, pp 1153-1159.

Jobling M.F., Huang X., Stewart L.R., Barnham K.J., Curtain C., Volitakis I., Perugini M., White A.R., Cherny R.A., Masters C.L., Barrow C.J., Collins S.J., Bush A.I., Cappai R. (2001). "Copper and Zinc binding modulates the aggregation and neurotoxic properties of the prion peptide PrP 106-126". *Biochem.* Vol 40, pp 8073-8084.

Jobling M.F., Stewart L.R., White A.R., McLean C., Friedhuber A., Maher F., Beyreuther K., Masters C.L., Barrow C.J., Collins S.J., Cappai R. (1999). "The hydrophobic core sequence modulates the neurotoxic and secondary structure properties of the prion peptide 106-126". *J. Neurochem.* Vol 73, pp1557-1565.

Jones E.M., Wu B., Surewicz K., Nadaud P.S., Helmus J.J., Chen S., Jaroniec C.P., Surewicz W.K. (2011). "Structural polymorphism in amyloids: new insights from studies with Y145Stop prion protein fibrils". *J Biol Chem.* Vol 286 No 49, pp 42777-42784.

Kourie J.I. (2001). "Mechanisms of prion-induced modifications in membrane transport properties: implications for signal transduction and neurotoxicity". *Chem. Biol. Interact.* Vol 138 No 1, pp 1-26.

Kourie J.I., Kenna B.L., Tew D., Jobling M.F., Curtain C.C., Masters C.L., Barnham K.J., Cappai R. (2003). "Copper modulation of ion channels of PrP[106-126] mutant prion peptide fragments". *J. Membr. Biol.* Vol 193 No 1, pp 35-45.

Kuwata K., Matumoto T., Cheng H., Nagayama K., James T.L., Roder H. (2003). "NMR-detected hydrogen exchange and molecular dynamics simulations provide structural insight into fibril formation of prion protein fragment 106-126". *Proc. Natl. Acad. Sci. USA* Vol 100 No 25, pp 14790-14795.

Laganowsky A., Liu C., Sawaya M.R., Whitelegge J.P., Park J., Zhao M., Pensalfini A., Soriaga A.B., Landau M., Teng P.K., Cascio D., Glabe C., Eisenberg D. (2012). "Atomic view of a toxic amyloid small oligomer". *Science* Vol 335, pp1228-1231.

Lee S.W., Mou Y., Lin S.Y., Chou F.C., Tseng W.H., Chen C.H., Lu C.Y., Yu S.S., Chan J.C. (2008). "Steric zipper of the amyloid fibrils formed by residues 109-122 of the Syrian hamster prion protein". *J. Mol Biol.* Vol 378 No 5, pp 1142-1154.

Liu D.C., Nocedal J. (1989). "On the limited memory method for large scale optimization". *Math. Programming B* Vol 45, pp 503-528.

Ma B.Y., Nussinov R. (2002). "Molecular dynamics simulations of alanine rich β-sheet oligomers: insight into amyloid formation". *Protein Science* Vol 11, pp 2335-2350.

Morley J.F., Brignull H.R., Weyers J.J., Morimoto R.I. (2002). "The threshold for polyglutamine-expansion protein aggregation and cellular toxicity is dynamic and influenced by aging in Caenorhabditis elegans". *Proc. Natl. Acad. Sci. USA* Vol 99 No 16, pp 10417-10422.

Nelson R., Sawaya M.R., Balbirnie M., Madsen A., Riekel C., Grothe R., Eisenberg D. (2005). "Structure of the cross-beta spine of amyloid-like fibrils". *Nature* Vol 435 No 7043, pp 773-778.

Nocedal J., Morales J. (2000). "Automatic preconditioning by limited memory quasi-Newton updating". *SIAM J. Opt.* Vol 10, pp 1079-1096.

Norstrom E.M., Mastrianni J.A. (2005). "The AGAAAAGA palindrome in PrP is required to generate a productive PrP^{Sc}-PrP^C complex that leads to prion propagation". *J. Biol. Chem.* Vol 280, pp 27236-27243.

Ogayar A., Snchez-Prez M. (1998). "Prions: an evolutionary perspective". *Internatl. Microbiol.* Vol 1, pp 183-190.

Orengo C.A., Brown N.P., Taylor W.R. (1992). "Fast structure alignment for protein databank searching". *Proteins: Struct. Funct, and Genetics* Vol 14, pp 139-167.

Pan K.M., Baldwin M., Nguyen J. (1993). "Conversion of α-helices into β-sheets features in the formation of the scrapie prion proteins". *Proc. Natl. Acad. Sci. USA* Vol 90, pp 10962-10966.

Pawar A.P., Dubay K.F. et al. (2005). "Prediction of aggregation-prone and aggregation-susceptible regions in proteins associated with neurodegenerative diseases". *J. Mol. Biol.* Vol 350 No 2, pp 379-392.

Reilly C.E. (2000). "Nonpathogenic prion protein (PrP^C) acts as a cellsurface signal transducer". *J. Neurol.* Vol 247, pp 819-820.

Sasaki K., Gaikwad J., Hashiguchi S., Kubota T., Sugimura K., Kremer W., Kalbitzer H.R., Akasaka K. (2008). "Reversible monomer-oligomer transition in human prion protein". *Prion* Vol 2 No 3, pp 118-122.

Sawaya M.R., Sambashivan S., Nelson R., Ivanova M.I., Sievers S.A., Apostol M.I., Thompson M.J., Balbirnie M., Wiltzius J.J., McFarlane H.T., Madsen A., Riekel C.,

Eisenberg D. (2007). "Atomic structures of amyloid cross-beta spines reveal varied steric zippers". *Nature* Vol 447 No 7143, pp 453-457.

Sunde M., Serpell L.C. et al. (1997). "Common core structure of amyloid fibrils by synchrotron X-ray diffraction". *J. Mol. Biol.* Vol 273 No 3, pp 729-739.

van der Spoel D., Lindahl E., Hess B., van Buuren A.R., Apol E., Meulenhoff P.J., Tieleman D.P., Sijbers A.L.T.M., Feenstra K.A., van Drunen R., Berendsen H.J.C (2010). *Gromacs User Manual version 4.5.4*, www.gromacs.org.

Wagoner V.A. (2010). "Computer simulation studies of self-assembly of fibril forming peptides with an Intermediate resolution protein model". *PhD thesis*, North Carolina State University, Raleigh, North Carolina.

Wagoner V.A., Cheon M., Chang I., Hall C.K. (2011). "Computer simulation study of amyloid fibril formation by palindromic sequences in prion peptides". *Proteins: Struct., Funct., and Bioinf.* Vol 79 No 7, pp 2132-2145.

Wegner C., Romer A., Schmalzbauer R., Lorenz H., Windl O., Kretzschmar H.A. (2002). "Mutant prion protein acquires resistance to protease in mouse neuroblastoma cells". *J. Gen. Virol.* Vol 83, pp 1237-1245.

Wormell R.L. (1954). *New fibres from proteins*. Academic Press, page 106.

Young M.M., Skillman A.G., Kuntz I.D. (1999). "A rapid method for exploring the protein structure universe". *Proteins* Vol 34, pp 317-332.

Zanuy D., Ma B., Nussinov R. (2003). "Short peptide amyloid organization: stabilities and conformations of the islet amyloid peptide NFGAIL". *Biophys. J.* Vol 84 No 3, pp 1884-1894.

Zhang J.P. (2011a). "Optimal molecular structures of prion AGAAAAGA palindrome amyloid fibrils formatted by simulated annealing". *J. Mol. Model.* Vol 17 No 1, pp 173-179.

Zhang J.P. (2011b). "Atomic-resolution structures of prion AGAAAAGA amyloid fibrils". In *Amyloids: composition, functions, and pathology* (editors: Irene P. Halcheck and Nancy R. Vernon), Hauppauge, N.Y.: Nova Science Publishers, 2011, ISBN: 9781621005384 (hardcover), Chapter 10: 177-186.

Zhang J.P. (2011c). *Derivative-free hybrid methods in global optimization and applications - in December 2010*, Academic Publishing, ISBN 9783845435800, Chapter 4.2: 92-95.

Zhang J.P. (2011d). *Practical global optimization computing methods in molecular modelling - for atom-resolution structures of amyloid fibrils*, Academic Publishing, ISBN 978-3-8465-2139-7, 192 pages.

Zhang J.P., Sun J., Wu C.Z. (2011a). "Optimal atomic-resolution structures of prion AGAAAAGA amyloid fibrils". *J. Theor. Biol.* Vol 279 No 1, pp 17-28.

Zhang J.P., Gao D.G., Yearwood J. (2011b). "A novel canonical dual computational approach for prion AGAAAAGA amyloid fibril molecular modeling". *J. Theor. Biol.* Vol 284 No 1, pp 149-157.

Zhang J.P., Varghese J.N., Epa V.C. (2006). "Studies on the conformational stability of the rabbit prion protein". *CSIRO Preventative Health National Research Flagship Science Retreat, Aitken Hill, Melbourne, 12-15 September 2006*, Poster in Excellence.

Zhang Z.Q., Chen H., Lai L.H. (2007). "Identification of amyloid fibril-forming segments based on structure and residue-based statistical potential". *Bioinf.* Vol 23, pp 2218-2225.

Zhu C., Byrd R.H., Nocedal J. (1997). "L-BFGS-B: Algorithm 778: L-BFGS-B, FORTRAN routines for large scale bound constrained optimization". *ACM Trans. Math. Softw.* Vol 23, pp 550-560.

Chapter 12

XAO Analysis – AO's and Their Populations in Crystal Fields

Kiyoaki Tanaka and Yasuyuki Takenaka

Additional information is available at the end of the chapter

1. Introduction

There are two ways to the analysis of electron density distribution (EDD) measured by X-ray diffraction. Multipole refinement (Stewart, 1969, 1973; Hansen & Coppens, 1978) expresses EDD as a linear combination of spherical harmonics. Using the determined functions, EDD is analyzed with Bader topological analysis (Bader, 1994). The concept of critical points of Bader theory defines the characters of chemical bonds. However, it does not give atomic orbital (AO) and its electron population, except for orbitals in high symmetry crystal fields where AO's are defined by theory. The other is based on quantum-mechanical orbital-models and gives the physical quantities just mentioned. The objective of this chapter is to introduce one of the second methods, the X-ray atomic orbital analysis (XAO) (Tanaka et al., 2008; hereinafter referred to as *I*), and its application to metal and rare-earth complexes, where advantages of XAO are revealed quite clear.

XAO is closely related to the electron population analysis (Stewart, 1969; Coppens et al., 1971). However, it was abandoned since the orthonormal relationships between orbital functions caused severe parameter interactions in conventional least-squares refinements. Later, the method incorporating the ortho-normal relationships between AO or MO (molecular orbital) was formulated by employing Lagrange's unknown multiplier method (Tanaka, 1988; hereinafter referred to as *II*).

The AO-based EDD refinement has been started from 3d-transition metal complexes with so high crystal symmetry that the AO's are known by the crystal field theory. Spin states of the metals in perovskites, $KCoF_3$ and $KMnF_3$ (Kijima et al., 1981, 1983) and $KFeF_3$ (Miyata et al., 1983) were determined to be high spin. On the other hand, mixed orbitals of $3d_{x^2-y^2}$ and $3d_{z^2}$ in Jahn-Teller distorted $KCuF_3$ were determined (Tanaka et al., 1979). The hybrid orbital of $3d_{z^2}$ and 4s of Cu^+ which made O-Cu-O straight bond in $CuAlO_2$ was also determined (Ishiguro et al., 1983). The investigations on $KNiF_3$ and $KCuF_3$ are, to our knowledge, the first ones that determined spin states and AO's by X-ray diffraction, respectively. The Cu^{2+} ion in $[Cu(daco)_2](NO_3)_2$ (daco: diazacyclooctane) is in the crystal field C_i. Five d-orbitals were determined with the least-squares method stated in *II*.

These investigations revealed that the anharmonic vibration (AHV) of atoms could not be ignored, since significant peaks still remained after removing the $3d$-peaks on difference density maps. They were attributed to the AHV of the metal atoms (Tanaka & Marumo, 1982; Ishiguro et al. 1983). For these investigations, a method proposed by Dawson et al., (1967), which had been applied only to high symmetry crystals, was made applicable to atoms in a general crystal field (Tanaka & Marumo, 1983).

Since the ratio of the number of bonding electrons to those in the unit cell becomes smaller as the atomic number increases, very accurate structure factors are necessary for X-ray EDD investigations of rare-earth compounds. Thus, the EDD analysis based on chemical-bond theories had not been done when we started the study on CeB_6. Actually, the ratio in CeB_6 is 1/88, which demands us to measure structure factors with the accuracy less than 1 %. Rare-earth crystals are usually very hard and good resulting in enhanced extinction and multiple diffraction (MD). Therefore, MD was investigated using the method by Tanaka & Saito (1975) in which an effective way to detect MD and correct for it were proposed. It introduced the time-lag between the relevant reflections, which usually do not occur at exactly the same time, to the method by Moon and Shull (1964). The effect of MD was demonstrated through the study of PtP_2 (Tanaka, et al., 1994) by measuring intensities avoiding MD and compared to those measured at the bisecting positions.

The frontier investigations aiming to measure 4f-EDD were done for CeB_6 (Sato, 1985) at 100 and 298 K and for nonaaqualanthanoids (Ln:La, Ce, Pr, Nd, Sm, Eu, Gd, Tb, Dy, Yb and Lu) (Chatterjee at al. 1988). Significant peaks were found around these rare-earth atoms on the residual density maps. However, the aspherical features of 4f-EDD were not analyzed quantitatively by the X-ray scattering factors calculated with AO's in a crystal field. The first 4f-EDD analysis of CeB_6 with the crystal field model but without spin-orbit interaction exhibited T_{1u} state was occupied (Tanaka et al. 1997). The ground state of the 4f-state was found to be $4f(j=5/2)\Gamma_8$ by the inelastic neutron scattering (Zirngiebl et al., 1984). When the spin-orbit interaction is taken into account, the total sum **J** of the orbital and spin angular momentum **L** and **S**, **J=L+S**, becomes a good quantum number. In the following discussion *J* values of *p*, *d*, *f* orbitals are attached as a subscript on the right-hand side of each orbital. The experiment does not contradict to our results. It was, however, investigated again at 100, 165, 230 and 298 K (Tanaka & Ōnuki, 2002) by introducing the spin-orbit interaction. Our experiments ascertained that 4f electron occupied $4f_{5/2}\Gamma_8$ orbitals. It revealed further that the electron population of them, $n(4f_{5/2}\Gamma_8)$, decreased on lowering the temperature. $4f_{5/2}\Gamma_8$ electrons are transferred to the B_6 moiety below room temperature. In addition to it just significant amount of $4f_{5/2}\Gamma_7$ electrons were found at 298 K, which was expected to be due to the thermal excitation. Thus, the electron population above room temperature became very interesting. Since the energy gap between the two $4f_{5/2}$ orbitals is about 500 K, the X-ray intensities were measured at 430 K expecting for $4f_{5/2}\Gamma_7$ to be more populated (Makita et al., 2007). The results were surprising. The $4f_{5/2}\Gamma_7$ is more populated than $4f_{5/2}\Gamma_8$, and $5d_{5/2}\Gamma_8$ is fully occupied at 430 K, which is expected to be at least a few ten thousands degrees higher than the $4f$ orbitals. The electron configurations of CeB₆ at 340 and 535 K were further investigated to confirm the results (Makita et al., 2008). The EDD of SmB₆ ($4f^5$) was studied (Funahashi, et al., 2010) below room temperature. $4f_{5/2}\Gamma_8$ and $4f_{5/2}\Gamma_7$ orbitals are fully or partially occupied. It is noted that $5d_{5/2}\Gamma_8$ occupation was also found. The physical properties such as electric resistivity are correlated to the electron configuration.

2. Basic formalism of XAO

2.1. AO's in crystal fields

The i-th AO, $\psi_{\alpha,i}(\mathbf{r})$ (i=1,2,...,M) of the α-th atom are expressed with N basisfunctions, $\psi_k(\mathbf{r})$,

$$\psi_{\alpha,i}(r) = \sum_{k}^{N} c_{ik}\psi_k(\mathbf{r}), \ i = 1,2,...,M, \tag{1}$$

where c_{ik} is a constant to be determined in the XAO analysis with the least-squares method incorporating orthonormal relationship between AO's (Tanaka, 1988). Matrices and vectors are written in upper-case letters and bold lower-case letters, and superscript ' means a transposed matrix or a row vector, respectively. The basis functions, $\psi_k(\mathbf{r})$ are listed in Table 1 of I. They are expressed using polar coordinate $(r,\ \theta,\ \phi)$ as a linear combination of a product of a radial function $R_{nl}(r)$ and a spherical harmonic $Y_{lm_l}(\theta,\phi)$ as,

$$\psi_k(\mathbf{r}) = \sum_{m_l=-l}^{l} d_{k,m_l} R_{nl}(r)Y_{lm_l}(\theta,\phi), \tag{2}$$

where n, l and m_l are principal, azimuthal and magnetic quantum numbers. The non-relativistic radial functions, $R_{nl}(r)$ (Mann, 1968) and relativistic functions calculated by HEX (Liberman et al., 1971) were mainly used for the XAO analysis. d_{k,m_l} 's are listed in Table 1 of I. Since approximate c_{ik}'s were necessary at the start of the non-linear least-squares calculation, they were calculated by taking the crystal field, which was calculated by placing a proper point charge on each atom, as a perturbation to the system (Kamimura, et al., 1969). For details, see eqs. (52)-(56) of I. The adjustable variables c_{ik}'s in the least-squares calculation are listed for all point group symmetries in Tables 4 and 5 in I.

2.2. Least-squares method incorporating ortho-normal condition

In a non-linear least-squares refinement, mathematical and physical relationships between parameters should be taken into account to avoid parameter interaction. The conventional one used in X-ray crystallography was improved to obtain AO/MO by taking into account the orthonormal relationship between wave functions (Tanaka, 1988),

$$\sum_{m}^{N}\sum_{n}^{N} c_{im}^{*} c_{jn} s_{mn} = \delta_{ij} \tag{3}$$

where $s_{mn} = \int \psi_{m}^{*} \psi_{n} d\mathbf{r}$ and N is the number of basis functions. δ_{ij} is Kronecker's δ. Assuming c_{im}'s are real, it is rewritten with (M,N) matrix $C=\{c_{im}\}$ and (N,N) matrix $S=\{s_{mn}\}$,

$$CSC' = I, \tag{4}$$

where I is the unit matrix of order M. Using Lagrange's unknown multiplier λ_{ij}, the value Q to be minimized in the least-squares method becomes,

$$Q = \mathbf{v}' M_{f}^{-1} \mathbf{v} - \sum_{i \leq}^{M} \sum_{j}^{M} \lambda_{ij} (\sum_{m}^{N} \sum_{n}^{N} c_{im} c_{jn} s_{mn} - \delta_{ij}) \tag{5}$$

For definition of $\mathbf{v}$ see eqs. (10a) to (10d) of II. Putting c_{im} as the sum of the true value c_{im}^{0} and its small shift Δc_{im}, Q is expressed by ignoring the terms higher than second order of Δc_{im},

$$Q = \mathbf{v}' M_{f}^{-1} \mathbf{v} - \sum_{i \leq}^{M} \sum_{j}^{M} \lambda_{ij} (\sum_{m}^{N} \sum_{n}^{N} (\Delta c_{im} c_{jn}^{0} + c_{im}^{0} c_{jn}) s_{mn} - \delta_{ij}), \tag{6}$$

Putting $\lambda_{ij}' = (1+\delta_{ij})\lambda_{ij}/2$, Q is differentiated to give,

$$\delta Q = 2\delta \mathbf{x}'(A' M_{f}^{-1} A \mathbf{x} - A' M_{f}^{-1} \mathbf{f}) - 2 \sum_{i}^{M} \sum_{j}^{M} \lambda_{ij}' \sum_{m}^{N} \sum_{n}^{N} \delta(\Delta c_{im} c_{jn}^{0} s_{mn}) = 0, \tag{7}$$

where $\mathbf{x}$ represents least-squares estimates of parameters. The first term is obtained following the usual procedure. For $\mathbf{x}$, A and M_f, see Hamilton (1964). After alligning Δc_{im}'s linearly and adding at the end of $\mathbf{x}$ for simplicity, λ_{ij} is expressed in terms of c_{im} by using the orthonormal relationship in eq. (3). Then the final secular equation is obtained,

$$(I - R)A' M_{f}^{-1} A \mathbf{x} = (I - R)A' M_{f}^{-1} \mathbf{f}, \tag{8}$$

where I is unit matrix of order P, total number of variables. The (P,P) matrix R is expressed in terms of c_{ik}'s. It makes the inverse matrix of $(I - R)A' M_{f}^{-1} A$ in (8) calculable. The explicit form of $R=\{r_{ij}\}$ under the limiting conditions (10) for i, j, k and l is,

$$r_{ij} = \{SC'C\}_{kl}, \tag{9}$$

$$P - MN + (m-1)N + 1 \leq i, j \leq P - MN + mN,$$
$$k = i - \{P - MN + (m-1)N\}, \ l = j - \{P - MN + (m-1)N\}, \tag{10}$$

where m runs from 1 to M. The other elements r_{ij} not defined here are zero.

2.3. Electron density and structure factors

The electron density $\rho_\alpha(\mathbf{r})$ of the αth atom centered at $\mathbf{r}_\alpha^{atom}$ is divided into that of the core and valence orbitals, $\rho_{\alpha,valence}(\mathbf{r})$. It is further expressed from eq. (1),

$$\rho_{\alpha,valence}(\mathbf{r}) = \sum_i n_{\alpha,i}\rho_{\alpha,i}(\kappa_i(\mathbf{r}-\mathbf{r}_\alpha^{atom})) = \sum_i n_{\alpha,i}\Psi_{\alpha,i}^*(\kappa_i(\mathbf{r}-\mathbf{r}_\alpha^{atom}))\Psi_{\alpha,i}(\kappa_i(\mathbf{r}-\mathbf{r}_\alpha^{atom})), \qquad (11)$$

where $n_{\alpha,i}$ is the number of electrons of $\Psi_{\alpha,i}$. κ_i expresses the expansion $(\kappa_i > 1)$ or contraction $(\kappa_i > 1)$ of i-th orbital (Coppens et al., 1979). Atom α in the asymmetric unit is translated to $\mathbf{r}_{\alpha\sigma}^{atom}$ by the σ-th crystal symmetry operation. Then the structure factor is,

$$F(\mathbf{k}) = \sum_\alpha \omega_\alpha \sum_\sigma f_{\alpha\sigma}(\mathbf{k})\exp[i\mathbf{k}\cdot\mathbf{r}_{\alpha\sigma}^{atom}]T_{\alpha\sigma}(\mathbf{k}), \qquad (12)$$

where $f_{\alpha\sigma}(\mathbf{k})$ is an atomic scattering factor and $\mathbf{k}$ is a scattering vector. The αth atom at $\mathbf{r}_{\alpha\sigma}^{atom}$ is displaced by atomic vibration to $\mathbf{r}_{\alpha\sigma}^{atom}+\mathbf{u}$. It perturbs the periodicity of the crystal and reduces diffracted intensity. Then the atomic displacement parameter (ADP),

$$T_{\alpha\sigma}^{atom}(k) = \langle\exp(i\mathbf{k}\cdot\mathbf{u})\rangle, \qquad (13)$$

is calculated as an average in space and time of exp(i**k**·**u**) by assuming each atom vibrates independently from each other. $f_{\alpha\sigma}$ is the sum of the scattering factors of core and valence electrons, $f_{\alpha\sigma}^{core}(\mathbf{k})$ and $f_{\alpha\sigma}^{valence}(\mathbf{k})$. $f_{\alpha\sigma}^{valence}(\mathbf{k})$ is expressed as,

$$f_{\alpha\sigma}^{valence}(\mathbf{k}) = \sum_i n_i f_{\alpha,\sigma,i}^{valence}(\mathbf{k}), \qquad (14)$$

where n_i is the number of electrons of the ith valence orbitals. $f_{\alpha,\sigma,i}^{valence}(\mathbf{k})$ is a Fourier transform of $\rho_{\alpha,\sigma,i}^{valence}(\mathbf{r})$. Final expression of $f_{\alpha,\sigma,i}^{valence}$ becomes,

$$f_{\alpha,\sigma,i}^{valence} = \sum_k \sum_{k'} c_{ik}^* c_{ik} \sum_{K=|l-l'|}^{l+l'} <j_K> \sum_{M=-K}^{K} i^K \sqrt{2(2K+1)}\Theta_K^M(\beta_\sigma)\exp(iM\gamma_\sigma)$$
$$\times \sum_{m_l=-l}^{l}\sum_{m_{l'}=-l'}^{l'} d_{k,m_l}^* d_{k',m_{l'}} c^K(lm_l, l'm_{l'}), \qquad (15)$$

where $K+l+l'$ is even. $M = m_l - m_{l'}$, and $|l-l'| \le K \le l+l'$. $\langle j_K \rangle$ is expressed as,

$$\langle j_K \rangle = \int R_{nl}(r)R_{n'l'}(r)j_K(kr)r^2 dr \qquad (16)$$

j_K is Bessel function of order K. For $(\beta_\sigma, \gamma_\sigma)$, see Appendix B of I.

2.4. Anharmonic vibration (AHV)

Gaussian probability density functions (p.d.f.) of atoms are Fourier transformed to give harmonic ADP. Gram-Charier (G-C) formalism is widely used to introduce AHV. However, it expresses aspherical EDDs both from orbital functions and AHV, as shown clearly by Mallinson et al. (1988). Thus, it may not be adequate for accurate EDD researches. The method based on Boltzmann statistics (Dawson et al. 1967, Willis, 1969) was employed,

$$T_{\alpha\sigma}^{atom}(\mathbf{k}) = \int \exp(-V(\mathbf{u})/k_BT)\exp(i\mathbf{k}\cdot\mathbf{u})d\mathbf{u} / \int \exp(-V(\mathbf{u})/k_BT)d\mathbf{u}, \qquad (17)$$

where k_B is Boltzmann constant, and T is absolute temperature. Assuming each atom vibrates independently and defining $u(u_1,u_2,u_3)$ on the Cartesian coordinates parallel to the principal axes of the thermal ellipsoid of harmonic ADP, potential $V(\mathbf{u})$ is expressed as,

$$V(\mathbf{u}) = V_0 + \frac{1}{2}\sum_{i=1}^{3}b_iu_i^2 + \sum_{i,j,k}{}'c_{ijk}u_iu_ju_k + \sum_{i,j,k,l}{}'q_{ijkl}u_iu_ju_ku_l + \ldots \qquad (18)$$

where i, j, k, l run from 1 to 3 and ' means they are permutable. There are 10 cubic and 15 quartic terms. Potential expansion terminates at the quadratic terms in the harmonic ADP.

By assuming the sum of the third and the fourth terms in (18) is much smaller than k_BT, b_i, c_{ijk} and q_{ijkl} are introduced into structure factor formalism using (12), (17) and (18).

3. Experimental

3.1. Multiple diffraction (MD)

MD occurs when two or more reflections are on the Ewald sphere. When the incident beam with wave length λ enters parallel to the unit vector $\mathbf{j}$, and a primary reflection, the intensity of which is measured, is at the reflecting position, the conditions for a secondary reflection, $h_1h_2h_3$, to occur become as follows (Tanaka & Saito, 1975),

$$\left| \mathbf{e}_3 \cdot (\mathbf{j}/\lambda + h_3\mathbf{b}_3) \right| \le 1/\lambda \qquad (19)$$

$$\left| \mathbf{e}_2 \cdot (\mathbf{j}/\lambda + h_2\mathbf{b}_2 + h_3\mathbf{b}_3) \right| \le \left[1/\lambda^2 - \left\{ \mathbf{e}_3 \cdot (\mathbf{j}/\lambda + h_3\mathbf{b}_3) \right\}^2 \right]^{1/2} \equiv r_3 \qquad (20)$$

$$\mathbf{e}_1 \cdot (\mathbf{j}/\lambda + h_1\mathbf{b}_1 + h_2\mathbf{b}_2 + h_3\mathbf{b}_3) = \pm\left[r_3^2 - \left\{ \mathbf{e}_2 \bullet (\mathbf{j}/\lambda + h_2\mathbf{b}_2 + h_3\mathbf{b}_3) \right\}^2 \right] \qquad (21)$$

$\mathbf{b}_1$, $\mathbf{b}_2$, $\mathbf{b}_3$ are reciprocal-lattice vectors when the primary reflection exactly fulfills the reflecting condition and $\mathbf{e}_1$, $\mathbf{e}_2$, $\mathbf{e}_3$ are the unit vectors defined as,

$$e_3 = \frac{b_1 \times b_2}{|b_1 \times b_2|}, e_2 = \frac{e_3 \times b_1}{|e_3 \times b_1|}, e_1 = \frac{b_1}{|b_1|}. \tag{22}$$

Secondary reflections only around the surface of the Ewald sphere are searched using eqs. (19) to (21). Integers h_3 are selected from (19) and integers h_2 are calculated from (20) for each h_3. Then the value of h_1 is evaluated from eq. (21) for each (h_2, h_3). The point (h_1, h_2, h_3) is judged to be a reciprocal lattice point if h_1 is an integer. The Ewald sphere has width due to the divergence of the incident beam, finite size and mosaicity of the crystal, and wavelength spread. It permits some margin from an integer to h_1.

The change of intensity by MD was first formulated by Moon & Shull (1964). Primary and secondary beams usually come on the Ewald sphere with an angle-lag $\Delta\varepsilon$. Using $\Delta\varepsilon$ The formalism was modified for a spherical crystal with a radius r (Tanaka & Saito, 1975),

$$\frac{\Delta I_1}{I_1} = \frac{\sqrt{\pi r^2}}{8} Q_{01} \sum_i \left[-g_{01;0i}\left(\frac{Q_{0i}}{Q_{01}}\right) - g_{01;1i}\left(\frac{Q_{1i}}{Q_{01}}\right) + g_{01;i1}\left(\frac{Q_{0i}}{Q_{01}}\right)\left(\frac{Q_{i1}}{Q_{01}}\right) \right]. \tag{23}$$

Subscripts '0', '1' and 'i' stand for the incident, primary and secondary beams. Q_{mn} is the integrated reflectivity per unit volume of a crystal for the m-th incident beam and n-th diffracted beam (m-n process). It is proportional to the squares of the structure factor. In i-1 process, for example, a secondary beam i acts as an incident beam and is diffracted along the direction of the primary beam 1, which enhances the intensity of the primary beam. $g_{ij;mn}$ is a function of $\Delta\varepsilon$, Lorentz and polarization factors of the simultaneous i-j and m-n processes. The polarization factor for a general multi-diffracted process was recently formulated (Tanaka et al., 2010). For $g_{ij;mn}$ see Tanaka & Saito (1975). Since $\Delta I_1/I_1$ depends on the ratio of Q_{mn}'s and $\Delta\varepsilon$ of sharp primary and secondary reflections, ΔI_1 fluctuates sharply. Eq. (23) was applied to diformohydrazide (Tanaka, 1978) and two peaks on the N-N bond in the difference density map became one peak at the middle of the bond.

The EDD analyses of compounds with metals and rare-earth elements need accurate measurements of structure factors, however MD affects the measured intensity seriously since Q_{mn} is large. However correction for it does not seem fruitful, since each primary reflection has many secondary reflections and MD depends sharply on crystal orientation and half-width of each Bragg peak. Thus it is better to avoid MD by rotating a crystal around the reciprocal vector of the primary reflection (ψ – rotation) so that secondary reflections move apart from the Ewald sphere and measure at ψ where $\Delta I_1/I_1$ is minimum. For actinoid compounds, MD cannot be avoided and the correction for it becomes necessary.

4. XAO analysis for crystals in the O_h crystal field

4.1. 3d, 4f and 5d orbitals in the O_h crystal field

In the present chapter EDD investigations on the first-transition metal complexes and rare earth compounds in O_h crystal field are mainly stated. Thus it is worth while to represent the energy level splitting of d- and f-states in Figs.1(a) and (b) (Funahashi, 2010). The spin-orbit splitting is neglected in a strong field model. For $(3d)^n$ systems strong field model is often employed. In $(4f)^n$ and $(5d)^n$ systems the two splittings are of the same order of magnitude. Weak field models are employed for CeB_6 and SmB_6. Their quantization axes point to the face-centre of the cubic unit cell, where no B atom exists. Note that the order of the levels in Fig. 1(a) is inverted, that is, t_{2g} orbitals lies lower than e_g, when ligands are on the quantization axes like $KCoF_3$ (section 4). In weak field $5d$- and $4f$-states split into $j=3/2$ and $j=5/2$ states, and into $j=5/2$ and $j=7/2$ states by the spin-orbit interaction. The O_h crystal field splits them further. Lobes of $4f_{5/2}\Gamma_8$ extend along <100> directions while those of $4f_{5/2}\Gamma_7$ orbitals along <110> and <111> as illustrated in Fig. 7 of Makita et al. (2007).

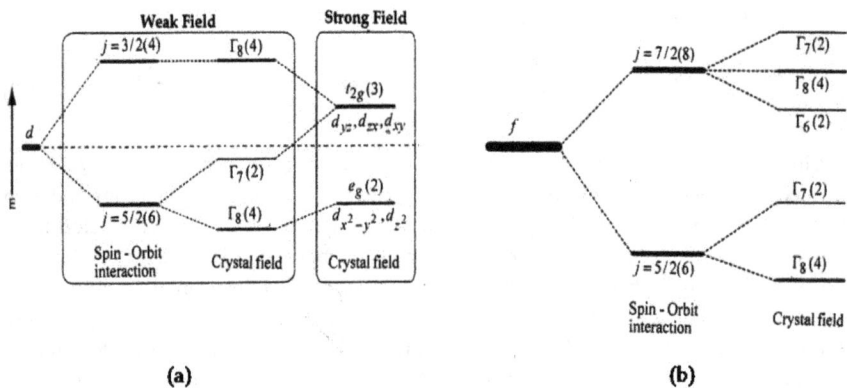

Figure 1. Energy-level splitting of (a) d- and (b) f-states in the O_h crystal field with the degeneracy of each state in parentheses. Each orbital of a strong field can have two electrons.

4.2. 3d-EDDand spin states in KCoF₃, KMnF₃ and KFeF₃

The peak due to 3d electrons was first reported for $[Co(NH_3)_6][Co(CN)_6]$ (Iwata & Saito, 1975) around the Co atom in $\overline{3}$ crystal field. Later electron distribution at 80 K was reported (Iwata, 1977). 3d-EDD in γ–Co_2SiO_4 (Marumo et al., 1977) and $CoAl_2O_4$ (Toriumi, et al., 1978) were also observed. The electron density around Co in $KCoF_3$ was first analysed quantitatively based on the $3d$-orbitals in O_h crystal field (Kijima et al., 1981). A crystal was shaped into a sphere of diameter 0.13 mm, and intensities were measured up to $2\theta=130°$with a four-circle diffractometer using $MoK\alpha$ radiation without avoiding MD. c_{ik}'s in (1) are fixed constants because of the high symmetry as listed in Tables 1(b) and 5(b) of I. κ in eq. (11) was not introduced to the investigations in sections 4 5.1 and 5.2. Scattering factors of $d_{x^2-y^2}$, d_{z^2}, d_{yz}, d_{zx} and d_{xy} orbitals are calculated with eq. (15). When electrons occupy the five d-orbitals equally, that is $(t_{2g})^{4.2}(e_g)^{2.8}$ for Co^{2+}, atoms have spherical EDD. Hereinafter the refinement with the spherical scattering factors and without assuming AHV is called as spherical-atom refinement, in which type I or II anisotropic extinction (Becker &

Coppens, 1974a, b, 1975) was assumed. The difference density map assuming the spherical-atom, high and low spins, $(t2g)^5(eg)^2$ and $(t2g)^6(eg)^1$, are shown in Figs. 2(a) to (c). Peaks remained after spherical-atom refinement were reduced and enhanced after high- and low-spin refinements, respectively. Large and almost no peaks in Fig. 2(c) and Fig. 2(b) demonstrate that Co^{2+} is in the high-spin state. Spin-state was first determined by X-ray diffraction in this study. In the similar way, the spin-states of 3d-transition metals in $KMnF_3$ (Kijima et al., 1983) and $KFeF_3$ (Miyata et al., 1983) were determined to be high-spin.

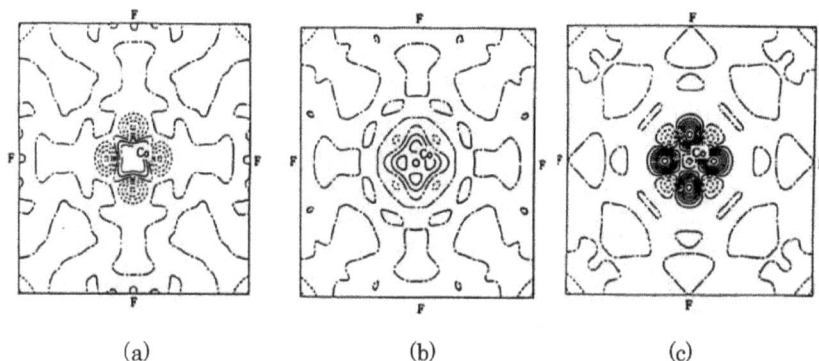

| (a) | (b) | (c) |

Figure 2. Difference density after refinements of (a) spherical-atom, (b) high-spin, (c) low- spin models. Contours at 0.2 eÅ^{-3}. Negative and zero in broken and dashed-dotted lines.

5. Experimental AO determination

When symmetries of crystal fields are low, c_{ik}'s in (1) become unknown variables, and it is necessary to determine them by the least-squares method keeping the orthonormal relationship between AO's. Independent variables, c_{ik}'s, and the relation between them are listed for 32 point group symmetries in Tables 4 and 5 of I.

5.1. Jahn-Teller distortion in KCuF$_3$ and mixed $d_{x^2-y^2}$ and d_{z^2} orbital

In KCuF$_3$ each F$^-$ ion between Cu^{2+} ions (3d^9) shifts from the centre by Jahn-Teller effect. It makes short Cu-F$_s$, medium Cu-F$_m$, parallel to c-axis, and long Cu-F$_l$ bonds, resulting in mmm point group symmetry for Cu^{2+}. Difference density after spherical-atom refinement exhibits non-equivalent four holes in Fig. 3 (b) (Tanaka et al. 1979). Table 5(b) of I indicates a mixing of $3d_{z^2}$ and $3d_{x^2-y^2}$ in the crystal field mmm,

$$\psi_G = \cos(\varphi/2)d_{z^2} + \sin(\varphi/2)d_{x^2-y^2},$$
$$\psi_E = \sin(\varphi/2)d_{z^2} - \cos(\varphi/2)d_{x^2-y^2}. \tag{24}$$

The peaks on $Cu\text{-}F_l$ in Fig. 3(a) correspond to the lone pair of ψ_G. Putting two and one electrons to ψ_G and ψ_E, $cos(\phi/2)$ became 0.964(18). 3d-peaks in Fig. 3 (a) and (b) were deleted in (c) and (d). It confirms orbitals in eq. (24) are quite reasonable. The orbital functions except the phase factor were determined for the first time from X-ray diffraction experiment. However, significant peaks still remained. The positive and negative peaks along <100> and <110> in Fig.3(c) indicate preferential and inhibitive vibration of Cu^{2+}, respectively. Therefore, AHV of Cu^{2+} was analysed (Tanaka & Marumo, 1982). The obtained potential in (18) corresponds to the peaks and removes them as seen in the residual density in Fig. 4. Our study on AHV started from this investigation. Since the AHV peaks appear in Fig. 3(c) and (d) after the d-electron peaks were removed, the aspherical peaks due to electron configuration and vibration of Cu^{2+} seem to be separated well.

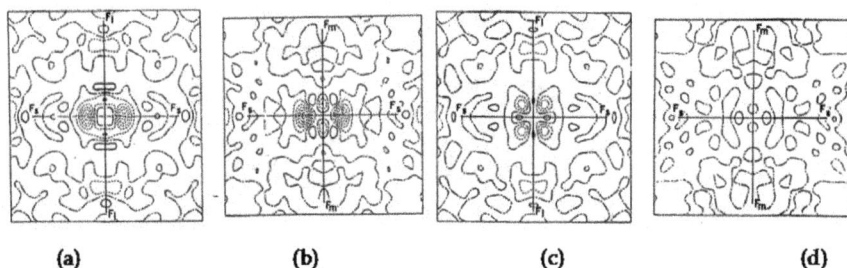

(a)   (b)   (c)   (d)

Figure 3. Difference density after spherical-atom refinement on (a) (001) and (b) (220) planes passing Cu^{2+} at (1/2,0,0). That after d-orbital analysis on (c) (001) and (d) (220). Contours are the same as those in Fig. 2.

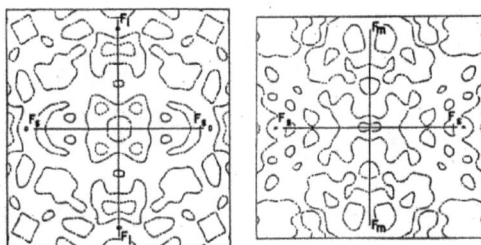

Figure 4. Difference density after AHV analysis. Contours are the same as those in Fig.2.

5.2. 3d orbitals in $\bar{1}$ crystal field of $Cu(daco)_2(NO_3)_2$

The Cu^{2+} ion ($3d^9$) of $Cu(daco)_2(NO_3)_2$(daco: 1,5-diazacyclooctane) is on the centre of symmetry and forms a coordination plane with the four N atoms in daco (Hoshino, et al., 1989). Difference densities on and perpendicular to the Cu-N$_4$ plane after the spherical-atom refinement are shown in Fig. 5(a) and (b). Negative four peaks in Fig. 5 (a) near Cu^{2+} correspond to a 3d-hole. However, the peaks do not locate on the Cu-N bonds. Since the crystal field of the Cu atom is C_i, d-orbitals of Cu^{2+} are expressed as a linear combination of $d_{x^2-y^2}, d_{z^2}, d_{yz}, d_{zx}$ and d_{xy} orbitals and all the 25 coefficients c_{ik} were refined with the method in *II*. Since approximate c_{ik}'s are necessary for the non-linear least-squares refinement, they are calculated putting a point charge on each atom (Kamimura et al., 1969). $\langle j_K \rangle_{3d}$ of Cu^{2+} were taken from International Tables for X-ray crystallography (1974, vol. IV). After each cycle of refinement, new set of coefficients were orthonormalized by the Löwdin's method (Löwdin, 1950). Refinement was converged and the coefficients c_{ik}'s, population n_i and κ_i are listed in Table 1. The number of significant c_{ik}'s is seven among 25 coefficients. Orbital 1 in Table 1 is the hole orbital. It is not a pure $d_{x^2-y^2}$ but a mixed orbital of $d_{x^2-y^2}$ and d_{zx} orbitals, $c_{14}=0.57(59)$ is not significant. The hole and large negative and positive peaks in Fig. 5 were removed in Fig. 6. Further AHV refinement removed the peaks in Fig.6, but that prior to the d-orbital analysis did not improve Fig. 5.

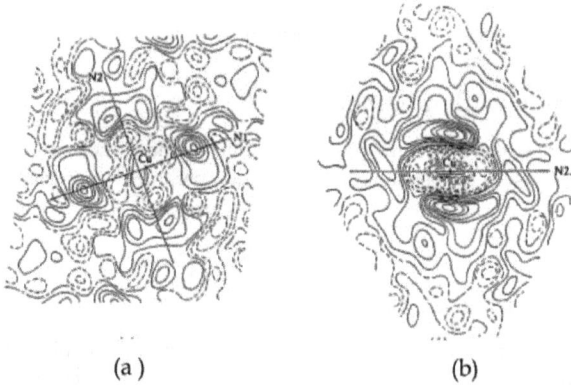

(a)   (b)

Figure 5. Difference density after spherical-atom refinement on (a) Cu-N₄ plane and (b) the plane perpendicular to it. Contours are the same as those in Fig. 2.

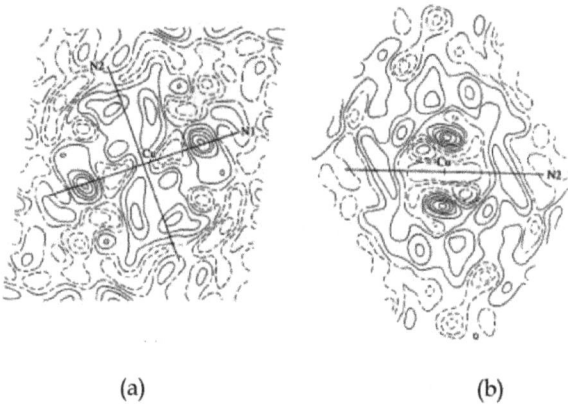

(a)   (b)

Figure 6. Difference density after 3d-orbital analysis. Contours are the same as in Fig. 2

When G-C formalism was applied to an iron complex, it removed the peaks equally well as the multipole refinement did (Mallinson, et al., 1988). Our AHV analysis is based on the classical thermodynamics and does not have such a problem, though the p.d.f. function $exp[-V(\mathbf{u})/k_BT]$ is not integrable (Scheringer, 1977) at a place far from the nucleus. However, we can apply it safely as long as it is applied within the region where the sum of the third and fourth terms in (18) is much smaller than k_BT. The condition is usually fulfilled in the investigations of EDD.

i	n_i	κ_i	$d_{x^2-y^2}$	d_{z^2}	d_{yz}	d_{zx}	d_{xy}
1	1.0	0.97(11)	**0.75(34)**	0.21(59)	0.23(216)	0.57(59)	0.10(79)
2	2.0	1.37(7)	0.08(28)	0.12(17)	−0.19(19)	0.10(26)	**−0.97(6)**
3	2.0	0.81(5)	**0.63(22)**	−0.15(52)	−0.44(44)	**−0.61(34)**	0.06(22)
4	2.0	0.71(4)	0.04(32)	**−0.93(13)**	−0.10(29)	0.34(41)	−0.06(10)
5	2.0	0.90(5)	0.15(76)	−0.22(45)	**0.84(43)**	−0.42(85)	**−0.22(15)**

Table 1. 3d-orbital parameters c_{ik}'s of Cu^{2+}. Significant ones in thick lines

6. 4f-EDD in rare-earth hexa-borides

4f-EDD analysis has become more and more important since many rare-earth compounds with interesting physical properties, such as high-temperature super conductors, have been found. The EDD of rare-earth hexa-borides were investigated since they are famous for their properties related to the Kondo effect.

6.1. XAO analysis of CeB₆

Ce and B atoms are at the body-centre and on the edges of the cubic unit cell with the B_6 regular octahedra at the corners as shown in Fig. 7. Ce is in an O_h crystal field and B atoms are on 4-fold axes. CeB₆ is a typical dense Kondo material with Kondo-temperature T_K=2 K. The magnetic moment decreases with that of temperature and vanishes below T_K. However, Kondo effect starts above room temperature (Ōnuki et al. 1984). Thus, change of EDD below room temperature is interesting although our diffractometer did not permit to measure below 2K. Intensities were measured at 100, 165, 230 and 298 K aiming to separate the two kinds of asphericity of EDD stated in 5.1 and 5.2 (Tanaka & Ōnuki, 2002). MD was avoided by ψ–scan with the program *IUANGL* (Tanaka et al., 1994). Crystal was shaped into a sphere with a radius 36 μm to reduce absorption, extinction and MD.

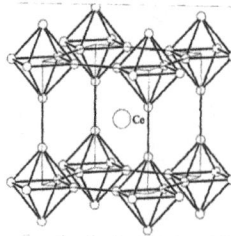

Figure 7. Crystal structure of CeB₆

Spin-orbit interaction was introduced into the XAO analysis. Four-fold degenerate $4f_{5/2}\Gamma_8$ orbitals were taken from Table 4 (b) of I as,

$$\Psi_1^{4f} = \sqrt{\frac{5}{6}}\psi_1 + \sqrt{\frac{1}{6}}\psi_5, \Psi_2^{4f} = \sqrt{\frac{1}{6}}\psi_2 + \sqrt{\frac{5}{6}}\psi_6, \Psi_3^{4f} = \psi_3, \Psi_4^{4f} = \psi_4, \tag{25}$$

where ψ_i (i=1 to 6) are basis functions in Table 1(c) of I. In the similar way $5p_{1/2}$, $5p_{3/2}$ and $4f_{5/2}\Gamma_7$ and $4f_{7/2}\Gamma_6, \Gamma_7, \Gamma_8$ orbitals were introduced into the XAO analysis.

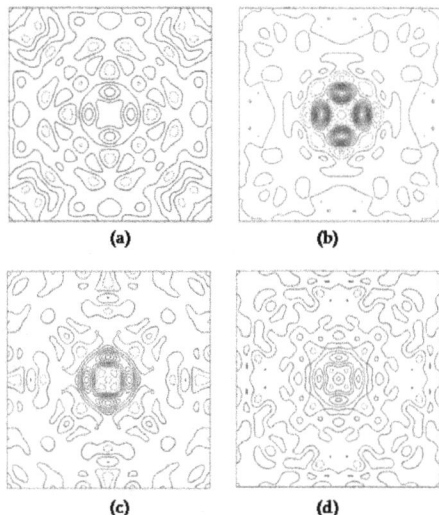

(a)

(b)

(c)

(d)

Figure 8. Difference density on (100) plane around Ce at (a) 100 K, (b) 165 K, (c) 230 K, (d) 298 K. Contours are the same as those in Fig. 2.

Temp. Ce	$n(\Gamma_8)^*$	$\kappa(\Gamma_8)$	$\kappa(5p_{3/2})$	q_{1111}	q_{1122}	B	$n(2p_x)$	$n(2p_z)$	$\kappa(2p_x)$	$\kappa(2p_z)$	Ce – B
100K	13(1)	179(8)	95(2)	47(19)	–131(48)		27(6)	103(8)	93(31)	92(12)	3.0416(3)
165K	15(2)	158(12)	98(2)	–14(6)	39(15)		52(4)	52(8)	85(16)	108(11)	3.0440(4)
230K	19(2)	157(11)	100(2)	0(4)	–1(11)		69(4)	16(8)	51(19)	130(18)	3.0459(7)
298K	21(2)	160(10)	100(2)	6(2)	–13(5)		67(4)	19(8)	54(17)	130(13)	3.0439(4)

Table 2. Orbital and AHV parameters. n(X), and κ(X) are a number of electrons and κ of the orbital X. n(x10^{-2}), κ (x10^{-2}), q_{iiij} (x10^{-19} JÅ^{-4}). $5p_{3/2,1/2}$ and $2s$ are fully occupied. $n(2p_x)=n(2p_y)$ and $\kappa(2p_x)=\kappa(2p_y)$. Unit of Ce-B is Å. *:n(Γ_7) is 4.2(3.6)x10^{-2} only at 298 K.

6.1.1. XAO analysis of CeB$_6$ below room temperature

After spherical-atom refinement for ions, Ce^{3+} and B$^{0.5-}$, the populations of the orbitals were refined keeping the sum of them equal to that of nuclear charges. When populations of them exceeded one/two or became negative, they were fixed to one/two or zero. In our program

QNTAO (Tanaka, 2000) each sub-shell is treated as an pseudo-atom sharing atomic parameters with the other sub-shells. It enables us to analyze the EDD of non-stoichiometric atoms, since valence electrons can be treated independently from the core electrons. Basis functions in Table 1 of *I* are automatically assigned to each AO labelled with $2l+1$ and the number of basis functions. Since high symmetry fixes c_{ik}'s in (1), the other orbital parameters and atomic parameters, including AHV were refined. Electron density around Ce on (001) after spherical-atom refinement is shown in Fig. 8. The peak heights around Ce increase from *0.6, 1.2 to 2.0 e$Å^{-3}$* on lowering the temperature to 165 K and reduce to *0.6 e$Å^{-3}$ at 100 K*. After XAO analysis they were removed, exhibiting they were due to the *4f*-electrons. Parameters of AO's and AHV are listed in Table 2. No electrons were found in $4f_{7/2}$ orbitals but B-2s and Ce-5p are always fully occupied. $n(4f_{5/2}\Gamma_7)$ has value 0.042(36) only at 298 K but $4f_{5/2}\Gamma_8$ is occupied at the four temperatures. It agrees with the previous experiments (Zirngiebl et al., 1984;Sato et al., 1984). Shorter Ce-B at 298 K than that at 230 K seems to be correlated to $4f_{5/2}\Gamma_7$. It extends along $\langle 111 \rangle$ or to the centre of B_6 octahedron making them extend to reduce electrostatic repulsion. Expanded $2p_x$ still has electrons. Total number of *4f*-elctrons mainly composed of $4f_{5/2}\Gamma_8$ are 0.98(11), 0.77(8), 0.61(7) and 0.52(6) at 298, 230, 165 and 100 K. They decrease with temperature. Table 2 tells $2p_{x,y}$ and *4f* electrons are transferred to $2p_z$, main contributor to the shortest covalent B-B bonds between B_6 octahedra (called as B-B$_{out}$).

Electron accumulation at *B-B$_{out}$* enhances q_{1111} and q_{1122} of Ce at 165 K. They change signs and become more larger at 100 K. Negative and positive q_{1111} at 165 K and 100K indicates that vibration of Ce at body-centre to the face-centre is favorable at 165 K, and becomes unfavorable at 100 K. This is the reason for the enhanced peak in Fig. 8(b) and reduced one in (a). Negative q_{1122} at 100 K indicates the attractive force to the edge centre increased by the accumulated $2p_z$ electrons. Since sum of the mean radii of B-2p and Ce-5p in free space, 2.2048+1.7947=3.9995 Å (Mann, 1968) is longer than Ce-B distance in Table 2, a slight expansion/contraction of them affects the potential seriously. Since the 5p orbitals are fully occupied, EDD's of them are spherical. B-$2p_x$ is closer than $2p_z$ to the spherically distributed Ce-$5p_{3/2}$ electrons. $\kappa(5p_{3/2})$ exhibits slight expansion on lowering the temperature. Expansion and contraction of $5p_{3/2}$ and $4f_{5/2}\Gamma_7$ in Table 2 reduces the potential of *4f* orbitals lying closer to the nucleus. However, it contradicts to the decrease of *4f* population. It indicates that the system resists against losing *4f* electrons from Ce by making the potential of Ce more stable. It reminds us of the Le Chatelier's law: the system resists against the change. Then why are *4f* electrons allowed to flow out of Ce in spite of the reduced potential energy of the Γ_8 state? The enhanced AHV at 165 and 100 K produced new ways of vibration along $\langle 100 \rangle$ and $\langle 110 \rangle$ directions increasing the ways to attain the energy of the system. Therefore, the enhanced AHV at lower temperatures means an increase in entropy. Since the electron transfer itself from Ce to (B-B)$_{out}$ increases entropy, it cannot be stopped.

The XAO analysis of CeB$_6$ was applied to the weak-field model. The crystal structure, EDD, electron populations, expansion/contraction parameters and AHV parameters were quite consistent with each other at different temperatures. Therefore, the 4f-EDD is concluded to be measured and analyzed successfully.

6.1.2. 4f population inversion and fully occupied 5d states at 430 K

The electron population at 298 K exhibited slightly occupied $4f_{5/2}\Gamma_7$. Since the energy gap between $4f_{5/2}\Gamma_8$ and Γ_7 was reported to be 530-560 K (Loewenhaupt et al., 1985; Zirngiebl et al., 1984), XAO analysis at 430 K was performed to observe more electrons in Γ_7 (Makita et al., 2007). For details of the high-temperature diffraction equipment, see PhD thesis of Makita (Makita, 2008). Scattering factors of Ce were evaluated from relativistic AO's calculated by the program *GRASP* (Dyall et al., 1989). Difference density around Ce on (001) after the spherical-atom refinement is shown in Fig. 9(a). The 4f-peaks closest to Ce elongate along <110> in contrast to those below room temperature in Fig. 8. They are surrounded by the four peaks of 0.65 eÅ^{-3}, which extend along <100>, and disappeared after the analysis of the 4f peaks. However, the population of $4f_{5/2}\Gamma_8$ and Γ_7 were 0.06(3) and 0.36(11). Higher temperature reverses the populations of $4f_{5/2}\Gamma_8$ and Γ_7. The peaks outside the 4f-peaks, called 5d peaks, remained almost unchanged. Accordingly $5d_{5/2}\Gamma_7$ and Γ_8 orbitals were further refined. The population of $5d_{5/2}\Gamma_8$, $n(5d_{5/2}\Gamma_8)$, exceeded 1.0 and was fixed. $n(4f_{5/2}\Gamma_8, \Gamma_7)$ and $n(5d_{5/2}\Gamma_8)$ were 0.06(2), 0.37(1) and 1.0, while $5d_{5/2}\Gamma_7$ was vacant. R factor was reduced from 1.25 to 1.16 %. The peaks outside the 4f peaks reduced slightly from 0.65 to 0.59 eÅ^{-3} in Fig. 10(b). Since 5d orbitals extend in a large area, in contrast to 4f orbitals, slight change of EDD is significant. Thus electron densities around 5d–peaks in Figs. 9 (a) and (b) were numerically integrated to give 3.41 and 1.57 electrons. XAO analysis explained 54 % of the 5d-peaks. Therefore $d_{5/2}\Gamma_8$ is concluded to be occupied, though peaks still remained. It may be ascribed to the inaccurate 5d-AO's (Claiser et al., 2004).

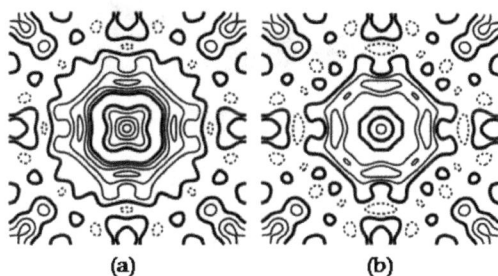

Figure 9. Differnece density around Ce at 430 K on (100) after (a) spherical-atom refinement and (b) final residual density. Contours at 0.2 eÅ^{-3}. Negative and zero contours in dotted and thick lines.

Why are 5d orbitals occupied? The energy level of the Ce-5d and B-2p calculated by Liberman et al., (1971) and Mann (1967) are -0.63 and -0.62 a.u. They are very close compared to the Ce-4f level at -0.75 a.u. Thus B-2p electrons are transferred to 5d levels predominantly according to the first-order perturbation theory. The radial distribution functions of the relevant AO's are illustrated in Fig. 10. Putting the origin of B at r=3.045 Å, 2p was drawn to the reverse side. 5d and 2p overlap well, and it is expected that 2p electrons which distribute all over the crystal through the network of B-B covalent bonds can be transferred most probably to 5d orbitals. Electrons traveling in the crystal are expected to stay at $5d_{5/2}\Gamma_8$ orbitals when they come close to Ce.

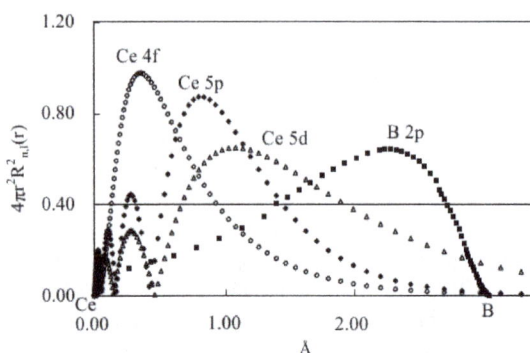

Figure 10. Radial distribution functions of Ca and B.

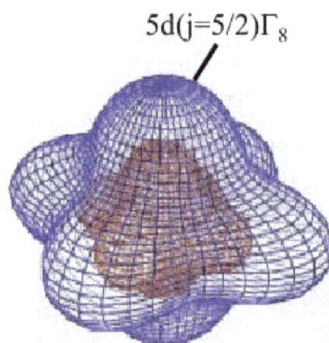

$$5d(j=5/2)\Gamma_8$$

Figure 11. $4f_{5/2}\Gamma_8$ (red) and $5d_{5/2}\Gamma_8$ obritals.

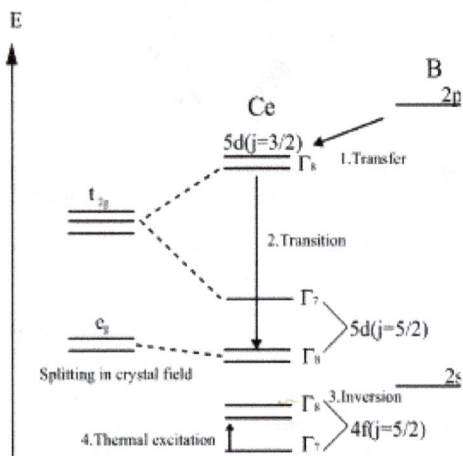

Figure 12. 5d- and 4f-energy levels expected from the electron populations.

Why does $4f_{5/2}\Gamma_8$-Γ_7 inversion occur? $5d_{5/2}\Gamma_8$ orbitals are located outside of $4f_{5/2}\Gamma_8$ having exactly the same symmetry (Fig. 11). Since $5d_{5/2}\Gamma_8$ is fully occupied, the potential of the $4f_{5/2}\Gamma_8$ orbital is enhanced. It is the reason for the population inversion of $4f_{5/2}\Gamma_8$ and Γ_7. The energy level diagram expected from the electron populations obtained by the XAO analysis is shown in Fig.12. Since the quantization axes are defined parallel to <100>, T_{2g} states locate higher than E_g. $5d_{3/2}\Gamma_8$ orbital closely related to T_{2g} is located highest among the 5d orbitals as shown in Fig. 12. From the populations obtained, 2p electrons seem to be transferred to $5d_{5/2}\Gamma_8$ orbitals, directly or first to $5d_{3/2}\Gamma_8$ and then to $5d_{5/2}\Gamma_8$.

6.1.3. Electron configuration at 340 and 535 K

In order to confirm the 5d-occupation in CeB$_6$ at 430 K, XAO analyses were done at 340 and 535 K. The difference densities, as well as electron configurations, are different from each other as shown in Fig. 13. Since the electrons are continually transferred between 2p and 5d orbitals, the crystal field seems to change with temperature resulting in different directions of the 4f and 5d peaks in Fig. 13. However, it is confirmed that 5d orbitals in CeB$_6$ are occupied above room temperature (Makita et al., 2008).

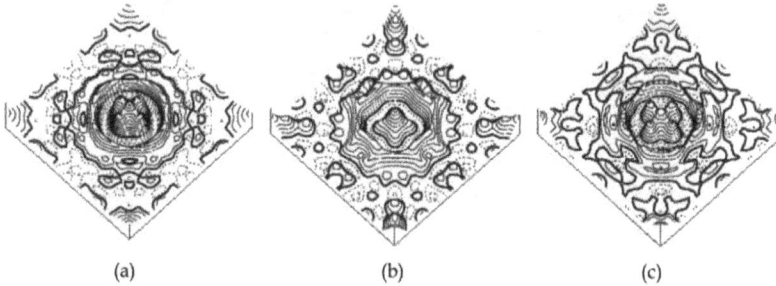

(a)   (b)   (c)

Figure 13. Bird-eye view of difference densities around Ce at (a) 340, (b) 430 and (c) 535 K. Contours as in Fig. 9 except the interval of 0.1 eÅ^{-3}. Frames are parallel to <100>axes.

6.2. XAO analysis of SmB$_6$

SmB$_6$ formally has five 4f electrons. In order to extend XAO analysis to multi 4f-electron system, EDD of SmB$_6$ was measured at 100, 165, 230 and 298 K (Funahashi, et al., 2010). It is interesting to see how physical properties as a Kondo insulator are explained by the XAO analysis. The shadow in the difference density at 230 K in Fig. 14 specifies roughly the area of 5d peaks. $5p_{3/2}$, $4f_{5/2}\Gamma_8$,Γ_7 and AHV were refined reducing the 4f peaks in Fig. 14(a). However, since 4f-peaks still remained along <100>, the populations of $4f_{7/2}\Gamma_6$,Γ_7,Γ_8 were refined but only Γ_6 orbital which extends along <100> more sharply than $4f_{5/2}\Gamma_8$ survived. Since peaks (peak A) remained in the 5d-area, $5d_{5/2}\Gamma_8$,Γ_7 and $5d_{3/2}\Gamma_8$ were added in the refinement and only $5d_{5/2}\Gamma_8$ orbital, which stems from e_g orbital of the storing field model had population. Fig. 14(b) exhibits that peak A is reduced from 0.43 to 0.17 eÅ^{-3} and 4f-peaks are almost deleted. Final parameters are listed in Table 3.

The electron configuration in Table 3 is correlated to the physical properties of SmB₆ as follows: (a) SmB₆ is a Kondo insulator. Its electric resistivity increases gradually like semiconductors below room temperature and begins to increase steeply below 30 K with a decrease in temperature. It also begins to increase like metals above room temperature (Ueda & Onuki, 1998). $2p$ and $5d_{3/2}\Gamma_8$ orbitals consist of the conduction band (Kimura et al., 1990). Since B-$2p_z$ extends along the edge, it does not overlap with $5d_{3/2}\Gamma_8$ effectively and does not seem to contribute to the band. The population of $2p_x$ in Table 3 as well as the resistivity steadily increases on lowering temperature. n($5d_{5/2}\Gamma_8$) also increases from 230 K. The increase of populations indicates that of localized electrons. It may be correlated to the increase of the resistivity. (b) $4f_{7/2}\Gamma_6$ are vacant only at 100 K, although they are occupied at the other three temperatures. It may be correlated to the band gap between the $4f$ states, which is reported to start developing between 150 and 100 K (Souma, et al., 2002). (c) Among the $5d$ orbitals only the $5d_{5/2}\Gamma_8$ are occupied. Since $5d_{5/2}\Gamma_8$ orbitals correspond to eg in the strong field model as illustrated in Fig. 1, it agrees with the band calculation of LaB₆ by Harima (1988), reporting that $5d$-e_g and $2p$ of B consist of the conduction band.

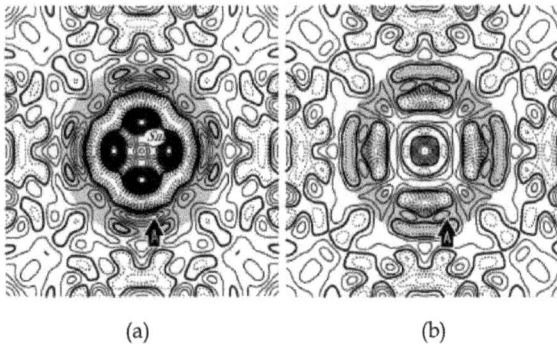

(a)   (b)

Figure 14. Difference density at 230 K after (a)spherical-atom refinement and (b) XAO analysis. Contours as in Fig. 13.

	$n(4f_{5/2}\Gamma_8)$	$n(4f_{5/2}\Gamma_7)$	$n(4f_{7/2}\Gamma_6)$	$n(5d_{5/2}\Gamma_8)$	$n(2p_x)$	$n(2p_z)$
100K	0.96(9)	0.48(12)	0.0	1.0	0.21(8)	0..45(17)
165K	1.0	0.67(13)	0.44(12)	0.75(24)	0.16(13)	0.55(39)
230K	1.0	0.61(13)	0.59(13)	0.56(16)	0.13(7)	0.66(28)
298K	0.77(3)	0.59(7)	0.47(6)	0.78(6)	0.07(3)	0.81(12)

Table 3. Temperature dependence of electron populations of Sm $4f$, $5d$ and B $2p$ orbitals.

7. Bright future for X-ray crystallography

EDD investigation was limited up to $3d$-transition-metal complexes. However, XAO analysis made EDD investigations of rare-earth compounds as well as non-stoichiometric ones possible. Its application to organic compounds can be attained when it is developed to X-ray molecular orbital analysis (XMO). Since the least-squares method stated in 2.2 was formulated for MO models, XMO analysis will be accomplished in a near future. When actinoid compounds become our targets, however, MD-effect is expected to be so much that the avoidance of it by ψ – rotation is impossible and the correction for it will be inevitable.

CeB_6 is a possible quantum-material to emit UV light when electrons in $5d_{5/2}\Gamma_8$ could be transferred to $4f_{5/2}\Gamma_8$, as the investigation at 430 K revealed. The d-f transition is a permitted one by quantum mechanics. Since the 5d-occupation is found in the ground state of CeB_6, some external force is necessary to make the transition occur. A electron populations in CeB_6 and SmB_6 found by the XAO analysis demostrate the importance of the EDD analysis based on quantum-mechanical orbitals.

As discussed in 5.3, the aspherical properties of EDD and AHV are separated better by the method based on classical Boltzmann statistics than the G-C method. However, recent development of neutron diffraction will make it possible to get intrinsic ADP's and use them as known parameters in X-ray EDD analyses. It will improve XAO analysis of rare-earth complexes and makes the XMO analysis surer and easier.

The accuracy of X-ray structure-factor measurements has been improved so much that every crystallographer will investigate EDD easily as a part of their X-ray structure analysis. The top-up operation with constant incident beam intensity at SR facilities has improved the accuracy of the structure factor measurements from 1 % to 0.1 %. Future of X-ray diffraction is bright.

Author details

Kiyoaki Tanaka and Yasuyuki Takenaka
Nagoya Institute of Technology, Hokkaido University of Education, Japan

8. References

Bader, R. F. W. (1994). Atoms in molecules A quantum theory, Clarendon Press, Oxford, Great Britain

Becker, P. J. & Coppens, P. (1974a). Extinction within the limit of Darwin transfer equations. I. General formalisms for primary and secondary extinction and their application to spherical crystals. *Acta Cryst.* A30, 129-147.

Becker, P. J. & Coppens, P. (1974b). Extinction within the limit of Darwin transfer equations. II. Refinement of extinction in spherical crystals of SrF_2 and LiF. *Acta Cryst.* A30, 148-153.

Becker, P. & Coppens, P. (1975). Extinction within the limit of Darwin transfer equations. III.Non-soherical crystals and anisotropy of extinction. *Acta Cryst.* A31, 417-425.

Chatterjee, A., Maslen, E. N. & Watson, K. J. (1988). Electron densities in crystals of nonaaqualanthanoid(III) tris(trifluoromethanesulfonates). *Acta Cryst.* B44, 386-395.

Claiser, N., Souhassou, M. & Lecomte, C. (2004). Problems in experimental charge modeling of rare *earth* atom complexes: the case of gadolinium. *J. Phys. Chem. Solids*, 65, 1927-1933.

Coppens, P., Guru Row, T. N.., Leung, P., Stevens, E. D., Becker, P. J. & Yang, Y. W. (1979). Net atomic charges and molecular dipole moments from spherical-atom X-ray refinements, and the relation between atomic charge and shape. *Acta Cryst.* A35, 63-72

Coppens, P., Willoughby, T. V. and Csonka L. N. (1971). Electron population analysis of accurate diffraction data. I. Formalisms and restrictions, *Acta Cryst.* A27, 248-256.

Dawson, B., Hurley, A. C. & Maslen V. W. (1967). Anharmonic vibration in fluorite structures. *Proc. Roy. Soc. London Ser. A,* 298, 255-263.

Dyall, K. D., Grant, I. P., Johnson, C. T., Parpia, F. A. & Plummer, E. P. (1989). GRASP* A general-purpose relativistic atomic structure program. *Comput. Phys. Commun.* 55, 425-456.

Funahashi, S. (2010). XAO(X-ray atomic orbital) analysis of the EDD of rare-earth compounds and their physical properties, PhD thesis, Grad. School of Sci. & Eng. Nagoya Inst. of Tech.

Funahashi, S., Tanaka, K. & Iga, F. (2010). X-ray atomic orbital analysis of 4f and 5d electron configuration of SmB_6 at 100, 165, 230 and 298 K. *Acta Cryst.* B66, 292-306.

Hamilton, W. C. (1964). *Statistics in physical science,* The Ronald press company, New, York.

Hansen, N. K. & Coppens, P. (1978). *Testing aspherical atom refinement on small-molecule data sets, Acta Cryst.* A34, 909-921.

Harima, H. (1988). *Theoretical study of the Fermi surface of the rare-earth compounds,* PhD thesis, Tohoku Univ. Japan.

Hoshino, N., Fukuda, Y., Sone, K., Tanaka, K. & Marumo, F. (1989). Single-Crystal X-ray and Optical Studies on Bis(1,5-diazacyclooctane)Copper(II) Nitrate, a thermochromic copper(II) complex involving hydrogen bonds between the ligand and the counter Ion, *Bull. Chem. Soc. Jpn.,* 62, 1822-1828.

Ishiguro, T., Ishizaa, N., Mizutani, M., Kato, M., Tanaka, K. & Marumo, F. (1983). Charge-density distribution in crystals of $CuAlO_2$ with d-s hybridization. *Acta Cryst.* B39, 564-569.

Iwata, M. (1977). X-ray determination of the electron density distribution in crystals of $[Co(NH_3)_6][Co(CN)_6]$, *Acta Cryst.* B33, 59-69.

Iwata, M. & Saito, Y. (1973), The crystal structure of Hexaamminecobalt(III) Hexacyanocobaltate(III): an accurate determination, *Acta Cryst.* B29, 822-832.

Kamimura, H., Sugano, S. & Tanabe, Y. (1969). *Ligand field theory and its application,* Tokyo, Syōkabō.

Kijima, N., Tanaka, K. & Marumo, F. (1981). Electron density distribution in crystals of potassium trifluorocobaltate(II), *Acta Cryst.* B37, 545-548

Kijima, N., Tanaka, K. & Marumo, F. (1983). Electron density distribution in crystals of $KMnF_3$ and $KNiF_3$, *Acta Cryst.* B37, 545-548

Kijima, K., Tanaka, K. & Marumo, F. (1983), Electron density distribution in crystals of potassium trifluorocobaltate(II). *Acta Cryst.* B39, 557-561

Kimura, S., Nanba, T., Kunii, S. & Kasuya, T. (1990). Interband optical spectra of rare-earth hexaborides. *J. Phys. Soc. Jpn,* 59 3388-3392.

Liberman, D. A., Cromer, D. T. & Waber, J. T. (1971). Relativistic self-consistent field program for atoms and ions. *Comput. Phys. Commun.* 2, 107-113.

Lowenhaupt, M., Carpenter, J. M. & Loong, C. –K. (1985). Magnetic excitations in CeB₆. *J. Magn. Magn. Mater.* 52, 245-249. (1971).

Löwdin, P. O. (1950). On the nonorthogonality problem connected with the use of atomic wave functions in the theory of molecules and crystals. *J. Chem. Phys.* 18, 365-375.

Makita, R. (2008). Investigation on the physical properties of heavy-atom compounds by X-ray analysis of electron density, PhD thesis. Grad. School of Mat. Sci. & Eng., Nagoya Inst. Tech.

Makita, R., Tanaka, K., Ōnuki, Y. & Tatewaki, H. (2007). Inversion of 4f-states in CeB₆ thermally excited at 430 K, *Acta Cryst.* B63, 683-692.

Makita, R., Tanaka, K., Ōnuki, Y. (2008). 5d and 4f configuration of CeB6 at 340 and 535 K, *Acta Cryst.* B64, 534-549.

Mallinson, P. R., Koritsanszky, T., Elkaim, E., Li, N. & Coppens, P. (1988). The Gram-Charlier and multipole expansions in accurate X-ray diffraction studies. Can they be distinguished? Acta Cryst. A44, 336-342.

Mann, J. B. (1968). Atomic structure calculations II. Hartree-Fock wavefunctions and radial expectation values: Hydrogen to Lawrencium. *Report LA3691. Los Alamos NatioFnal Laboratory*, New Mexico, USA.

Marumo, F., Isobe, M. & Akimoto, S. (1977). Electron-density distribution in crystals of γ-Fe₂SiO₄ and γ-Co₂SiO₄. *Acta Cryst.* B33, 713-716.

Miyata, N., Tanaka, K. & Marumo, F. (1983). Electron density distribution in crystals of iron(II) potassium trifluoride, *Acta Cryst.* B39, 561-564.

Moon, R. M. & Shull, C. G. (1964). The effect of simultaneous reflections on single-crystal neutron diffraction intensities. *Acta Cryst.* 17, 805-812.

Ōnuki, Y., Furukawa, Y. & Komatsubara, T. (1984). Dense Kondo behavior in CeₓLa₁₋ₓAl₂. *J. Phys. Soc. Jpn.*, 53, 2734-2740.

Sato, S. (1985). Aspherical charge distribution in a crystal of CeB₆. *J. Magn. Magn. Mater.* 52, 310-312.

Scheringer, C. (1977). On the interpretation of anisotropic temperature factors. III. Anharmonic motions. *Acta Cryst.* A33, 879-884.

Souma, S., Kumigashira, H., Ito, T., Takahashi, T. & Kunii, S. (2992). Direct observation of pseudogap of SmB6 using ultrahigh-resolution photoemission spectroscopy. *Physica B*, 312-313, 329-330.

Stewart, R. F. (1969).Generalized XRay scattering factors, *J. Chem. Phys.* 51, 4569-4577.

Stewart, R. F. (1969). Electron population analysis with generalized xray scattering factors: higher multipoles. *J. Chem. Phys.* 51, 4569-4577.

Tanaka, K. (1978). Electron density distribution in crystals of diformohydrazide, *Acta Cryst.* B34, 2487-2494.

Tanaka, K. (1988). X-ray analysis of wavefunctions by the least-squares method incorporating orthonormality. I. General formalism, *Acta Cryst.* A44, 1002-1008.

Tanaka, K. (1993). X-ray analysis of wavefunctions by the least-squares method incorporating orthonormality. II. Ground state of the Cu²⁺ ion of bis(1,5-diazacyclooctane)copper(II) nitrate in a low-symmetry crystal field. *Acta Cryst.* B49, 1001-1010.

Tanaka, K., Kato, Y. & Ōnuki, Y. (1997). 4f-electron density distribution in crystalso of CeB₆ at 165 K and its analysis based on the crystal field theory. *Acta Cryst.* B53, 143-152.

Tanaka, K., Konishi, M. & Marumo, F. (1979). Electron-density distribution in crystals of KCuF₃ with Jahn-Teller distortion. *Acta Cryst.* B35, 1303-1308.

Tanaka, K., Kumazawa, S., Tsubokawa, M., Maruno, S. & Shirotani, I. (1994). The multiple-diffraction effect in accurate structure factor measurements of PtP₂ crystals. *Acta Cryst.* A50, 246-252.

Tanaka, K., Makita, R., Funahashi, S., Komori, T. & Zaw Win (2008). X-ray atomic orbital analysis. I. Quantum-mechanical and crystallographic framework of the method. *Acta Cryst.* A64, 437-449.

Tanaka, K. & Marumo, F. (1982). Electron-density distribution and anharmonic vibration in crystals of potassium trifluorocuprate(II). *Acta Cryst.* B38, 1422-1427.

Tanaka, K. & Marumo, F. (1983). Willis formalism of anharmonic temperature factors for a general potential and its application in the least-squares method. *Acta Cryst.* A39, 631-641.

Tanaka, K. & Ōnuki, Y. (2002). Observation of the 4f-electron transfer from Ce to B₆ in the Kondo crystal CeB₆ and its mechanism by multi-temperature X-ray diffraction. *Acta Cryst.* B58, 423-436.

Tanaka, K. & Saito Y. (1975). Simultaneous reflection: Its detection and correction for intensity perturbation. *Acta Cryst.* A31, 841-845.

Tanaka, K., Takenaka, Y., Funahashi, S., Sakakura, T. & Komori, T. (2010). A general expression of the polarization factor for multi-diffraction processes, *Acta Cryst.* A66, 438-440.

Tanaka, K., Tsubokawa, M., Maruno, S. & Shirotani, I. (1994), The multiple-diffraction effect in accurate structure-factor measurements of PtP₂ crystals. *Acta Cryst.* A50, 246-252.

Toriumi, K., Ozima, M., Akaogi, M. & Saito, Y. (1978). Electron-density distribution in crystals of CoAl₂O₄. *Acta Cryst.* B34, 1093-1096.

Ueda, K. & Onuki, Y. (1998). *Physics of heavy electron systems*, p. 224, Shokabo, Tokyo.

Willis, B. T. M. (1969). Lattice vibrations and the accurate determination of structure factors for the elastic scattering of X-rays and neutrons. Acta Cryst. A25, 277-300.

Zirngiebl, E., Hillerbrands, B., Blumenröder, S., Güntherrodt, G., Lowenhaupt, M., Carpenter, J. M., Winzer, L. & Fisl, Z.,(1984). Crystal-field excitation in CeB₆ studied by Raman and neutron spectroscopy. *Phys. Rev. B*. 30, 4052-4054.

Permissions

The contributors of this book come from diverse backgrounds, making this book a truly international effort. This book will bring forth new frontiers with its revolutionizing research information and detailed analysis of the nascent developments around the world.

We would like to thank Jason B. Benedict, for lending his expertise to make the book truly unique. He has played a crucial role in the development of this book. Without his invaluable contribution this book wouldn't have been possible. He has made vital efforts to compile up to date information on the varied aspects of this subject to make this book a valuable addition to the collection of many professionals and students.

This book was conceptualized with the vision of imparting up-to-date information and advanced data in this field. To ensure the same, a matchless editorial board was set up. Every individual on the board went through rigorous rounds of assessment to prove their worth. After which they invested a large part of their time researching and compiling the most relevant data for our readers. Conferences and sessions were held from time to time between the editorial board and the contributing authors to present the data in the most comprehensible form. The editorial team has worked tirelessly to provide valuable and valid information to help people across the globe.

Every chapter published in this book has been scrutinized by our experts. Their significance has been extensively debated. The topics covered herein carry significant findings which will fuel the growth of the discipline. They may even be implemented as practical applications or may be referred to as a beginning point for another development. Chapters in this book were first published by InTech; hereby published with permission under the Creative Commons Attribution License or equivalent.

The editorial board has been involved in producing this book since its inception. They have spent rigorous hours researching and exploring the diverse topics which have resulted in the successful publishing of this book. They have passed on their knowledge of decades through this book. To expedite this challenging task, the publisher supported the team at every step. A small team of assistant editors was also appointed to further simplify the editing procedure and attain best results for the readers.

Our editorial team has been hand-picked from every corner of the world. Their multi-ethnicity adds dynamic inputs to the discussions which result in innovative outcomes. These outcomes are then further discussed with the researchers and contributors who give their valuable feedback and opinion regarding the same. The feedback is then collaborated with the researches and they are edited in a comprehensive manner to aid the understanding of the subject.

Apart from the editorial board, the designing team has also invested a significant amount of their time in understanding the subject and creating the most relevant covers. They scrutinized every image to scout for the most suitable representation of the subject and create an appropriate cover for the book.

The publishing team has been involved in this book since its early stages. They were actively engaged in every process, be it collecting the data, connecting with the contributors or procuring relevant information. The team has been an ardent support to the editorial, designing and production team. Their endless efforts to recruit the best for this project, has resulted in the accomplishment of this book. They are a veteran in the field of academics and their pool of knowledge is as vast as their experience in printing. Their expertise and guidance has proved useful at every step. Their uncompromising quality standards have made this book an exceptional effort. Their encouragement from time to time has been an inspiration for everyone.

The publisher and the editorial board hope that this book will prove to be a valuable piece of knowledge for researchers, students, practitioners and scholars across the globe.

List of Contributors

Bart Kahr and Alexander G. Shtukenberg
Department of Chemistry, New York University, New York City, USA

Kouhei Okitsu
Nano-Engineering Research Center, Institute of Engineering Innovation, Graduate School of Engineering, The University of Tokyo, 2-11-16 Yayoi, Bunkyo-ku, Tokyo 113-8656, Japan

Yasuhiko Imai and Yoshitaka Yoda
Japan Synchrotron Radiation Research Institute, SPring-8, 1-1-1 Kouto, Mikazuki-cho, Sayo-gun, Hyogo 679-5198, Japan

Nickolay Y. Chirgadze
Canada Campbell Family Cancer Research Institute, Ontario Cancer Institute, Princess Margaret Hospital, University Health Network, Toronto, Ontario, Canada
Department of Pharmacology and Toxicology, University of Toronto, Toronto, Ontario, Canada

Gera Kisselman, Wei Qiu, Vladimir Romanov, Christine M. Thompson and Robert Lam
Canada Campbell Family Cancer Research Institute, Ontario Cancer Institute, Princess Margaret Hospital, University Health Network, Toronto, Ontario, Canada

Kevin P. Battaile
Hauptman-Woodward Medical Research Institute, IMCA-CAT, Advanced Photon Source, Argonne National Laboratory, Argonne, Illinois, USA

Emil F. Pai
Canada Campbell Family Cancer Research Institute, Ontario Cancer Institute, Princess Margaret Hospital, University Health Network, Toronto, Ontario, Canada
Departments of Biochemistry, Molecular Genetics, and Medical Biophysics, University of Toronto, Toronto, Ontario, Canada

Legrand Vincent
LUNAM Université - Université de Nantes - Ecole Centrale Nantes, Institut de Recherche en Génie Civil et Mécanique (GeM), France

Henrich H. Paradies, Hendrik Reichelt and Chester A. Faunce
The University of Salford, Joule Physics Laboratory, Manchester, United Kingdom

Peter Quitschau
Fachhochschule Südwestfalen, University of Applied Sciences, Biotechnology & Physical Chemistry, Iserlohn, Germany

Kurt Zimmermann
The Symbio Herborn Group Inc., Institute for Microecology, Herborn, Germany

Dinesh G. (Dan) Patel and Jason B. Benedict
Department of Chemistry, State University of New York at Buffalo, Buffalo, New York, USA

Rakesh Gudavarthy
Missouri University of Science and Technology, Rolla, Missouri, USA Current address: Intel Corporation, Hillsboro, OR, USA

Elizabeth A. Kulp
Missouri University of Science and Technology, Rolla, MO, USA

Xin Ding, Matti Tuikka and Matti Haukka
Department of Chemistry, University of Eastern Finland, Joensuu Campus, Joensuu, Finland

Nader Noroozi Pesyan
Urmia University, Iran

Kazimierz Stróż
Faculty of Informatics and Materials Science, University of Silesia, Katowice, Poland

Jiapu Zhang
The University of Ballarat, Australia

Kiyoaki Tanaka and Yasuyuki Takenaka
Nagoya Institute of Technology, Hokkaido University of Education, Japan